SCIENCE EDUCATION IN CONTEXT

Science Education in Context

An International Examination of the Influence of Context on Science Curricula Development and Implementation

Edited by

Richard K. Coll
University of Waikato, Hamilton, New Zealand

and

Neil Taylor
University of New England, Amidale, Australia

SENSE PUBLISHERS
ROTTERDAM / TAIPEI

A C.I.P. record for this book is available from the Library of Congress.

ISBN 978-90-8790-247-6 (paperback)
ISBN 978-90-8790-248-3 (hardback)
ISBN 978-90-8790-249-0 (e-book)

Published by: Sense Publishers,
P.O. Box 21858, 3001 AW Rotterdam, The Netherlands
http://www.sensepublishers.com

Printed on acid-free paper

This book was developed with the assistance of a Dean's Initiative Grant from the Faculty of The Professions at the University of New England.

TABLE OF CONTENTS

COLL AND TAYLOR

SCIENCE EDUCATION IN CONTEXT: AN OVERVIEW AND SOME OBSERVATIONS

INTRODUCTION

This book presents an international perspective of the influence of educational context on science education. By this we mean *the context in which the teaching and learning takes place*, rather than the use of a context-based approach to learning and teaching (Pilot & Bulte, 2006). The focus is on the interactions between curriculum development and implementation in non-Western and non-English-speaking contexts (i.e., outside the UK, USA, Australia, NZ, etc.).

There has been much written about the problems and issues associated with the development of curriculum, especially in science education. A complication in curriculum development and implementation is the mode of development, with the use of a centre-periphery model being common (McGee, 1997). Even with wide consultation, Bell, Jones and Carr (1995) comment that a key feature of curriculum development is the inevitable tension that arises between stakeholder groups such as: government, 'industry', curriculum developers, teachers, school authorities, teachers and students. Each of these cohorts holds certain views about what is important and what should (or should not) be included in any curriculum statement or policy document. It also is important to understand what the term curriculum means (Hume & Coll, 2005). Here we distinguish between the formal or official curriculum (i.e., that as represented in formal policy documents), and the implemented or enacted curriculum (i.e., the interpretation of official policy statements and subsequently delivered by teachers).

Effective science curriculum is of considerable importance internationally because there is increasing pressure by governments to produce technically skilled people from the compulsory and tertiary education sectors; people capable of participating actively in the so-called 'knowledge economy'. Such pressure is often acute in 'developing' or non-Western countries, which oftentimes have economies based on the rather fickle (especially post-September 11) tourism industry, or relatively low-value agricultural produce. Factors such as isolation from major markets like Europe and the USA further exacerbates problems with economic competition - as do powerful trading blocks and the apparent inability of World Trade Organization to open markets to free competition for agricultural produce. The World Bank, in particular, routinely talks of a greater need for skilled workers in developing countries (and indeed in Western countries) for those who wish to participate in the knowledge economy (see, e.g., World Bank, 1996, 2006), and to shift away from commodities to value-added products.

There are similar calls for more technically skilled workers even in more developed wealthy countries. However, arguably of greater interest here is a general need to produce a scientifically-literate populace (Coll & Taylor, 2004; Laugksch, 2000). Miller (2000) sees scientific literacy as being able to read about, understand and participate in a public policy debate involving science or technology, which many authors see as an important outcome of the school system (see Laugksch, 2000 and references therein). We live in the most scientifically and technologically advanced age ever. Incredible advances have been made in science, in medical science, technology and engineering, and information and communication technologies. Advances in education and science education have strived to incorporate these advances in science education (Fensham, 1988; Gilli, 2000). Most countries have national curricula in science, and many have witnessed the development of new national curriculum statements in related subjects such as technology (e.g. Jones, 1997; Sade & Coll, 2003). However, many authors (e.g., Eisenhart, Finkel & Marion, 1996; Jenkins, 1990; Preece & Baxter, 2000; Shamos, 1996) express alarm about the prevalence of the public's beliefs that are not in accord with scientific views, seeing this lack of agreement as an unfortunate indictment of the compulsory science education sector.

THEORETICAL BASIS TO THE BOOK

Recent research seeking to understand science education has been developed based upon the notion of social cognition and sociocultural views of learning (Vygotsky, 1986; Wenger, 1991). According to social cognition theory learning occurs in a variety of contexts or situations (Hennessy, 1993; Hennessy & Murphy, 1999). Sociocultural views of learning derived from this place considerable emphasis on the *social* component within the particular context or situation in which learning occurs. Educational research has in the past been grounded in the assumption that it is possible (indeed desirable) to investigate an individual removed from their social context. Many early learning theories sought to achieve this in order to obtain data 'uncontaminated' by subjective influence of researchers. Other approaches assume that cultural and social issues can be incorporated as additional variables once the basic form of mental functioning in an individual is isolated and understood. However, according to Wertsch (1991):

> The basic tenet of a sociocultural approach to mind is that human mental functioning is inherently situated in social interactional, cultural, institutional, and historical context. Such a tenet contrasts with approaches that assume, implicitly or explicitly, that it is possible to examine mental processes such as thinking or memory independently of the sociocultural setting in which individuals and groups function. (p. 86)

A key feature of the book is then a focus on the 'situatedness' of learning, and its influence on curriculum. Each author writes about a particular educational issue.

Many first provide an overview with some detail of the background about science education in the particular educational context, before exploring the influence of the key sociological and contextual factors on the issue of interest, and in some cases research projects they have been intimately involved with.

ABOUT THE CONTRIBUTORS

An important and distinguishing feature of the book is that it draws upon the experiences and research of local experts from an extremely diverse cohort across the world. Each chapter is concerned with some aspect of science curriculum (interpreted broadly to include technology, environmental and mathematics curriculum) in a specific educational context. In some cases the chapter is in the form of a story or narrative, in other case it draws from particular research inquiry conducted by the author and his or her colleagues. The book addresses topics such as: curriculum development; research or evaluation of an implemented curriculum; discussion of pressures driving curriculum reform or implementation of new curricula (e.g., technology or 'new mathematics'); the influence of political, cultural, societal or religious mores on education; governmental or ministerial drives for curriculum reform; economic or other pressures driving curriculum reform; the influence of external assessment regimes on curriculum; and so on.

The editors have made a conscious effort to remain 'hands off' when editing, and to allow the contributors own voice to be heard. This book represents their story, not ours.

REFERENCES AND BIBLIOGRAPHY

Bell, B., Jones, A., & Carr, M. (1995). The development of the recent national New Zealand science curriculum. *Studies in Science Education, 26*, 73-105.

Coll, R.K., & Taylor, N. (2004). Probing scientists' beliefs: How open-minded are modern scientists? *International Journal of Science Education, 26*(6), 757-778.

Duit, R. (2000). *Bibliography: Student's alternative frameworks and science education* (5th ed.). Kiel, Germany: University of Kiel.

Eisenhart, M., Finkel, F., & Marion, S.F. (1996). Creating the conditions for scientific literacy: a re-examination. *American Educational Research Journal, 33*, 261-295.

Fensham, P.J. (1988). Familiar but different: Some dilemmas and new directions in science education. In P.J. Fensham (Ed.), *Development and dilemmas in science education* (pp. 1-26). London: Falmer.

Gilli, D. (2000). Science education and economic developments: Trends relationships and research agenda. *Studies in Science Education, 35*, 27-58.

Hennessy, S. (1993). Situated cognition and cognitive apprenticeship: Implications for classroom learning. *Studies in Science Education, 22*, 1-41.

Hennessy, S., & Murphy, P. (1999). The potential for collaborative problem solving in design and technology. *International Journal of Technology and Design Education, 9*, 1-36.

Hume, A., & Coll, R.K. (2005, July). *The impact of a new assessment regime on science learning*. Paper presented at the 34th Annual Conference of the Australasian Science Education Research Association. Hamilton, New Zealand.

Jenkins, E. (1990). Scientific literacy and school science education. *School Science Review, 71*, 256.

Jones, A. (1997). Recent research in learning technological concepts and processes. *International Journal of Technology Design Education, 7*(1-2), 83-96.

Laugksch, R.C. (2000). Scientific literacy: A conceptual overview. *Science Education, 84*, 71-94.

Lave, J. (1991). Situated learning in communities of practice. In L.B. Resnick, J.M. Levine & S.D. Teasly (Eds.), *Perspectives on socially shared cognition* (pp. 63-82). Washington, DC: American Psychological Association.

McGee, C. (1997). *Teachers and curriculum-decision-making*. Palmerston North, New Zealand: Dunmore.

Miller, J. (2000). The development of civic scientific literacy in the United States. In D. Kumar & D. Chubin (Eds.), *Science, technology and society: A source book on research and practice* (pp. 21-47). Dordrecht: Kluwer.

Pilot, A., Bulte, M.W. (2006). What do you 'need to know'? Context-based education. *International Journal of Science Education*, *28*(9), 953-956.

Preece, F.W., & Baxter, J.H. (2000). Scepticism and gullibility: The superstitious and pseudo-scientific beliefs of secondary school students. *International Journal of Science Education*, *22*(11), 1147-1156.

Sade, D., & Coll, R.K. (2003). Technology and technology education: Views of some Solomon Island primary teachers and curriculum development officers. *International Journal of Science and Mathematics Education*, *1*(1), 87-114.

Shamos, M. (1996). The myth of scientific literacy. *Liberal Education*, *82*(3), 44-49.

Vygotsky, L. (1986). *Thought and language*. Translated by A. Kozulin. Cambridge, MA: MIT Press.

Wertsch, J.V. (1991). A sociocultural approach to socially shared cognition. In L.B. Resnick, J.M. Levine & S.D. Teasly (Eds.), *Perspectives on socially shared cognition* (pp. 85-100). Washington, DC: American Psychological Association.

World Bank. (1996). *Pacific island economies: Building a resilient economic base for the twenty-first century*. Washington, DC: World Bank.

World Bank. (2006). *Education for the knowledge economy*. Retrieved 10 November 2006, from http://web.worldbank.org/WBSITE/EXTERNAL/TOPICS/ EXTEDUCATION/,,contentMDK:20161496~menuPK:540092~pagePK:148956~piPK:216618~theSitePK:282386,00.html

Yates, G.C.R., & Chandler, M. (2000). Where have all the skeptics gone? Patterns of new age beliefs and anti-scientific attitudes in pre-service primary teachers. *Research in Science Education*, *30*(4), 377-397.

AFFILIATIONS

Richard K. Coll
Centre for Science & Technology Education Research
University of Waikato
New Zealand

Neil Taylor
School of Education
University of New England
Australia

PACIFIC

ATWEH, BERNARDO AND BALAGTAS

CAPACITY BUILDING AS A COLLABORATIVE MODEL IN INTERNATIONAL DEVELOPMENT PROJECTS: LESSONS FOR THE PHILIPPINES

INTRODUCTION

The role of professional development for the successful implementation of educational reform has been highlighted by different authors. For example, Hargreaves (1994) asserts that "significant change in curriculum, assessment or any other domain is unlikely to be successful unless serious attention is also paid to teacher development and the principles of professional judgment and discretion contained within it" (p. 242). Similarly, Sprinthall, Reiman and Thies-Sprinthall (1996) affirm that the "massive failures of the [many] national curriculum projects of the 1960s" (p. 666) have raised interest in (re)investigating and (re)theorizing the teachers' role in educational change. Arguably, these issues are perhaps more problematic in educational reforms in less affluent countries where research knowledge and resources are often very limited. Educational reform in many less affluent countries has modeled itself after reform movements in the USA (Standards) and UK (outcome based education). Further, the increasing international aid projects designed to assist less industrialized countries in implementing educational reform has the potential of escalating the uncritical transfer of priorities, agendas and practices from industrialized to less industrialized countries.

In science education, the past 100 years have witnessed a rapid increase of international contacts and collaborations between academics around the globe in the form of conferences, publications, courses for international students, exchanges of curricula and professional development programs, a multitude of cross-national comparative studies and, of course, consultancy projects. At the same time that the trends of globalization might be providing increasing opportunities for our academic work, consultancies and publications, it is also leading to an ever-increasing gap between the haves and have-nots, between the rich and the poor. This has created a dilemma for socially responsible academics that face a tension between their willingness to assist their counterparts from developing nations and their care not to indirectly impose their priorities, assumptions and values on them.

Richard K. Coll, Neil Taylor (eds.), Science Education in Context: An International Examination of the Influence of Context on Science Curricula Development and Implementation, 3–15.

This chapter discusses a model of collaborative capacity building as a model for professional development of science educators in a less industrialized country, the Philippines, towards the implementation of the new Teacher Education Curriculum in their pre-service teacher training courses. It also discusses issues on the use of a capacity building model as a means of international collaboration between donor and recipient countries in foreign educational aid projects with particular attention to minimizing the uncritical transfer of knowledge and values. First, we will discuss the context of the project. In the following section we will discuss the concept of capacity building as understood in the design and implementation of the project. The chapter concludes with some lessons that we have learnt about capacity building and collaboration in international projects.

The Context of the Project

In 2004, the Commission on Higher Education (CHED) in the Philippines issued a new set of guidelines on what ought to be taught in the teacher education curriculum. Bernardo (in press) describes three important changes in the new teacher education curriculum – including the science education component. First, the curriculum incorporates an enhanced set of General Education requirements that could allow teacher education students to develop their language and communication, quantitative, and other cognitive competencies required for higher learning. Second, the Professional Education courses are designed to be integrated. The theories, principles and methods of teaching should be taught in a manner so that students explore how they relate to problems of teaching and learning in a real setting where there are diverse types of students, environments, and curricular objectives. This feature of the curriculum expands the experiential component of teacher education which begins in the second year through field observations of actual classes and culminates in a practice teaching requirement in the last year. Third, the Content Specialization courses for both the elementary and secondary education programs are drastically increased in the new curriculum. There is no specialization required for elementary teachers but they have to take a total of 57 units, in the same way as secondary teacher trainees with a major area of study, and to possess adequate content knowledge in all subject areas in the basic education curriculum.

Cognizant of the new features of the new curriculum that teachers need to implement, CHED linked with an Australian university[i] to help design a capacity building project that would enable the teacher educators to develop the skills required for the implementation of the new curriculum. The project was conceived to be completed in five stages: (1) Networking electronically between the Facilitators[ii]; (2) a one-week Designing of the Capacity Building Workshop between five Filipino and two Australian Facilitators held in Brisbane in September 2005; (3) a two-week Capacity Building Program for 46 faculty members of tertiary education institutions and universities from 15 regions in the Philippines held in Manila October 2005; (4) the implementation of Professional

Development Activities in the various regions using action research conducted by the 45 participants between November 2005 and January 2006; and (5) a Writing Workshop in Manila held in March 2006, involving participants from the 15 regions for the writing and editing of a book (Atweh, Balagtas, Bernardo, Ferido & Macpherson, in press) as the final documentation of the project.

THE PROJECT AS A CAPACITY BUILDING ACTIVITY

Capacity building, a term of increasing use in educational literature during the last few decades, remains a contested term because it is used in many contexts and means different things to different people. Often it is used to justify and argue for different policies and outcomes. Seddon (2001) demonstrates how the construct has been used by a wide cross-section of contexts all the way from writers within the transformative to the neo-conservative economic rationalist paradigms. This ambiguity about the term is not without its price. In a book published by Oxfam (Eade, 1997), the author quotes a report from the Community Development Resource Association of South Africa that states:

> Our lack of adequate theory of capacity building reduces our own capacity to engage in the practice. We lack the theory because we are not thinking through what we see before us. And we are avoiding thinking things through because to face the obvious will be to radically transform our practice. We are avoiding genuine accountability. (p. 1)

It is beyond the scope of this chapter to provide a comprehensive critique of the capacity building discourses. However, we will use some of the principles often associated with capacity building to reflect on effective and sustainable professional development models in educational settings in less industrialized nations.

Capacity Building as Professional Development

Here, we will use and expand a model suggested by Harris and Lambert (2003) to discuss capacity building as a particular type of professional development. In their book, *Building Leadership Capacity for School Improvement*, the authors present a two-dimensional model to compare and reflect upon the different professional development activities. The first dimension relates to what they call the *breadth of involvement* in the programs. At one extreme lie the professional development activities that target individuals, or multiple individuals, who work alone within the institution, for example, an individual or group of teachers or principals. At the other extreme are programs that involve the whole school community including parents, students, teachers, administrators and possibly education departments. The second dimension discussed by Harris and Lambert relates to the *skilful involvement* of the program. On one extreme are the professional development activities that develop low level skills, such as writing lesson plans according to set specifications, writing of outcomes/objectives, use of computer software or developing behavioral management techniques. At the other extreme are the professional development activities that aim at improving learning outcomes of students or reducing student disengagement from education, or adapting the curriculum to take into account the needs of the disadvantaged, such as students from low socioeconomic families or indigenous backgrounds.

Here we will add another dimension to this model to differentiate between usual professional development and capacity building. This dimension relates to the *level of involvement* by the participants in the activity. Many professional development activities are designed with minimal teacher input. In some hierarchal organizations, teachers are directed to attend scheduled seminars that may or may not develop skills and learnings that they need, or perceive that they need. At other times, teachers, individually or collectively, can identify needs or problems in their own practice and either collaborate with each other to satisfy those needs or seek external support to assist in the satisfaction of these needs. Likewise, they may have full control on issues of timing, duration, depth and form of their professional development activities.

In this context we will use capacity building to refer to professional development activities that: a) are collaborative in a sense they involve a variety of stakeholders in a practice, b) aim to develop high order abilities and skills to assure independence and the ability to learn, and c) are participatory in the sense that the participants can take an active role in their own development.

The discussions during Stage 2 of the project among the facilitators revealed that the model of professional development used by many previous international collaborative projects in the Philippines was the cascading model of training (Bernardo, 1999; Talisayon et al., 1998). While the details may vary from one project to another, the basic design was training a small group of teachers at the national or central level, using modules, usually prepared by external 'experts', with the expectation that they will go back to their regions and schools and train selected others in the use of similar modules, and so on.

Such models of professional development are often referred to as 'training of trainers'. In spite of their wide usage and perceived efficiency, there is a significant amount of evidence in the international literature and research in education that casts doubt on the long term efficacy of such approaches for professional development. For example, Nguyen (1992) evaluated one such program for mathematics and science teachers and found that less than half the program goals for teacher development were met, and it was the higher level goals such as the development of higher order thinking skills and integration of values that were often not addressed. Bernardo, Clemeña and Prudente (2000) surveyed Filipino teachers and concluded that the general sentiment was that the learning from these types of professional development activities was not very relevant to their needs. Among the many findings of the survey, the teachers reported that: a) only a handful of teachers from their schools get to participate in such professional development programs, b) they are not usually consulted about the focus or content of such professional development programs, and c) they usually attend only when they are required by their superiors to do so. This was among other reports suggesting that the cascading professional development programs do not embody the important features of capacity building.

On the other hand, there are reports of international collaborative projects on teacher development in the Philippines that are perceived to be successful. Balagtas and Ferido (in press) reported that a collaborative project of the Philippine and Australian governments in 1996 called Project in Basic Education (PROBE) was perceived to be an effective model of professional development. PROBE involved: a) a large number of teachers and administrators, b) a very intensive and interactive program that was focused on developing critical and creative thinking skills, and c) allowing teachers to create various curriculum resource materials. The participants in this program were able to continue building on their professional knowledge and skills even after the program was completed, and to share their new knowledge and skills with other teachers and teacher educators. Similar positive results are found in locally initiated professional development programs that involve teachers in long-term intensive programs that address the teachers' fundamental beliefs and attitudes about the nature of teaching and learning (e.g., Pattuinan, 1994; Reteracion, 1996). Such programs succeeded in developing the teachers' higher level thinking and problem-solving skills, higher level knowledge of the content area, and instructional flexibility and creativity. It is worth noting that these programs seem to embody the key dimensions of capacity building *of* professional development activities as discussed here.

The 2005 collaborative project reported in this chapter was first conceptualized by its developers as a cascading model of professional development that was typically implemented in international collaborative professional development programs. However, we, the authors of this chapter, together with the other facilitators of the program, saw the need to move away from this model, and to move towards a model that would embody the key dimensions of capacity building as professional development. To express the importance of breadth of involvement, skilful involvement and level of involvement of the participants in the capacity building program, we adopted the metaphor of *ripples* to describe our conceptualization of our work in this project. Ripples unsettle the *status quo* to bring much needed nourishment and air to the depth of our day to day lives and practices. They might be caused by an input from outside; but they also start by creatures from the depth struggling for fresh air. Further, they take different shapes depending on our local contexts. Unlike a waterfall that cascades from one source, ripples have different sources and a wider reach. Finally, ripples interact with each other to form larger ripples and, under the right conditions, may even become waves. Using this metaphor avoided the usual dichotomy between the *top down* and *bottom up* models of professional development which we saw as problematic.

Our thinking behind this capacity building project was that it should become a vehicle for starting ripples of change in the different regions of the country. We are conscious that curriculum reform and professional development does not happen overnight. It does not happen simply as a result of a change in policy from above. We envisaged our Capacity Building Workshop as a small contribution from the outside that undoubtedly will be reduced in its effectiveness over time. What we intended was to build the capacities of a group of educators to start their own ripples and share their learnings with others.

Capacity Building and the Professionalization of Academics

As Atweh (2003) argued, in many Western countries teachers do not reach a full professional status. Arguably, this situation applies to teacher educators in many less industrialized countries as well. Noddings (1992) notes that there are "bright spots, however, in this bleak picture. Reform movements are pressing for changes designed to professionalize teaching" (p. 206). What is of interest to us here are Noddings' identification of several components of a profession, three of which are directly relevant to the discussion in this chapter.

First, professionals are acknowledged as having special knowledge in their field of expertise. This knowledge is essential for professionals to be effective in the provision of their service and in improving their practice. Evidence from the UK points to the fact that teacher change is most effective when teachers have explicit knowledge about their context and can articulate their personal theories on students' learning (Askew, Brown, Rhodes, Johnson & Wiliam, 1997). This theoretical knowledge is an essential criterion that is often used to differentiate between a technician and a professional. In this project, we understood capacity building to include both theoretical knowledge and practical skills. The various activities in the Capacity Building Workshop included some theoretical presentations by the facilitators and some invited speakers as well as practical activities on writing components of the syllabus and project plans. All sessions attempted to demonstrate praxis between theory and practice.

Second, professionals work within an atmosphere of collegiality. They work with others to produce professional knowledge and to improve their practice and the conditions of their practice. There is much evidence from research on school change in Australia that shows that more effective changes result when all the teachers in a school cooperate to change their practice in their individual classes (Baird & Northfield, 1992). In designing this project we employed the model of collaborative working teams within the different regions designing professional development activities that took the local environment and resources into consideration. Such collaborations avoided the shortcomings of 'one size fits all' approaches to curriculum reform and professional development.

Third, professionals enjoy a degree of autonomy. There is some evidence that indicates teachers' sense of autonomy is correlated with their enthusiasm about their work and even their performance (Noddings, 1992). It can also be argued that changes can be more permanent and effective if initiated and controlled by the teachers themselves. Teachers do not resist reform when they choose aspects of their practice that need to be improved, and they control the direction of change. However, it is also true that total autonomy is not possible or desirable. There are other people who have an interest and a right to be involved in educational decisions. They too must be involved in educational change. In the context of this project, the development of the regional professional development activities has contributed to the adoption of the reform agenda by the participating teacher educators.

Capacity Building as Collaborative

Conscious of Hargreaves (1994) warning about the uncritical usage of the term collaboration (that at times can be nominal and coercive), here we document aspects of this project that show the genuine collaborative nature of the design and implementation. At different stages of its design and implementation, this project involved collaboration between different groups of people at different levels of authority. The foundation for this collaboration was the fact that the project was directly related to the priorities and agenda of both parties. Prior to this project, the CHED had identified professional development of teacher education staff for the new curricula as a high priority and had already allocated funds towards such activities in the different regions in the country.

The first instance of collaboration was the development of the funding proposal that identified the overall aims and structure of the project. This was a collaborative effort between staff from CHED responsible for the implementation of the new curriculum, and professional and academic personnel from the collaborating university. Due to distance between the partners, email and teleconferencing were utilized to negotiate the proposal. An important component of such collaboration was that the different parties had a chance to share their visions, principles and experiences in previous projects as well as negotiate details such as timelines, scope of the project, and the required resources.

Second, realizing the limitation of collaboration at a distance, the design of the project also included an intensive face-to-face collaborative planning week as the first component of the funded project (Stage 2). The main aim of this stage was to design the Capacity Building Workshop to be conducted in Stage 3. The plans for the planning week began with the facilitators exchanging information about themselves and for the purpose of establishing a good working relationship among the group members. Establishing sound personal relationships was important for developing the working relationship that the group was able to develop. Further, intensive collaboration was conducted between the participants in developing the agenda for the face-to-face meeting in Brisbane. The Australian facilitators drew up the first draft of the timetable for the five days as well as their expectations of the outcomes. The Filipino facilitators had slightly different expectations of the five days and of the mode of operation. Based on the needs of the various participants, the timetable of Stage 2 was revised several times prior to the actual meeting. Each group was free to express their concerns and listened to each others views. At times, a compromise was difficult to achieve, however, the sprit of good will, trust, and respect for the knowledge base of each participant resulted in a commonly agreeable agenda and fruitful discussion.

Third, the spirit of collaboration has continued through the conduct of the Capacity Building Workshop in the Philippines (Stage 3 above). All seven facilitators took part in the activities and met regularly to make day-to-day decisions on the conduct of the sessions. Symbolic of this collaborative contribution, we adopted the practice in all project documents of avoiding distinctions between the Australian and Filipino facilitators and listed the names alphabetically. However, this collaborative contribution was not always equal in time. The majority of sessions were led by the Australian facilitators. This was a reflection of the funding rules of the granting organization determining the amount of funding available to support their involvement in the project in contrast to the limited support that the Filipino facilitators have received. Unfortunately, this is not an isolated instance of the unequal resources available to the different international collaborative partners.

Other activities in the project were also conducted collaboratively between the facilitators and the others involved in the implementation of their regional professional development activities (Stage 4). The three representatives from each region had drawn up plans for their professional development activities that involved working with the local CHED offices, their own institutional administrators, regional Philippine Association for Teacher Educators (PAFTE) groups, and other relevant interested parties. Often they have crossed the boundaries of traditional demarcations, spheres of interest and responsibilities, and were able to negotiate their plans to suit different stakeholders' agendas.

The last aspect of collaboration was the practice of critical friends used at different times during the project. Regional groups' output from the workshop activities and their plans for the professional development activities were subjected to critical comments, not from the facilitators, but from their peers in other groups. Carefully planned introduction of this practice has assured the development of trust between the participants and sense of openness to being assisted by others where advice is taken as a form of critique of ideas rather than a personal assault. This feedback from critical friends was cultivated in a writing workshop (Stage 5) where each draft chapter was refereed by at least two or three other authors and facilitators, leading to an improved quality of chapters.

LEARNINGS FROM THE PROJECT

Learning about Capacity Building

From our experience as facilitators in this project, we make the following observations about capacity building as means of professional development and international collaboration. First, a primary concern for any country in developing professional development programs to implement educational change, is how to assure the widest benefit at a minimal cost. Rightly so, we may add. However, in designing this project we raise two other considerations. Here we argue that the criteria of depth of expertise and long term benefit need to be taken into account in planning for change. The old Chinese proverb is appropriate: "Give a man a fish, and you feed him for a day. Teach a man [sic] to fish, and you feed him for life". In this project we aimed not only to develop knowledge about the new curriculum and how to implement it as a recipe, but to develop the expertise of at least 46 participants to design professional development activities on their own as well, as conduct research and engage in academic publications. Arguably, such benefits not only meet the immediate needs of educators in the Philippines to implement the current curriculum but also develop long term expertise towards autonomy and self-reliance. We should note here that the criteria of depth of expertise and sustainable benefit are not necessarily in conflict with the criteria of efficiency, but should be taken as part of its definition.

Second, as Atweh (2003) argues, not only professional development of teachers is important for effective curriculum reform, but this professional development must also cater for the different teacher's needs. These available professional development activities should match the expectations of the educational system from the teachers. If teachers are expected to be professionals and accountable, then developing professional development activities to strengthen their professionalism must be in place. To apply the theory of Habermas, in order for teachers to conduct their practice, they need technical knowledge (e.g., they need a huge repertoire of teaching and classroom management methods, knowledge of content, use of materials, resources and technologies, evaluation and assessment techniques) and practical knowledge (e.g., teachers need to understand the social context of the school and of their students in order to cater for their particular needs and to develop their teaching and learning theories in order to reflect on their practice). Since technical knowledge is open to "the tyranny of regulations" and practical knowledge is open to "deceit and false consciousness" (Grundy, 1987, p. 17), a different type of knowledge is required for decision-making. Grundy argues that "the emancipatory interest gives rise to autonomous, responsible action based upon prudent decisions informed by a certain kind of knowledge" (p. 18). While control and understanding are the motivating factors of the previous knowledge-constituted interests, empowerment, that is, "the ability of individuals and groups to take control of their own lives in autonomous and responsible ways" (p. 19), is the motivation for emancipatory knowledge. Professional development activities that are based on capacity building are more attuned with these latter needs and provide teachers with a sense of autonomy and a sense of agency and control over their work.

Lastly, academics around the world are under increasing pressure to perform and excel in all areas of administration, teaching, research and service to the profession and community. Here we argue that combining two or more of such agendas is prudent, as well as beneficial to the individual academic and the educational system. In this project, by combining action research and professional development, we have attempted to have different impacts on the life of Filipino academics.

Learning about International Collaborations

From our experience, we make the following observations about collaboration. First, collaboration does take time. Several emails and lengthy meetings were spent to plan and implement this project. In the busy schedule of all participants and the large components of the project, at times it appeared that it might have been more efficient for one coordinator to make decisions. At times, this was done. However, participants should be aware of the potential conflict between efficiency and developing a sense of ownership by all participants. Even at times where efficiency in time is essential, participants should allow for dissent with and contestation of each other views and decisions. Only through dedication to the aims of the project and respect for each others' views can collaboration be successful.

Second, collaboration does not necessary imply an equal amount or the same type of contribution. As Grundy (1998) reminds us, parity of esteem should be the guiding principle by which collaboration should be judged. In developing projects of this scale, each participant has their own expertise and knowledge, which is often complementary to the others' contributions. While some overseas facilitators might have had more access to literature about some theories of professional development, the local facilitators' knowledge of the local context is invaluable to the success of such a project. The process of collaboration among the facilitators could be described as a mutual Vygotskian scaffolding process, where each facilitator is a *more knowledgeable other*, helping each other facilitator to attain higher levels of understanding regarding specific aspects of the program, but drawing from different types and forms of knowledge.

Third, collaboration involves risk-taking. Both groups of facilitators were dealing with unknown territory in this project. The project employed different models of professional development familiar in the Filipino context. Similarly, the context of higher education in the Philippines was at best partially understood by the Australian facilitators. We were conscious that we were sailing in uncharted waters with little experience in this particular context to guide us. However, the sense of community, the trust in, respect for, and faith in each other was our only guarantee for success. In the risky undertaking, each facilitator had to trust the other facilitators' knowledge, experience, and whatever else they had to bring into the collaborative risk. The Filipino facilitators had to trust the Australian facilitators' experience and success with similar capacity building activities in Australia; while the Australian facilitators had to trust the Filipino facilitators' understanding and efficacy in teacher development activities in the Philippine context. At the same time, each facilitator had to trust his or her one respective capacity to see the project through. This spirit of trust, respect and faith in each other is a good alternative model to conventional 'transfer of technology' frameworks that characterize most international development projects from more developed countries to less developed ones.

Lastly, collaboration challenges the traditional boundaries and hierarchies that systems may develop and impose. Curriculum reform is best achieved through collaborative endeavors between the final implementers of change (i.e., the lecturers themselves) and their administrators and people responsible for development of policy and those who control the resources. One of the sources of success in this project has been the ability of the participants to negotiate their plans with different stakeholders. For example, we saw the tension that arises from the emphasis on emancipatory interests, as the facilitators had to constantly explain, convince, and negotiate with other project stakeholders about the adoption of capacity building as our professional development model where, as discussed above, the original proposal developed as training of trainers. The most difficult feature of capacity building was to convince these stakeholders of related features that did not fit the established institutional practices. The other participants who conducted the 'rippling' activities in their respective regions likewise encountered similar tensions and difficulties when they attempted to do things differently from the established practices. Negotiation was a key feature of the implementation of the capacity building activities. At times such negotiation was lengthy and frustrating, and it was more successful in some cases than in others. However, the lessons learnt were undoubtedly profound. One lesson shared by the different participants in this project was that if teachers are expected to think differently about their role in curriculum reform, they have to be thought about differently by the educational system. Teachers are the most important link in reforming education.

CONCLUSIONS

In this chapter we discussed our learning from a professional development activity for the implementation of the new teacher education course in the Philippines, designed and implemented in collaboration between local and international stakeholders. In significant ways, the project was a new experience for many of the participants. First, it adopted the model of capacity building in contrast to train the trainer models more familiar in the Philippines. Second, it was designed to reflect genuine and critical collaboration between the different participants. In this chapter, we reflected upon our expectations of and experiences in both features.

In terms of capacity building, we argue that professional development activities in international collaborative projects in science education should aim at the development of higher order capacities that enable the participants to be agents of reform in contrast to merely developing skills to implement programs developed by others. In this project the participating academics have developed capacities and theoretical knowledge not only to understand the basics tenets of the new curriculum but also to design, conduct and document their own projects for the implementation of the reform. Further, capacity building challenges not only the way that academics in less industrialized countries see themselves, but also how they are seen by the system. At times these new capacities imply a certain level of conflict between old and new thinking that needs to be carefully negotiated.

Further, our experience demonstrated the importance of being critical about claims of collaboration between different participants with different experiences and aspirations. This holds true particularly in the area of international collaborations in science education. The different participants should be aware of their differing agendas and capacities to contribute to the project; they should develop a sense of trust and take risks to negotiate their differences rather than merely accommodate those differences.

NOTES

[i] Project jointly funded by Commission of Higher Education of the Philippines and AusAid in Australia and managed by the Queensland University of Technology.

[ii] Bill Atweh, Marilyn Balagtas, Allan Bernardo, Merlea Cabalquinto, Marlene Ferido, Ian Macpherson, and June Salana.

REFERENCES AND BIBLIOGRAPHY

Askew, M., Brown, M., Rhodes, V., Johnson, D., & Wiliam, D. (1997). *Effective teachers of numeracy: Report of a study carried out by the teacher training authority 1995-6.* London: School of Education, King's College.

Atweh, B. (2003). Understanding for changing and changing for understanding: Praxis between practice and theory through action research in mathematics education. In P. Valero & R. Zevenbergen (Eds.), *Researching the socio-political dimensions of mathematics education: Issues of power in theory and methodology* (pp. 187-206). Dordrecht: Kluwer.

Atweh, B., Balagtas, M.U., Bernardo, A.B.I., Ferido, M.B., & Macpherson, I. (Eds.). (in press). *Ripples of change: A journey of teacher education reform in the Philippines.* Pasig City, the Philippines: Commission on Higher Education.

Baird, J., & Northfield, J. (Eds.). (1992). *Learning from the PEEL experience.* Melbourne: Monash University Press.

Balagtas, M.U., & Ferido, M.B. (in press). Building teacher educators' capacity in student assessment: Where have we been and where are we heading? In B. Atweh, M.U. Balagtas, A.B.I Bernardo, M.B. Ferido & I. Macpherson, I. (Eds.). *Ripples of change: A journey of teacher education reform in the Philippines.* Pasig City, the Philippines: Commission on Higher Education.

Bernardo, A.B.I. (1999). The striving and the struggle for teacher development: The contexts, issues, trends, and opportunities in in-service teacher training in the Philippines. In Y. Tabata & L. Griek (Eds.), *Ensuring opportunities for the professional development of teachers* (pp. 141-158). Higashi-Hiroshima City, Japan: Hiroshima University UNESCO-APEID Associated Centre.

Bernardo, A.B.I. (in press). Contrasting the old and the new Philippine pre-service teacher education curriculum. In B. Atweh, M.U. Balagtas, A.B.I. Bernardo, M.B. Ferido & I. Macpherson (Eds.), *Ripples of change: A journey of teacher education reform in the Philippines.* Pasig City, The Philippines: Commission on Higher Education.

Bernardo, A.B.I., Clemeña. M.S., & Prudente, M.S. (2000). *The contexts and practices of science and mathematics education in the Philippines: Foundations of responsive science and mathematics teacher education programs.* Manila: JICA & Lasallian Institute for Development and Educational Research.

Eade, D. (1997). Capacity-building: An approach to people-centred development. Oxford, UK: Oxfam Publications.

Grundy, S. (1987). *Curriculum: Product or praxis*. London: Falmer.

Grundy, S. (1998). Research partnerships: Principles and possibilities. In B. Atweh, S. Kemmis & P. Weeks (Eds.), *Action research in practice: Partnerships for social justice in education* (pp. 37-46). London: Routledge.

Hargreaves, A. (1994). *Changing teachers, changing times: Teachers' work and culture in the postmodern age*. London: Cassell.

Harris, A., & Lambert, L. (2003). *Building leadership capacity for school improvement*. London: Open University Press.

Nguyen, E. (1992). *An evaluation of the SEDP mass training program for Mathematics III in selected regions of Luzon*. Unpublished PhD Dissertation, University of the Philippines, College of Education, Quezon City, The Philippines.

Noddings, N. (1992). Professionalization and mathematics teaching. In D.A. Grouws (Ed.), *Handbook of research on mathematics teaching and learning* (pp. 197-208). New York: Macmillan and National Council for Teachers of Mathematics.

Pattuinan, D. (1994). *Effects of a constructivist professional development program on the problem solving skills of secondary school mathematics teachers.* Unpublished PhD Dissertation, University of the Philippines, College of Education, Quezon City, the Philippines.

Reteracion, N. (1996). *Effects of the constructivist model of learning on teachers' conceptions of science, teaching, learning and classroom practices*. Unpublished PhD Dissertation, University of the Philippines, College of Education, Quezon City, the Philippines.

Seddon, T. (2001). Exploring capacity-building: From functionalist to political analysis. *Australian and New Zealand Journal of Vocational Education*, *9*(2), 61-86.

Sprinthall, N.A., Reiman, A.J., & Thies-Sprinthall, L. (1996). Teacher professional development. In J. Sikula (Ed.), *Handbook of research on teacher education* (pp. 666-703). New York: Macmillan.

Talisayon, V., Ulep, S., Cajilig, N., De Guzman, F., Fajardo, A., Ferido, M., Gallos, F., Galvez, E., Magno, M., Mendoza, A., Reyes, R., & Tan, M. (1998). Materials and methods in basic education and in-service teacher training in science and mathematics (1960-1998). In E.B. Ogena & F.G. Brawner (Eds.), *Science education in the Philippines: Challenges for development* (pp. 109-150). Taguig, Metro Manila: NAST, SEI, UP-CIDS.

AFFILIATIONS

Bill Atweh
Science & Mathematics Education Centre
Curtin University of Technology
Australia

Allan B.I. Bernardo
College of Education
De la Salle University-Manila
The Philippines

Marilyn Balagtas
College of Education
Philippines Normal University
The Philippines

ROMBO

SCIENCE EDUCATION AND CURRICULUM REFORM IN PAPUA NEW GUINEA

INTRODUCTION

The Papua New Guinea (PNG) education system considers science as a core subject to be taught in schools, particularly in the primary and secondary schools (Department of Education [DoE], 2004). While there are debates pertaining to the development and dissemination of relevant science curriculum from the PNG perspective, Wilson (n.d.) strongly argues that the teaching of science is difficult for teachers in many PNG schools, observing that teachers "blame lack of equipment, materials and their own limited background in science for this" (p. iii). Yet, grappling with the basics of science and how to best disseminate the theories and philosophies pertaining to science is vital, particularly for countries of the developing world such as PNG (DoE, 2004). Hence, the Government of PNG has embarked on a major curriculum reform, in which, both primary and secondary science is a major consideration.

Education reform in PNG began in 1993 with the support of donor agencies such as the World Bank, IMF and Aus-AID, among others. This was driven by a perceived need to improve the country's economy through the development of an educated workforce, and to raise the literacy rate, since PNG has the lowest literacy rate in the Asia-Pacific region (Webster, 2005). In collaboration with aid agencies, the PNG Department of Education has been proactive in improving the standard of education through the development of a curriculum appropriate and relevant to meet the contextual needs of Papua New Guineans in a technologically developing world (DoE, 2003a). It is considered by the DoE that coordinated changes under the education reforms represent a paradigm shift in the nature of the primary and secondary science education curriculum – from a rigid, behaviorist, teacher-centered approach to a curriculum based on the ecological context of the learner. This shift is underpinned by both curriculum changes and teacher education initiatives under the auspices of the Department with the support of aid agencies (DoE, 2003a). Primary and secondary teachers have been progressively introduced to changes in the science curriculum in the hope of meeting the Government's policy and education reform agenda. These changes, including their historical and theoretical base to the development of the science curriculum, its implementation, and what is intended to achieve for the future of PNG are explored in this chapter.

Richard K. Coll, Neil Taylor (eds.), Science Education in Context: An International Examination of the Influence of Context on Science Curricula Development and Implementation, 17–31.

RELEVANCE AND BASIS OF THE CURRICULUM REFORM IN PAPUA NEW GUINEA: AN ECOLOGICAL SYSTEMS THEORY PERSPECTIVE

The theoretical framework that supported the child-centered and/or outcomes-based curriculum reform in PNG is Bronfenbrenner's Ecological Systems Theory (Bronfenbrenner, 1979, 1989, 1993). In defining his theory, Bronfenbrenner (1979) stated that:

> The ecology of human development involves the scientific study of the progressive, mutual accommodation between an active, growing human being and the changing properties of the immediate settings in which the developing person lives, as this process is affected by relations between these settings, and by the larger contexts in which the settings are embedded. (p. 21)

Viewing the theory from a developmental psychology perspective, a child's development is influenced by the surrounding environment and the social contexts the child interacts with. The theory suggests that human behavior emanates from the function of the human person, and the environment in which the person exists (Damon & Lerner, 2006, 1997). For instance, the way children in Papua New Guinea are viewed, treated and included in the learning process in the classroom contexts play a major part in their learning achievements. Bronfenbrenner points out that the vital component of this theory is the degree of interaction that is evident within a layer-like structure (Ceci, Baker-Sennett & Bronfenbrenner, 1994). The interactions can be bidirectional or multidirectional in that the individual person and the specific context may influence each other.

From an ecological systems perspective, the environment is viewed as a configuration of structures from a child's development stages to maturity. Thomas (2005) notes that the ecological structures or layers are organized hierarchically with increasing complexity as the child develops and moves from the immediate context like a family home, school or village setting to the wider society. Each of these structures imposes its own degree of interaction and each one is relatively dependent upon the previous state of functioning of the individual in the social context (Bronfenbrenner, 1979, 1989, 1993; Thomas, 2005).

Bronfenbrenner conceptualized ecological systems theory in terms of successive layers that influence a developing person. These nested structures or successive layers are called the *microsystem*, the *mesosystem*, the *exosystem* and *macrosystem*. From the perspective of PNG education curriculum reform, the context is the school culture which is the microsystem, and the children in the school are seen as inseparable from this context. This is because 'school culture' or 'the way things are done here' (Prosser & Warburton, 1999) influences how children are viewed and treated.

Hence, in summary Bronfenbrenner proposed that human development is influenced by factors operating at different 'systems levels' within a broad ecological structure. These different levels exert reciprocal influences on each other (Baltes, Lindenberger & Staudinger, 2006; Magnusson & Stattin, 2006). For instance, negative attitudes toward science exhibited by teachers when teaching can have repercussions for the achievement of the outcomes-based education in PNG schools. On the other hand, the child's state or condition can also influence the perceptions of a child held by teachers (Bronfenbrenner & Morris, 2006).

Microsystem Level

The first system level, the microsystem contains all factors within a child's immediate environment (Bronfenbrenner, 1979, 1989, 1993). This is the innermost level and is viewed as the context in which a child interacts with the immediate environment such as a home, school, and peer group which ultimately fosters human development. At the microsystem level, an analysis of Bronfenbrenner's ecological theory identifies activities, roles, and interrelationships as three fundamental factors that promote a child's development at the immediate context. Thomas (2005) defines these factors:

> The term *activities* refers to what people are doing. *Roles* are the actions expected of people holding a position in society, such as positions of parent, infant, sibling, teacher, friend, coach, or the like. *Interpersonal* relations are the ways people treat one another, as shown by what they say and do as a result of them being together [original emphasis]. (p. 351)

Hence, the science curriculum in PNG had to be embedded in a child-entered and/or outcomes-oriented philosophy, which ultimately must reflect the child's learning and assimilation capacity. In other words Bronfenbrenner's conceptualization of the microsystem also underpins the individualization of context-based instructions and practices like the science curriculum differentiation in PNG classrooms. This is to meet an individual child's specific learning needs and social relationships with peers and capable adults: "In effect, how the child grows up is also strongly affected by what is said or done to the child - or in the child's presence - by parents, siblings, other relatives, teachers, coaches, club leaders, and the like" (Thomas, 2005, p. 352).

Mesosystem Level

The mesosystem encompasses interrelations of two or more settings in which the developing person actively participates. For example, in the case of a child the mesosystem consists of relations between home, school, neighborhood and peer groups (Bronfenbrenner, 1979; Bronfenbrenner & Ceci, 2005; Bronfenbrenner &

Morris, 2006). According to the DoE (2005b), family members' beliefs about certain curriculum outcomes, and the family's relationships with the school, may affect the learning outcomes. Similarly, how a child relates to his or her peers in the classroom setting may affect relationships outside the classroom situation (Kugelmass, 2004). For example, in inclusive settings, negative attitudes exhibited in the classroom situation during the science curriculum learning process can affect social relations such as playing games or experiential learning.

Exosystem Level

The exosystem level does not involve the developing person as an active participant, but is a structure in which events occur that affect, or are affected by, what is happening in the setting containing the developing person (Bronfenbrenner, 1979). The service delivery agency responsible for child support programs such as teacher aid is an example of an exosystem setting. How the agency is organized can affect the acquisition of intended curriculum learning outcomes. For instance, in PNG, there is an agency called the Callan Services Resource Center that plays a role in the dissemination and implementation of an inclusive curriculum, in the hope of giving equal learning opportunities to all children. That means the linkages and the interrelationships that exist at the exosystem level can influence effective teaching practices. The degree of interaction at the exosystem level with teachers responsible for implementing the intended curriculum and associated policies, may impact upon the other organizational levels (Peck, Hayden, Wandschneider, Peterson & Richarz, 1989). These could be considered as the school's features and practices which in turn can affect the learning experiences of all children in lower or upper primary schools.

Macrosystem Level

The macrosystem represents the broader perspective of school culture in which the other ecological systems (microsystem, mesosystem and exosystem) are embedded (Bronfenbrenner, 1979; Bronfenbrenner & Ceci, 2005; Tudge, Gray & Hogan, 1997). Bronfenbrenner observes that the macrosystem is thus the wider cultural milieu from which the different ecological systems, practices, and cultures arise. It consists of "the array of attitudes, practices, and convictions shared throughout society at large" (Thomas, 2005, p. 355).

At the macrosystem level, within the wider school culture, other subcultures such as the students, teachers, parents, and boards of trustees also exist, and these influence each other through the decisions made by these individuals and their individual and collective actions and practices (Prosser & Warburton, 1999). Examination of the features and practices of the social milieu suggest they appear to be much the same - but they employ different approaches and strategies to meet needs of individuals within the different levels of the ecological systems (Bronfenbrenner, 1979, 1989, 1993). For instance, children who struggle to achieve science curriculum outcomes are generally viewed through a 'deficiency' lens and are characterized by a perception of shortcomings and difficulties associated with teaching children with learning problems. According to Mitchell (2005), the way such children are included within the educational contexts may differ from one culture or subculture to another. The PNG education curriculum contains a 'practical life skills' based curriculum as an alternative, intended to provide alternative education opportunities.

A Reality Check for the Papua New Guinea Education System

Education in PNG has been dominated by imported models of education, based on those of colonial powers, predominantly the British and latterly Australian education systems (Avalos, 1993; Maha & Flaherty, 2003). The original education system was derived from a behaviorist philosophy with little recognition of local conditions or learners' needs. Teachers were expected to develop structured, preempted teaching programs and plans, with aims and objectives that had to be achieved at the end of each lesson. These could not deviate from the prescribed national curriculum goals and objectives. This rigid approach of curriculum development and implementation was strictly monitored by school inspectors who sought to ensure teachers implemented the prescribed national curriculum. The science curriculum was regarded as a major or core subject to be taught across all grades (Grades 1-12 before reform and Grades 3-12 after reform). Discovery and investigative learning were not promoted by education authorities, including curriculum developers and the implementing authorities like school inspectors and provincial and national education officers.

Hence, teachers as implementers of the curriculum were bound by national, centrally-prescribed curriculum aims, goals and objectives which had to be taught under the watchful eyes of the school inspectors who acted as auditors vested with the power to hire and fire. Depending on the extent to which the prescribed curriculum was implemented, teachers were either promoted or demoted, which obviously affected salaries and other entitlements. This often created animosity between teachers and the Ministry, but the hierarchical, authoritarian, leadership style in PNG education, meant that authorities were considered so powerful, teachers did not say much, living in fear of losing their jobs.

CONTEXTUALIZING CURRICULUM DEVELOPMENT AND IMPLEMENTATION IN PAPUA NEW GUINEA: SOME EMERGING TRENDS AND FEATURES

The PNG national curriculum underwent reform beginning in 1993 (DoE, 2002a, 2003a). The curriculum is now more 'liberal' in nature and has been developed based on the 'ecological systems theory' described above (Bronfenbrenner, 1979,

1989, 1993). The 'person' as an active and able learner and the 'educational context', nowadays play a major part in the way the curriculum is viewed, developed and implemented in PNG schools and classrooms. The liberal nature of the curriculum strives to help the child achieve both in terms of theory and/or academic content, and practical life skills, which might be useful for further study opportunities or for settling into village or community life (DoE 2004). In other words, the curriculum now concentrates on two perspectives - an 'academic pathway' and a 'practical life skills pathway' (these two curriculum pathways are explained further below). The reformed curriculum is considered more appropriate for PNG's educational context because many students cannot proceed to tertiary studies due to limited spaces at tertiary institutions.

Curriculum before the Education Reforms

As noted above, prior to the education reforms of 1993, the curriculum in the primary and secondary education sectors in PNG was based on behaviorist theory (DoE, 2002a). Teachers were guided by specifically developed aims and objectives to be achieved at the end of every lesson. The basic classroom teaching practices such as teaching programs, classroom timetables, and the lesson plans were strictly structured, and had to be aligned with prescribed aims and objectives of the centrally-developed curriculum framework. Teachers had little choice in making changes or alterations to meet individual children's learning needs. According to Guy (1999), this was coupled with other duties such as extra class supervision, home-work marking, and classroom management, resulting in high workloads for PNG teachers.

The science curriculum also was strictly structured, and there were few opportunities for children to engage in self-discovery or experiments in practical classes (Boeha & Robins, 2000). Teachers often taught from prescribed teachers' guides, while students depended entirely on answers available in student resource books or text books developed by the Department of Education. This promoted rote learning, rather than higher order thinking skills (DoE, 2004). In so doing, this approach of teaching and learning treated students as 'recipients of knowledge and skills', rather than original 'thinkers, discoverers and explorers of new knowledge, skills and ideas'.

At the same time, the previous science curriculum, like other curriculum learning areas, was heavily influenced by the rigorous, centrally-formulated and directed examination system in PNG. Having done science as a core subject at both primary and secondary school, students were required to sit for examinations at the end of Grades 6, 10 and 12. Teachers were highly focused, and concentrated their teaching practices towards preparing students to pass the examinations rather than teaching science as a holistic and self-exploratory subject. This practice also meant that teachers could not explore ideas beyond what was centrally prescribed, as this might disadvantage their students in the final examinations. A common view in PNG at the time was that if fewer students passed the exams from the total number of students in a particular grade, then the education authorities such as inspectors and the surrounding community would be unhappy, and conclude that this was the fault of the teacher. In other words, it was viewed that the teacher had not taught the students well, and this would affect his or her chances of promotion and salary increments.

Curriculum Reform in Papua New Guinea

The Government of PNG, through the Department of Education, embarked on an ambitious curriculum reform program (DoE, 2002a, 2002b, 2003a, 2003b, 2004). Based on a structural adjustment program, the curriculum reforms aimed to cover nine years in primary education, which included three years in the elementary (early childhood) years of learning taught in the vernacular (Paraide, 2002). In fact, the first three years of primary schooling consists of curriculum programs aimed at introducing children to basic knowledge and skills and their cultural underpinnings, and at the end of Year 2 a bridging program to English was conducted. Alderson (1984) suggests that children acquire skills and knowledge better and more quickly when they are introduced to a second language such as English, from the vernacular. However, although English is the medium of instruction in latter years of schooling, core subjects such as science are taught based on students' cultural backgrounds, before moving on to the wider 'scientific world'. This approach is considered useful and relevant for the children in PNG, in that it enables them to acquire culturally-based scientific knowledge, skills and ideas.

Outcomes-Based Curriculum in PNG Education System

The education reforms in PNG have gone a step further by overhauling the primary and secondary education curriculum with the support of the Australian Government (AusAID) (DoE, 2005a). The main purpose of these reforms was to develop a curriculum relevant for all Papua New Guineans. The previous education system resulted in many school leavers being seen as 'failures', and an outcomes-based curriculum was developed in an attempt to provide opportunities for all children to excel and succeed in life, whether this be via an 'academic pathway' or through a 'practical life skills pathway'. Either way, it was considered that all students should be given at least some basic knowledge, ideas and skills to live independent lives: "The most significant aspect of the new syllabuses is that they describe student learning in terms of learning outcomes and indicators and not in terms of objectives as was the practice previously" (DoE, 2006, p. 1). However, while the intentions of the outcomes-based education are genuine and laudable, Norman (2006) cautions that curriculum implementers have to be made aware of what constitutes an outcomes-based curriculum, and its major associated principles and/or tenets. This outcomes-based curriculum is a very new concept for both the education authorities in the existing PNG education system and the teachers at large as curriculum implementers. Therefore, an understanding of what constitutes outcomes-based education is vital in the hope of providing the major tenets and issues associated with it.

Outcomes-based Education

While many commentators have defined and/or explained what constitutes outcomes-based education, Spady (1994, 1998) is the principal authority in this field, and his significant contributions have been adopted in Papua New Guinea. According to Spady (1994), outcomes-based education is defined as:

> Focusing and organizing everything in the education system around what is essential for all students to be able to do successfully at the end of their learning experiences. This means starting with a clear picture of what is important for students to be able to do, then organizing the curriculum, instruction and assessment to make sure that learning ultimately happens. (p. 1)

Within the Papua New Guinea education context, it is the role of the Curriculum Development Division (CDD) and the Curriculum Reform Implementation Programme (CRIP) to ensure that the outcomes-based curriculum is relevant and appropriate for the local needs in PNG. However, from the PNG perspective, Norman (2006) points out that if the country is to achieve authentic outcomes-based education, the 'outcomes' must provide a "...clear, observable demonstration of student learning that occur[s] after a significant set of learning experiences" (p. 46). This suggests that the outcomes-based curriculum should be learner-centered in nature, and allow the child to demonstrate what he or she can do. It also takes into consideration what is going on within the surrounding ecological systems, so that the immediate surrounding environment is congruent with the teaching and learning processes. In this case, the learner and the environment are considered to be inseparable, and thus the associated activities and actions are geared towards giving greater opportunities for the good of the learner as a developing person in the school or classroom contexts in PNG.

SCIENCE CURRICULUM AND OUTCOMES-BASED EDUCATION IN PAPUA NEW GUINEA

In the light of outcomes-based education in Papua New Guinea, the science curriculum is best understood in terms of real life situations (DoE, 2003b; Wilson, n.d.). Science is perceived as how an individual person's life is influenced in terms of their associations in villages, communities, and at the provincial and national levels. According to the DoE (2003b), there are four levels of education structure which have different but related structures needed to achieve science outcomes before a child proceeds to the tertiary sector. For instance, elementary education covers culture and community, lower primary emphasizes environmental studies, upper primary attempts to integrate ideas covered in lower primary and lower secondary. Lower secondary is based on environmental science and applied science, and upper secondary on biology, chemistry, physics and applied science. In all of this, science strands and substrands must be linked to other subject areas, and the local cultural underpinnings in the hope of creating relevant units of work, whereby the teaching and learning activities, assessment, tasks and programs are linked in a thematic form of curriculum development and implementation across a given grade.

As noted above, the science curriculum is focused on two pathways the academic pathway and the practical life skills pathway. This is an exciting development for teachers as the curriculum implementers if they achieve the objectives of these two pathways using the rather basic resources available in local communities rather than resorting to expensive and sophisticated laboratories, although these are not totally done away with. At the same time, each of these pathways aim to achieve "science outcomes described as knowledge, skills and attitudes along with concepts, contexts, topics or issues based on the nature of the course" (DoE, 2003b, p. 3).

The academic pathway and the practical life skills pathway in outcomes-based science curriculum are student-centered in approach and the teacher takes a facilitating and/or guiding role, and students are given every opportunity to actively acquire and construct knowledge, skills and attitudes. This is thus a constructivist-based approach to teaching and learning where students are given opportunities to work out things for themselves to achieve the intended curriculum outcomes. According to the DoE (2003b), among others, in essence this approach has five phases of teaching and learning: engaging, exploring, explaining, elaborating, and evaluating. Each of these five phases of teaching and learning approach is described below.

Engaging

In this phase:

- Teachers engage students in activities that capture their interests and stimulate curiosity;
- Students raise questions;
- Teachers verify students' prior understanding of the topic; and
- Students compare ideas.

Exploring

In this phase students take on hands-on activities where they:

- Experience the phenomenon or concept;
- Explore the questions they have raised;
- Test their ideas; and
- Solve problems.

Explaining

Only after students have had opportunities to explore, do they have opportunities to:

- Compare their ideas with scientific explanations;
- Use scientific terminology; and
- Construct explanations that can be justified using information collected.

Elaborating

In this phase, students have opportunities to:

- Apply what they have learnt to new contexts; and
- Develop a deeper understanding of the problem or phenomenon as they discuss and compare ideas.

Evaluating

In this phase, students and the teacher:

- Look for evidence of changes in students' ideas, beliefs and skills; and
- Evaluate what students know and can do (adapted from DoE, 2003b, pp. 8-9).

Some Lessons Learnt from the Science Curriculum Reform, Development and Implementation Process

The education system in Papua New Guinea has gone through both structural as well as curriculum changes in an attempt to meet changing societal demands (Solon & Solon, 2006). In regard to the latter, the previous education system was so structured and rigid that it provided few options for children (DoE, 2006). However, with the introduction of an outcomes-based curriculum in PNG, there is now at least the potential to address the learning needs of *all* children. Specifically in the case of the science curriculum, the curriculum is more liberal and flexible in that there are more options for children. While this may sound superficially convincing, some three fundamental lessons have been learnt from the introduction of an outcomes-based science curriculum in PNG, and these require further thoughtful consideration.

First, it is clear that although the outcomes-based science curriculum was developed as part of education reforms designed to meet local needs, it was the work of an Australian contract company as the funding came from the Australian Aid for International Development (AusAID). Hence, the content of the outcomes-based science curriculum is still rather 'academic' in nature, and classroom teachers, school administrators and even provincial and national education authorities are finding it difficult to understand, assimilate and disseminate such content to students. Due to the apparent academic nature of the outcomes-based science curriculum, staff had to be trained in order to understand the content and implementation strategies. Another issue is that teachers and school administrators were not trained prior to the introduction of the outcomes-based curriculum. This placed the teachers and school administrators in a situation of needing to be trained by those who were conversant with outcomes-based curriculum. Although a selected number teachers have now been trained, this has been in the form of on-the-job training for many, with others going to regional education centers at their own expense. This has put considerable pressure on staff in terms of personal cost, workload, family obligations and associated factors.

Second, the shift from a behaviorist approach to science curriculum development and implementation is a necessary and higher step towards achieving equality and equity in terms of educational opportunities for all Papua New Guineans. The outcomes-based science curriculum encourages all students in the PNG education system to acquire some basic scientific knowledge, skills and attitudes necessary to progress in life, either through the academic pathway or the practical life skills pathway. Both of these pathways are necessary for every child to feel the worth and meaning of an authentic education culture. However, while there is perceived goodness in the outcomes-based science curriculum in PNG schools, as noted above the retraining of the personnel who are to implement the science curriculum is still to be fully achieved. This means teachers, school administrators, provincial and national education officers have to be informed through a consultative process about the effective implementation of an outcomes-based education in Papua New Guinea.

Third, "In Papua New Guinea, much attention is given to addressing the issue of making learning relevant to social needs" (Norman, 2006, p. 54). This suggests that in line with ecological systems theory (Bronfenbrenner, 1979, 1989, 1993), the science curriculum in PNG should focus on the child as a socially-developing person, on the real life situations in villages and surrounding communities, and on social settings of the learner. The learner here is perceived as being immersed within these social structures, and thus needs to understand the knowledge, skills and attitudes taught using local models, language, materials and experiences. The advantage of this kind of learning approach is the involvement of community leaders, parents, extended family members and other genuine and knowledgeable capable adults in the teaching and learning process. From the PNG perspective, this is seen to allow contextualizing the understanding of science within the local communities as students learn better when they are introduced to new ideas and difficult concepts using familiar things they are already aware of, and exposed to. This supports the basic principle of 'going from known to the unknown'. In so doing, the students and the surrounding communities take responsibility and ownership of the teaching and learning process in the PNG education system for the good of the children and their future.

Aspirations and Considerations for the Papua New Guinea Education System and Reformed Science Curriculum

The science curriculum development and implementation accompanying the education reforms has shed light in terms of paving a way for the future educational initiatives in Papua New Guinea. As of 2007, the PNG science curriculum has been overhauled considerably in the hope of addressing educational local needs. This has been achieved to an extent, and the basic science curriculum for elementary (early childhood), lower and upper primary, and lower and upper secondary schools were near completion at the time of writing. However, there is much yet to be achieved in terms of development of complementary local-based resource materials to help teachers meet the learning outcomes envisaged in the outcomes-based science curriculum. In order to achieve this, there are two basic curriculum aspirations and considerations that need to be pursued.

First and foremost, is to ensure that the existing science curriculum is user-friendly. This means the curriculum implementers have to feel at ease, take ownership, and responsibility for the effective dissemination of the curriculum. One way to achieve this is by training and awareness programs for the curriculum implementers emphasizing the benefits of the new curriculum. This approach is vital because the outcomes-based curriculum was developed by Australian experts, with some assistance from local education officers. Though the rhetoric is that the curriculum is PNG context-based, the ideas and examples are derived from Australian perspectives. In other words, their preconceived philosophies, theories and practices have influenced the curriculum development for the PNG education system. Hence, PNG experts should be allowed to take ownership of, and responsibility for, curriculum implementation so that when overseas experts leave at the end of their contracts, there is continuity and sustainability in terms of curriculum development and implementation.

The second aspiration and/or consideration is to strengthen the academic and practical life skills curriculum pathways described earlier so that all PNG children can benefit from the education system. However, this must be underpinned by the universal primary education (UPE) policy initiative where all children must be allowed to complete the primary education cycle before proceeding to secondary schools and the tertiary education sector (Webster, 2000, 2005). This might then pave the way for all PNG children to gain at least some formal education to enhance their lives, and help them make better informed decisions and act as 'educated citizens'. Whether or not children who have gone through the formal education system continue on to further education or go back into village life is not the issue in the end. The issue is to provide some basic science education to all citizens so that Papua New Guineans can live independent and productive lives in today's economically- and technologically-driven society.

SUMMARY

The education reforms in PNG have resulted in a curriculum that is more flexible and has the potential to better meet learners' needs. The science curriculum has been recently overhauled to meet local needs and this has, so far, been achieved through an ambitious program of education reform. However, prior to the education reforms, the primary and secondary education sector, and the general education system in Papua New Guinea, was primarily driven by an objectivist and/or behaviorist philosophy. Teachers were told to develop structured/preempted teaching aims and objectives which had to be achieved at the end of the lesson and were not to deviate from the centrally prescribed curriculum. This rigid approach of curriculum development and implementation was supported through the work of school inspectors who ensured that teachers implemented what was nationally prescribed, and any alterations or deviations were a breach of the intended curriculum principles and thus were seldom allowed. However, the reformed science curriculum is more flexible in nature, and was developed and implemented along the outcomes-based and/or an ecological systems theory based approach. The science curriculum now embraces 'the child' as a developing person who has the potential to learn and excel in life, and sees the educational context and/or environment as an influential factor in the teaching and learning process. However, in order to achieve the good intentions of the education reform, there is an urgent need for appropriate substantial professional development, school cultural re-organization and/or modification, and educational research to facilitate curriculum implementation, and to enhance effective teaching and learning in the context of PNG classrooms.

REFERENCES AND BIBLIOGRAPHY

Alderson, C.J. (1984). Reading in a foreign language: A reading or language problem. In C.J. Alderson & A.H. Urquhart (Eds.), *Reading in a foreign language* (pp. 122-135). London: Longman.

Avalos, B. (1993). Ideology, policy and educational change in Papua New Guinea. *Comparative Education, 29*(3), 275-292.

Baltes, P.B., Lindenberger, U., & Staudinger, U.M. (2006). Life span theory in developmental psychology. In W. Damon & R.M. Lerner (Eds.), *Handbook of child psychology: Theoretical models of human development* (6th ed., vol. 1, pp. 569-664). Mahwah, NJ: Wiley.

Boeha, B., & Robins, J. (2000). *Issues and trends in national development searchlight on Papua New Guinea: A thematic compendium of weekly articles from the Post Courier in the series 'Search Light with NRI'*. NRI special publication No. 30. Port Moresby: National Research Institute.

Bronfenbrenner, U. (1979). *The ecology of human development: Experiments by nature and design.* Cambridge, MA: Havard University Press.

Bronfenbrenner, U. (1989). Ecology systems theory. In R.I. Vasta (Ed.), *Annals of child development: Six theories of child development* (vol. 6, pp. 187-249). Greenwich, CT: JAI Press.

Bronfenbrenner, U. (1993). The ecology of cognitive development: Research models and fugitive findings. In R.H.Wozniak & K.W. Fischers (Eds.), *Development in context: Acting and thinking in specific environments* (pp. 3-44). Hillsdale, NJ: Erlbaum.

Bronfenbrenner, U., & Ceci, S.J. (2005). Heredity, environment and the question 'How': A first approximation. In U. Bronfenbrenner (Ed.), *Making human beings human: Bioecological perspectives on human development.* Thousand Oaks, CA: Sage.

Bronfenbrenner, U., & Morris, P.A. (2006). The bioecological model of human development. In W. Damon & R. M. Lerner (Eds.), *Handbook of child psychology: Theoretical models of human development* (6th ed., vol. 1, pp. 793-828). Mahwah, NJ: Wiley.

Ceci, S.J., Baker-Sennett, J.G., & Bronfenbrenner, U. (1994). Psychometric and everyday intelligence: Synonyms, antonyms and anonyms. In M. Ruther & D.F. Hay (Eds.), *Development through life: A handbook for clinicians* (pp. 239-259). London: Blackwell.

Damon, W., & Lerner, R.M. (Eds.). (1997). *Handbook of child psychology: Theoretical models of human development* (vol. 1, 5th ed.). New York: Wiley.

Damon, W., & Lerner, R.M. (Eds.). (2006). *Handbook of child psychology: Theoretical models of human development* (vol. 1, 6th ed.). New York: Wiley.

Department of Education. (2002a). *National curriculum statement for Papua New Guinea.* Waigani, Papua New Guinea: DoE.

Department of Education. (2002b). *The state of education in Papua New Guinea.* Papua New Guinea: Education Reform Facility and Monitoring Unit, DoE.

Department of Education. (2003a). *The state of education in Papua New Guinea. Updated.* Waigani, Papua New Guinea: Education Reform Facilitation and Monitoring Unit, DoE.

Department of Education. (2003b). *Science teachers' guide: Upper primary.* Waigani, Papua New Guinea: DoE.

Department of Education. (2004). *A achieving a better future: A national plan for education 2005-2014.* Waigani, Papua New Guinea: DoE.

Department of Education. (2005a). *Unit 1 philosophy of curriculum reform: Study guide: In-service units to support the implementation of the primary reform curriculum.* Waigani, Papua New Guinea: DoE.

Department of Education. (2005b). *Unit 6 learning and teaching for outcomes: Study guide: In-service units to support the implementation of the primary reform curriculum.* Waigani, Papua New Guinea: DoE

Department of Education. (2006). *Science teachers' guide: Upper primary.* Waigani, Papua New Guinea: DoE.

Guy, R. (1999). *Last in line: A teachers' work value study - DER Report Number 75.* Waigani, Papua New Guinea: National Research Institute.

Kugelmass, J.W. (2004). *The inclusive school: Sustaining equity and standards.* New York: Teachers' College Press.

Magnusson, D., & Stattin, H. (2006). The person in context: A holistic-interactionistic approach. In W. Damon & R.M. Lerner (Eds.), *Handbook of child psychology: Theoretical models of human development* (6th ed., pp. 400-464). Mahwah, NJ: Wiley.

Maha, A.C., & Flaherty, T.A. (Eds.). (2003). *Education for 21st century in Papua New Guinea and the South Pacific.* Goroka: University of Goroka.

Mitchell, D. (Ed.). (2005). *Contextualizing inclusive education: Evaluating old and new international perspectives.* New York: Routledge.

Norman, P. (2006). Outcomes-based education: A PNG perspective. *Contemporary PNG Studies DWU Research Journal, 5,* 45-57.

Paraide, P.M. (2002). *Discovering our heritage: An early assessment of lower primary learning in the reform curriculum. DER report number 76.* Port Moresby, Papua New Guinea: National Research Institute.

Peck, C.A., Hayden, L., Wandschneider, M., Peterson, K., & Richarz, S. (1989). Development of integrated preschools: A qualitative inquiry into sources of resistance among parents, administrators and teachers. *Journal of Early Intervention, 13,* 353-364.

Prosser, J., & Warburton, T. (1999). Visual sociology and school culture. In J. Prosser (Ed.), *School culture* (pp. 82-97). London: Paul Chapman.

Solon, M., & Solon, E. (2006). Implementing curriculum reforms in urban Madang schools: Issues and challenges. A PNG perspective. *Contemporary PNG Studies DWU Research Journal, 5,* 15-28.

Spady, W. (1994). Outcomes-based education: Critical issues and answers. *American Association of School Administrators*. Arlington, VA: AASA.

Spady, W. (1998). Paradigm lost: Reclaiming America's educational future. *American Association of School Administrators*, Arlington, VA: AASA.

Thomas, R.M. (2005). *Comparing theories of child development* (6th ed.). Belmont, CA: Thomas Wadsworth.

Tudge, J., Gray, J.T., & Hogan, D.M. (1997). Ecological perspectives in human development: A comparison of Gibson and Bronfenbrenner. In J. Tudge, M.J. Shanahan & J. Valsiner (Eds.), *Comparisons in human development: Understanding time and context* (pp. 72-105). Cambridge, UK: Cambridge University Press.

Webster, T. (2000). *Globalisation of education policies: The extent of external influences on contemporary Universal Primary Education policies in Papua New Guinea.* Port Moresby, Papua New Guinea: UPNG Press.

Webster, T. (2005, 11 February). Education for all: Where is PNG? Accelerating girls' education in PNG. *Post Courier*, pp. 2.

Wilson, M. (n.d.). *Science achievement of grade 12 students: The PNG perspective. Research report No. 63.* Waigani, Papua New Guinea: National Research Institute.

AFFILIATION

John Longo Rombo
Education Consultant
International Education Agency
Papua New Guinea

EDUCATION FOR SUSTAINABLE AGRICULTURE AND NATURAL RESOURCE MANAGEMENT ON BOUGAINVILLE

BACKGROUND: THE BOUGAINVILLE CRISIS

Bougainville is the northernmost island of the Solomon Islands archipelago in the southwest Pacific. For historical reasons it is part of the state of Papua New Guinea (PNG). In the region, conflicts over the management of natural resources are common sources of tension. Many of these are in response to problems associated with past and continuing colonial relations. For example, an armed rebellion over the Panguna mine in central Bougainville began as a land dispute following the eviction of traditional landowners, and escalated because ensuing mining activities polluted valued land and water systems. Mining, forest clearing, and logging are often linked to environmental degradation from unsustainable practices by transnational corporations operating across the Pacific region (Jenkins & Jenkins, 2005; Lockwood, 2004).

The Bougainville crisis, which stemmed from the conflict over the Panguna copper mine, became a protracted armed struggle that led to the deaths of thousands of people (Havini, 2004; Regan, 1998). It continued for almost a decade from 1989 to 1998 during which more than 15,000 people lost their lives in a civil war involving the Papua New Guinea Defense Force (PNGDF), the Bougainville Revolutionary Army (BRA) and various Bougainville Resistance Forces (BRF) (Joint Standing Committee on Foreign Affairs, Defense & Trade, 1999; Regan, 2001). Fighting also took place sporadically at different times and locations across Bougainville after the PNGDF withdrew in 1990 (O'Callaghan, 2002). This fighting was mainly between internal factions that formed in opposition to the BRA (Howley, 2002; Regan, 1998). To make matters worse, the PNG government imposed an economic blockade on Bougainville between 1990 and 1995, which affected the provision of essential services. The island of Buka on the northern end of Bougainville was affected less by this blockade (Dorney, 1998). Nevertheless, the only education and health services available in most parts of Bougainville for almost a decade were those devised by the islanders themselves (Zale, 2004). This chapter is about the lack of basic education that eventuated for many people as a result of the Bougainville crisis and how to contribute towards rectifying the problem.

Bougainville has since moved away from conflict and towards peace and reconciliation (Böge & Garasu, 2004; Carl & Garasu, 2002). This transition to peace led to the signing of the Bougainville Peace Agreement in 2001, which has

Richard K. Coll, Neil Taylor (eds.), Science Education in Context: An International Examination of the Influence of Context on Science Curricula Development and Implementation, 33–43.

been "remarkably successful" (Regan, 2002, p. 114). The agreement included goals to collect weapons, to disarm militant groups and set up a separate Bougainville administration (Böge & Garasu, 2004). Bougainville is currently an autonomous region within PNG under the guidelines of a new constitution. After 10 to 15 years of autonomy, a referendum will determine whether it becomes an independent nation or remains part of the state of PNG. In the meantime, the Autonomous Bougainville Government (ABG) has to contend with many social, economic and political challenges while the region recovers from armed conflict.

Residual violence, which persists at the community level in Bougainville, is exacerbated by posttraumatic stress, underemployment, and alcohol/drug abuse among both ex-combatants and civilians alike - who observed, participated in or experienced physical assault, killing, rape, torture and the destruction of property during the crisis (Hakena, Ninnes & Jenkins, 2006; Sirivi & Havini, 2004). The speed of recovery depends on the population overcoming these problems and returning to productive, normalized lives. The peace-building efforts of local Non-Government Organizations (NGOs) such as Leitana Nehan Women's Development Agency (LNWDA) and the Bougainville Inter-Church Women's Forum (BICWF) have continued to work with the community towards the sustaining of peace (Böge & Garasu 2004).

In spite of the efforts of local NGOs, certain key issues remain in Bougainville that could reignite internal conflict. Among these is the major issue of a low level of functional literacy and numeracy in the population. Illiteracy and innumeracy are consequences of a lack of basic education skills for both males and females who were unable to attend school due to the disruptive effects of the crisis. This chapter examines one means of alleviating this situation for some of these people, with a view of reducing an identifiable 'basic-education gap' in Bougainville. Local people describe this educational shortfall as 'a time bomb waiting to go off' and refer to those who were disadvantaged as the 'lost generation'. This gap in basic education is widely accepted as posing a significant challenge to socio-economic recovery and continued peace on Bougainville (see Flinders University Research News, 2006). Local people, leaders and the international community acknowledge the existence of the 'lost generation' of Bougainville and are concerned about the potential problems this poses. General fears are that frustration associated with lack of employment opportunities, delinquency and the effects of post-traumatic stress could push disaffected youth, especially (but not exclusively) illiterate men, into crime, cargo cults and militancy. The major concern is that this could destabilize Bougainville.

EDUCATIONAL ISSUES IN BOUGAINVILLE

Filling a Basic Education Gap

Information collected during five visits to Bougainville between 2003 and 2006 are synthesized here to outline the education problem faced by the lost generation. This problem affects people who missed out on basic primary school education during the crisis, and this chapter particularly focuses on those interested in agriculture and natural resource management (NRM) as a livelihood activity. Data collected from primary informants, NGO workers and government agents, by way of personal observations, conversations and interviews are compiled to produce the case study presented below. While most of the information summarized here was related in English, some accounts were interpreted from the national vernacular, PNG Pidgin.

Education for the Lost Generation

Significant numbers of the lost generation are ex-combatants, but not exclusively, as the majority of women who missed out on school were non-combatants. Most of the people who would have attended school during the crisis were born between 1977 and 1992. All are above the age of 15 years, the majority of whom are now adults. They include thousands of people, and could number as many as 30,000 according to estimates by NGO staff, leaders, government agents and locals. This number comprises all those for whom formal school education was disrupted during the crisis, and there is a sizable internal population that is less literate and numerate than the average adult over the age of 35 years in Bougainville. A significant proportion of this disadvantaged cohort either never finished or never commenced primary school, and therefore missed out on gaining a basic education to Year 8. They are now between the ages of 15 and 25 years, and categorized as 'the youth of Bougainville'. The aspirations, intentions and welfare of this group are the main focus of this chapter.

The opportunity exists for the lost generation to return to school, however, this option does create particular problems for students as well as teachers. Generally, those who missed out on education during the crisis are reluctant to return to a formal school setting because they are ashamed because they are much older than other students in class. Furthermore, teachers generally find mature students, particularly young men, problematic to control and teach in class. The ex-combatants among them also have special psychosocial needs for which the schools are unable to cater.

Certain avenues are available for people who have completed Year 8 (the end of primary school in PNG) to return to study, either at one of the five high schools, via Flexible Open Distance Education (FODE) run by the PNG Ministry of Education or by enrolling in the Certificate in Tertiary and Community Studies (CTCS) offered by the Open College of the University of PNG through the Buka Open Campus (BOC). The same cannot be said for those who neither commenced nor completed primary school. Although a number of NGOs provide adult literacy programs, and have done so since the crisis ended in 1998, there is a shortfall in student engagement. This is attributed to the difficulty of motivating adults to enroll in functional literacy courses.

What Kind of Basic Education?

In answer to the question, 'What kind of education would you like to engage in?', various aspects of agriculture and the management of natural resources featured significantly. Bougainville is predominantly an agricultural society. As in the rest of PNG, the majority of people are involved in agricultural or related activities. In terms of livelihood interests, the following areas were well represented in responses: growing and processing cash crops (e.g., cocoa, coconuts, vanilla); growing and marketing garden produce and diversifying food gardens to include fruit trees; growing flowers (e.g., orchids & heliconias) for the cut-flower market; small-scale timber milling; poultry keeping and pig husbandry; fishing and harvesting of marine resources (e.g., sea cucumbers, mud-crabs, reef fish); and eco-tourism (e.g., coral reefs, forests, beaches, bird watching, hiking, diving, game fishing, snorkeling and boating). In relation to science, the environment, soils, global warming and sustainable energy were specifically mentioned as being important subjects to learn.

In addition to the specifics outlined above, there is a great desire among this group to be able to read and write in Pidgin as well as in English, in order to be able to read newspapers and gain the necessary literacy and numeracy skills to participate in everyday life. Without these basic skills these people feel isolated, unable to contribute fully in the community and restricted in the extent to which they can participate, earn a living and fend for their families. Nevertheless, in spite of not having the benefits of a basic education, many acknowledge having learned many skills that helped them survive during the crisis including growing and gathering food, catching fish, using indigenous herbal medicines, devising alternative technologies and supplying their labor in various ways to cooperate with others in the community. Notwithstanding their disadvantages, they perform traditional dances, sing, produce art and craft, attend church and are involved in religious activities, vote in elections and volunteer in community activities. Therefore, not being able to read and write, particularly in English, has not handicapped this group entirely. However, they feel disadvantaged because of being unable to complete primary school, and hence have requested a basic education program, one that improves literacy and numeracy skills and revolves around their livelihoods in agriculture and the management of natural systems.

Negotiating for a Unique Learning Experience in Bougainville

Conversations with prospective students also revealed a desire to learn about subjects of global interest in combination with or in addition to the agricultural, NRM and science topics. These areas - including globalization, computers, the Internet, communications, English language, sustainable development, technology, the market, marketing, business, bookkeeping and microfinance - are recognized as important components of global learning. They are also seen as an integral part of any course that focuses on working with agricultural smallholdings (gardens, plantations, farms) or natural ecosystems (mangroves, reefs, forests, grasslands, islands, atolls, rivers).

Another frequent request from this group emphasized the necessity for the program to be developed with local input and significant negotiation of the content with prospective students. A related factor is a reluctance to import a curriculum from outside Bougainville, say from Australia or New Zealand, although incorporating course components from Vudal University in neighboring East New Britain is viewed as acceptable. Prospective students question whether the colonial influences of an imported program could have adverse effects on the future of Bougainville.

A third point of some significance is the insistence that functional literacy be part of the course. This is related to concerns about existing literacy programs being deemed dreary because they focus on irrelevant subjects when trying to teach people to read and write.

In short, the target group is asking for an indigenous program that is free from outside colonial influences. The content, structure, design and curriculum must be 'home-grown' to suit local needs and aspirations, and incorporate local knowledge as well as enhance self-reliance. Finally, the 'big ask' is for the students to learn functional and critical literacy as part of this practical course of study.

DISCUSSION

What Will This Program of Study Entail?

Those who engage with the Basic Education Program (BEP) may arguably need to first gain rudimentary literacy and numeracy skills before they are able to use computers or surf the Internet. However, from a different perspective, it is preferable to address the desired global areas of interest (i.e., the list above) simultaneously as students engage in agricultural and NRM activities. If so, course materials, primers, tutorial assistance and extension services, will have to be designed carefully to integrate basic literacy and numeracy skills into the activity based projects that are focusing on content-based areas such as science, business and communications.

There is also the question of how to assess what students learn and to know when basic literacy and numeracy levels are reached. A plausible suggestion is to begin with oral forms of assessment and action-based project outcomes, while students learn from doing as they participate in agricultural and NRM activities. Subsequently, educators could build on the strong oral tradition that exists in Melanesia. It is commonplace in Bougainville, as in many indigenous societies across Melanesia, for people to narrate stories, and debate and argue issues passionately (Sillitoe, 1998). Accordingly, students could be trained to report orally on their learning through a *viva voce* style interaction with tutors and among peers. A verbal approach such as this could be reinforced by simple written reports, which become more sophisticated as functional literacy and numeracy levels improve. These written reports are expected to gradually evolve towards meeting acceptable basic literacy standards. Students who achieve an acceptable level of proficiency finally receive a basic literacy certificate in their field of study - Agriculture and Natural Resource Management. Therefore, the program outlined for the lost generation in Bougainville involves the integration of authentic learning and assessment strategies (oral and written) as a way of engaging students in real life (agricultural and NRM) activities and issues (Brady & Kennedy, 2005). This would prove more motivating, meaningful and useful than passive, transmitted knowledge, often imported, which has little impact on or relevance to students' lives in a rural, postconflict society.

The program is envisaged to run mixed-mode, with components offered via distance education and supplemented with specially crafted projects and primers. It would require the support of roving extension facilitators and leadership of qualified adult education tutors. Government agricultural extension officers could assist with field projects and NGOs involved in agriculture and NRM could play an enabling role. The course could be reinforced with field schools based on demonstration sites selected from the students' own projects. Hence the program would be field oriented.

There is, however, an additional dimension to the challenge of designing a course for the lost generation. It is essential to recognize that a majority of the target group may still suffer from trauma, and hence require psychosocial support (Perry, 2006; Summerfield, 1991). This calls for training in peace education, conflict resolution skills and trauma counseling for extension facilitators, tutors and students alike because the situation is far from normal given the violent history of Bougainville. Planners who ignore the effects of the crisis would be increasing the risk of the program failing.

Institutional Support and Delivery

The BEP could be run in partnership with the Open College of the University of PNG, which currently offers the CTCS through the BOC. The CTCS caters for students who have completed Years 8 or Year 10 at school. It combines preliminary and foundation studies in literacy, numeracy and community development, and it offers a variety of elective subjects including agriculture, which can contribute to the development of life skills. The Open College and the government in Bougainville are collaborating with Vudal University in East New Britain to offer agricultural programs through the BOC at the diploma level. This would allow a pathway, for at least some students, from the BEP to university via the CTCS followed by a Diploma in Agriculture.

It is practical to run the BEP via decentralized extension centers situated in the 13 districts of Bougainville, established as subsidiaries of the BOC. A senior staff member located at the BOC could coordinate and administer the BEP activities taking place via these centers. Each extension centre would have a small team of local staff. One model is to have three staff at an extension center. A district team would comprise a qualified adult education tutor - to organize and supervise projects, learning materials for substantive areas, literacy and numeracy, workshops, field-schools, supervise extension staff and be responsible for assessment. In addition, each center would need at least two roving extension facilitators, preferably one male and one female, who support students with projects and learning tasks. Ultimately, the number of facilitators needed on the ground would depend on student demand.

The setting up of extension centers requires the sourcing of funds for infrastructure development including information technology (IT) capacity. As employment opportunities are few and the cash economy is relatively small in Bougainville, the question of how students pay for their studies is open to much debate in a society where education is available on a user-pays basis. A popular argument suggests the BEP should be free to encourage the lost generation to enroll, so as to simultaneously avert future problems and solve a major impediment to socioeconomic recovery. Even if the program is free and offered as an essential service, it needs funding. One proposition is to ask the company responsible for Panguna, Bougainville Copper Limited (BCL) to 'foot the bill' for the BEP because the crisis was caused by questionable practices in setting up and running the mine. This proposition is contentious, but many who hold the company responsible for the problem view it as appropriate that BCL should pay for the educational needs of the lost generation.

What Kind of Education Is This?

The education program described in this chapter involves the use of participatory processes to engage students in negotiating (agriculture & NRM) content and learning outcomes while designing a unique course of study that addresses educational disadvantage. Exploring participatory and empowering methods promoted by Paulo Freire is one way to approach the designing of this action-oriented program. Freire advocated involving the students themselves as co-investigators in a dialogic process in order to develop generative themes that are meaningful and relevant to them in their own world (Freire, 1996). It would be appropriate to draw on these ideas in the context of building critical literacy while concurrently engaging in active learning. This would, of course, also involve setting up dialectic processes to challenge participant's critical reasoning and motivation.

The fact that there are many educational benefits from student-centered and experiential learning is not new (see, Dewey, 1937). It is therefore widely accepted that students learn effectively when hands-on, practical, project-based approaches are employed, especially in their local settings. More recently, researchers have shown that active involvement has positive effects on student attitudes to science and that students are easily motivated to learn in the local environments with which they are more familiar or which are more meaningful (Barnhardt & Kawagley, 2005; Palmer, 2004). Also, culturally-specific and indigenous ways of knowing and learning are recognized to assist students in acquiring and utilizing knowledge in environmental studies by creating meaningful field-based experiences (Feinstein, 2004). The benefits of learning from an action-oriented and issues-based program of study easily extends to land, water, natural resources, food and agriculture (Hodson, 2003). Subsequently, the implications of such a program address several key principles of a constructivist approach to learning and teaching - "learner-centered experiences and activities", "occasions for students to interact and work collaboratively", and a means of "providing assistance for novice learners to develop expertise" (Krause, Bochner & Duchesne, 2003, p. 158).

After Piaget, a constructivist approach to learning and teaching is appropriate in developing this education program for the lost generation. It focuses on student-centered, experiential and discovery learning with students being encouraged to participate actively in their own learning, thus enabling them to make sense of their learning experiences (Brunner, 1966; De Vries, 1997). It is an education program in which the knowledge, skills, values and experiences that students bring to learning tasks are valued as each student's prior knowledge is regarded as important and unique, therefore taken into consideration in order to link these with more complex and related learning via new ideas (Krause et al., 2003). Furthermore, in line with a social constructivist approach to cognitive development (Vygotsky, 1986), learning in the program envisaged for Bougainville is 'scaffolded' through social interaction with students' peers and community experts in the relevant field.

In a different sense, the development of the program in Bougainville relates directly to the application of a constructivist-based research paradigm, where what is real, useful and meaningful from the perspective of taking some kind of action to change a situation (the provision of basic education through active learning in agriculture and NRM to address a disadvantage), is derived from a process whereby the community, in this case the target group, is involved actively (Lincoln & Guba, 2000).

CONCLUSION

The disadvantage faced by the lost generation in Bougainville is attributable to the political and economic actions taken by dominant states during the colonial period, and hence is a consequence of imperialism (Galtung, 1979; Said, 1993). In other words, the target group was deprived of a basic education due to the Bougainville crisis, which in turn resulted from a conflict between mining interests, including Australian and PNG interests, and local landowners (Clear & Hutchinson, 2006; Havini, 2004). It involved a secessionist movement against the state of Papua New Guinea that manifested in locals resisting negative socioeconomic and environmental impacts of the Panguna copper mine (Ballentine & Nitzschke, 2003). The case of the Panguna mine is an example of a negative response by local people to an inappropriate model of economic development (Jenkins, 1999). War as a negative response also has costly disabling effects that cause losses in human development and the potential for people to participate fully at all levels of social and economic life (Clear & Hutchinson, 2006). Hence these problems experienced in Bougainville may be described as products of mal-development processes, related to external structures that have had adverse impacts on internal groups due to economic and cultural dominance, thereby creating underdevelopment (Youngman, 2000).

The process of planning and implementing education programs in post-conflict situations is significant. Wondem Asres Degu (2005) warns that old conflicts often reignite or new types of conflict develop in situations where education solutions are imposed and not readily accepted by the various parties and/or the actors responsible for implementing the education solutions. In the case of the lost generation in Bougainville, the target group's interest in engaging with the problem and finding a suitable solution would drive the agenda for change. The affected group desires to actively remedy their lack of a basic education, which they see both as a social need and an essential human right, so that problems associated with poverty and disadvantage disappear as a consequence of actions taken. In addition, to avoid dependency, the group opts to remain independent, refusing external aid or intervention and choosing instead to work towards self-reliance. Self-reliance is a means of avoiding the dependency trap and decoupling from colonial influences (Bahaba, 1990). Therefore, the participants as co-learners wish to determine a future that is based on their own "efforts, aspirations and intentions", and not those dictated to them by others, especially from outside (Rogers, 1992, pp. 148-155). They have come across as knowing what they want; motivated, willing and capable of making decisions, setting goals, mapping out desired outcomes and acting on their own behalf to achieve desired ends even though the availability of funds is limited. This is genuine 'people-centered development' (Burkey, 1993). Hence, the lost generation appear committed to what Saovana-Spriggs (2003, pp. 210-213) calls a "whole of community" approach to recovery in Bougainville, from within and autonomously.

The BEP described here serves multiple purposes. In the first instance, the program informs and empowers the lost generation to work more efficiently in the agricultural and NRM sectors, particularly within the informal economy. Secondly, it is a means of achieving literacy and numeracy for disadvantaged youth while they are involved in livelihood activities, thereby creating a learning community by enhancing their motivation to learn. Ultimately, the BEP offers the disadvantaged youth an alternative means of improving their circumstances, which encourages them to contribute to a peaceful future in Bougainville while simultaneously providing an avenue for them to move away from drug and alcohol abuse, militancy and joining destructive cargo cults. Furthermore, it provides a bridge for students who wish to progress further in their education by preparing them for more advanced studies. The BEP is an active-learning program that engages dislocated youth from among the lost generation, providing them with a way out of disadvantage and underdevelopment and a means of contributing to socio-economic recovery in Bougainville.

REFERENCES AND BIBLIOGRAPHY

Bahba, H. (1990). The other question: Difference, discrimination and the discourse of colonialism, In R. Ferguson (Ed.), *Out there: Marginalisation and contemporary cultures* (pp. 71-89). Cambridge, MA: MIT Press.

Ballentine, K., & Nitzschke, H. (2003). *Beyond greed and grievance: Policy lessons from studies in the political economy of armed conflict.* Program on Economic Agendas in Civil Wars (EACW), International Peace Academy, New York. Retrieved 13 September 2006, from http://www.ipacademy.org/Publications/Publications.htm

Barnhardt, R., & Kawagley, A.O. (2005). Indigenous knowledge systems and Alaska native ways of knowing. *Anthropology and Education Quarterly, 36*(1), 8-23.

Böge, V., & Garasu, L. (2004). Papua New Guinea: A success story of post-conflict peacebuilding in Bougainville. In A. Heijmans, N. Simmonds & H. van de Veen (Eds.), *Searching for peace in Asia Pacific: An overview of conflict prevention and peacebuilding activities* (pp. 564-579). London: Lynne Rienner.

Brady, L., & Kennedy, K. (2005). *Celebrating student achievement: Assessing and reporting* (2nd ed.). French's Forest, NSW: Pearson Education Australia.

Brunner, J. (1966). *Towards a theory of instruction.* Cambridge, MA: The Belknap Press of Harvard University.

Burkey, S. (1993). *People first: A guide to self-reliant participatory rural development.* London: Zed Books.

Carl, A., & Garasu, L. Sr. (Eds.). (2002). *Weaving consensus: The PNG-Bougainville peace process, conciliation resources* (Accord, Issue 12), Retrieved 21 September 2006, from http://www.c-r.org/our-work/accord/png-bougainville/index.php

Clear, M., & Hutchinson, F. (2006). Learning from each other: A theoretical and applied overview of the relationship between disability studies and peace studies. *The Review of Disability Studies: An International Journal, 2*(2), 88-99.

De Vries, R. (1997). Piaget's social theory. *Educational Researcher, 26*(2), 4-17.

Dewey, J. (1937). *Experience and education.* New York: Macmillan.

Dorney, S. (1998). *Politics and mercenaries, and the Bougainville crisis: The Sandline Affair.* Sydney: ABC Books.

Feinstein, B.C. (2004). Learning and transformation in the context of Hawaiian traditional ecological knowledge. *Adult Education Quarterly, 54*(2), 105-120.

Flinders University Research News (9 March 2006). *Bougainville: Beginning again*, Retrieved 25 June 2007, from http://www.flinders.edu.au/?news=102

Freire, P. (1996) *Pedagogy of the oppressed.* London: Penguin Books.

Galtung, J. (1979). A structural theory of imperialism. In G. Modelski (Ed.), *Transnational corporations and world order: Readings in international political economy* (pp. 155-171). San Francisco: W.H. Freeman & Company.

Hakena, H., Ninnes, P., & Jenkins, B. (Eds.). (2006). *NGOS and post-conflict recovery: The Leitana Nehan Women's Development Agency, Bougainville.* Canberra: Asia Pacific Press.

Havini, M.T. (2004). Introduction. In J.T. Sirivi & M.T. Havini (Eds.), *As mothers of the land: The birth of the Bougainville Women for Peace and Freedom.* Canberra: Pandanus Books.

Hodson, D. (2003). Time for action: Science education for an alternative future. *International Journal of Science Education, 25*(6), 645-670.

Howley, P. (2002). *Breaking spears and mending hearts: Peacemakers and restorative justice in Bougainville.* London: Zed Books & The Federation Press.

Jenkins, B. (1999). Equitable mining and appropriate development for a secure peace. In G. Harris, N. Ahai & R. Spence (Eds.), *Building peace in Bougainville* (pp. 18-23). UNE Armidale: The Centre for Peace Studies, University of New England, NSW Australia and the National Research Institute, Waigani, Papua New Guinea (Special Publication No. 27).

Jenkins, K., & Jenkins B. (2005). Education for sustainable development and the question of balance: Lessons from the Pacific. *Current Issues in Comparative Education, 7*(2), 114-129.

Joint Standing Committee on Foreign Affairs, Defense and Trade. (1999). *Bougainville: The peace process and beyond - Report No 90 - to both Houses of Parliament: Tabled 27 September 1999*. Retrieved 21 September 2006, from http://www.aph.gov.au/house/committee/jfadt/bougainville/BVrepindx.htm

Krause, K., Bochner, S., & Duchesne, S. (2003). *Educational psychology for learning and teaching*. Victoria, Australia: Thomson Learning.

Lincoln, Y.S., & Guba, E.G. (2000). Pragmatic controversies, contradictions and emerging confluences. In N.K. Denzin, & Y.S. Lincoln, (Eds.), *Handbook of qualitative research* (2nd ed., pp. 163-188). London: Sage.

Lockwood, V.S. (2004). The global imperative and Pacific Island Societies. In V.S. Lockwood (Ed.), *Globalisation and culture change in the Pacific Islands* (pp. 1-39). Saddle River, NJ: Pearson/Prentice Hall.

O'Callaghan, M. (2002). The origin of the conflict. In A. Carl & L. Garasu (Eds.), *Weaving consensus: The PNG-Bougainville peace process*. London: Conciliation Resources. (Accord, Issue 12), Retrieved 21 September 2006, from http://www.c-r.org/our-work/accord/png-bougainville/index.php

Palmer, D. (2004). Situational interest and the attitudes towards science of primary teacher education students. *International Journal of Science Education, 26*(7), 895-908.

Perry, B.D. (2006). Fear and learning: Trauma related factors in adult education process, *New Directions for Adult and Continuing Education*, (110), 21–27.

Regan, A. (1998). Current developments in the Pacific. *The Journal of Pacific History*, *33*(3), 269-284.

Regan, A. (2001). Why a neutral peace monitoring force: The Bougainville conflict and the peace process. In M. Wehner & D. Denoon (Eds.), *Without a gun: Australia's experiences monitoring peace in Bougainville 1997-2001* (pp. 2-18). Canberra: Pandanus Books & the Research School of Pacific and Asian Studies, Australian National University.

Regan, A. (2002). The Bougainville political settlement and the prospects of sustainable peace. *Pacific Economic Bulletin, 17*(1), 114-129.

Rogers, A. (1992). *Adults learning for development*. London: Cassell.

Said, E.W. (1993). *Culture and imperialism*. New York: A.A. Knopf.

Saovana-Spriggs, R. (2003). Bougainville women's role in conflict resolution in the Bougainville peace process. In S. Dinnen with A. Jowitt & T. Newton Cain (Eds.), *A kind of mending: Restorative justice in the Pacific Islands* (pp. 195-214). Canberra: Pandanus Books.

Sillitoe, P. (1998). *An introduction to the anthropology of Melanesia: Culture and tradition*. Cambridge, UK: Cambridge University Press.

Sirivi, J.T., & Havini, M.T. (Eds.). (2004). *As mothers of the land: The birth of the Bougainville Women for Peace and Freedom*. Canberra: Pandanus Books.

Summerfield, D. (1991). The psychosocial effects of conflict in the Third World. *Development in Practice*, *1*(3), 159–173.

Vygotsky, L.S. (1986). *Thought and language* (trans. A. Kozulin). Cambridge, MA: MIT Press.

Wondem Asres Degu (2005). Reforming Education. In G. Junne & W. Verkoren (Eds.), *Postconflict development: Meeting new challenges* (pp. 129-145). London: Lynne Rienner.

Youngman, F. (2000). *The political economy of adult education*. London: Zed Books.

Zale, D. 2004, Human rights abuses and survival behind the blockade. In J. Sirivi & M. Havini (Eds.), *As mothers of the land: The birth of the Bougainville Women for Peace and Freedom* (pp. 47-50). Canberra: Pandanus Books.

AFFILIATION

Bert Jenkins
School of Professional Development and Leadership
University of New England
Australia

SADE

TECHNOLOGY EDUCATION DEVELOPMENT IN THE SOLOMON ISLANDS

TECHNOLOGY AND TECHNOLOGY EDUCATION

The Nature of Technology

Although attempting to define technology is a recent phenomenon, technology itself is as old as the human race, and has been part and parcel of human life throughout the struggle for survival and existence. People have been using technology ever since the existence of the human race to manipulate and seek control over their environment (Dyrenfurth, 1984). Adams (1993) explains that the word technology is derived from the Greek words *tekhne,* which means art or skill, and *logia,* which means science or study. Lux (1983) adds that "technology is a praxiological knowledge," (p. 1); in other words, it is knowledge of practice, and involves the practicing of an art or skill.

Over the past decades many scholars have been involved in studying technology in an attempt to understand it better. Hansen and Froelich (1994) observe that philosophers, sociologists, historians, and teachers have all continued to study this subject in search of a specific definition, yet an accepted definition remains elusive. Some authors see technology as the end product of human engineering endeavors. Johnson (1989), for example, defines technology as the "application of knowledge, tools, and skills to solve practical problems" (p. 2). Such authors view technology as the work of engineers, inventors and technologists when they apply knowledge, sometimes from scientific theories or other sources of knowledge and sometimes from experience, to create and commercialize devices and systems to meet human goals (Naughton, 1993). Other authors view technology as the result of human influence and control. Mitcham (1994) argues that technology is more than engineering and incorporates philosophy, since all technology has been developed by people to satisfy human needs and wants. Cutcliff (1981) agrees with Mitcham and states that:

> Technology is a social process in which abstract economic, cultural, and social values, shape, develop and implement specific artifacts and techniques that emerge from the distinct technical problem-solving activity called engineering which is embedded in that process. (p. 36)

Several authors have indicated that technology and technological practice is very much a part of society. Pacey (1993), for example, sees technology and technological practice in three dimensions: a cultural aspect, an organizational

Richard K. Coll, Neil Taylor (eds.), Science Education in Context: An International Examination of the Influence of Context on Science Curricula Development and Implementation, 45–53.

aspect, and a technical aspect. According to Pacey's model, technology is not only technical in nature, but also influenced by culture and the role of people and organizations in societies, and vice versa. Custer (1995) affirms that to better understand the term technology, the concepts of technology as artifact, knowledge, activity or process, and volition with both the human and social dimensions, have to be considered when conceptualizing technology.

Technology Education Curricula

Technology education is a subject that has been recently developed and is now gaining recognition internationally by many educators. It is difficult to deny the significance and influence that technology has on people and society; as a consequence many educators have sought to identify its roots and the rights of technology education as a separate discipline. Several authors differentiate technology education from science education - highlighting differences in philosophical and historical roots in order to assert technology's right to a place in the general education system (Gardner, 1997; Gradwell, 1999).

Patrina (1998) argues that technology education is a multidimensional discipline rather than a monodiscipline subject when considering the historico-political terrain in which it belongs. The adoption of a multidisciplinary approach to technology education is one way of avoiding the restricted views of different perspectives. Williams (1992) adds that "technology strives to go outside of itself while many traditional disciplines are much more introspective" (p. 84). In other words, technology does not protect its own knowledge within a confined academic orthodoxy, but rather encourages lateral thinking in solving practical problems (Williams, 1992). These disparate views reflect the nature of technology education as that of a broad base that includes personal development, both vocational and intellectual. A balance must be maintained between theory and practice, between method and product.

Technology is the major determining feature of the world in which we live; therefore, it is important that it is studied in schools (Jones, 2001). Technology education is now included as an essential key learning area in the curricula in many countries. However, technology education curriculum frameworks differ from country to country due to the different views of educators. This trend is evident in Black's (1994) description of the five perspectives of technology education that have been adopted by most countries to inform the development of a framework for technology education curricula. The five perspectives are: technology as craft skills; technology as design and making artifacts; technology as science; technology as the application of scientific principles; and technology as practical capability. McComick (1993) adds that the development of technology education differs across countries, and is influenced very much by various existing traditions forming interest groups that determine what is to be taught as technology in schools.

Teachers' Perceptions of Technology and Technology Education

Teachers' perceptions of technology and technology education have been examined and explored by researchers such as Jarvis and Rennie (1996), Jones and Carr (1992), Moreland (2003), Rennie (1987), and Symington (1987). Not surprisingly, these studies reveal that teachers hold a diverse range of views on technology and technology education. Symington (1987), for example, found that most primary teachers view technology as modern and sophisticated hardware, like computers, lasers and kidney transplants, while other teachers associate technology with a much broader definition. Rennie (1987) suggests that science teachers perceive science and technology to be related. Some teachers, for example, talked about technology as the application of scientific knowledge and others considered technology education to be the use of machines, making things to fulfill a need, or investigating how things work (Jarvis & Rennie, 1996). According to Rennie (1987), teachers considered technology to be artifacts, services, processes and techniques, problem-solving, improvements in the quality of life, and environment control.

Understanding teachers' perceptions of a subject is seen to be important in technology curriculum development and implementation, and teachers' perceptions of technology significantly affect what and how they teach (Jones & Carr, 1992). Teachers' concepts of technology are influenced by many variables such as background experiences, subject subculture, perceived need for change, teacher support, and personal attitudes towards change (Jones, 2001, 2005). The subject subculture in particular is seen to be strongly influential (Jones, 2001; Jones & Moreland, 2004).

Teacher Professional Development

Bell (2005) points out that an effective professional development program should empower teachers to try out something new by themselves, even well after the professional development process finishes. Jones (2003) says that long-term teacher development that is based on an ongoing process is crucial for effecting change in a classroom practice (see also Bell, 1993; Bell, 2005; Jones & Compton, 1998). This implies a culture of teacher development that sees *continual* professional growth and change as most important for effective teaching practice.

The teachers' positive response to the changes is another clue that indicates the anticipated effectiveness of a professional development program. Teacher response to changes required by any new curriculum is not simple, but rather a complex process. Bell (1993) states that unless the teachers themselves recognize the need for change and see it as preferable to their current situation, no change is likely to occur. However, long-term professional development programs can effect changes in all levels affecting the curriculum. Jones (2003) points out that long-term ongoing professional development in technology education in New Zealand has enabled technology education teachers to have a much broader view of technology and technology education consistent with the intentions of the New Zealand technology curriculum. He goes on to say that time is the crucial factor for developing a technology education culture in teaching technology education in New Zealand.

Teacher development, in itself, is a form of teacher support. Effective teacher development enhances the teacher's confidence in adjusting to new situations. Jones (2003) points out that teacher development is necessary to enhance teacher confidence in implementing a new curriculum. Other factors of teacher support in professional development include teachers talking to each other, school-based support, and including administrative support (Bell, 1993; Bell, 2005; Bell & Gilbert, 1996). Talking to each other about teaching practice is important, both in providing a feeling of support for each teacher and in helping to develop consistency of concepts. It also helps teachers feel more in control of their own development (Bell, 2005). A professional development that empowers teachers to interact and provide support for each other is to some degree indicating its effectiveness. A similar approach was also adopted for a research inquiry into the professional development in technology education in the Solomon Islands.

Thus the development of appropriate concepts in teachers is crucial for a successful implementation of curriculum proposals. In this chapter I consider technology education in the Solomon Islands, and in particular what Solomon Island teachers think of as technology and technology education. I conclude with reflection on the implications of their views on the implementation of technology education in the nation.

SOLOMON ISLAND TEACHERS' VIEWS OF TECHNOLOGY AND TECHNOLOGY EDUCATION

We have investigated Solomon Islands teachers' views on technology and technology education (Sade, unpublished results; Sade & Coll, 2003) through a sociocultural lens (Vygotsky, 1986). We wanted to know technology education teachers' existing views of technology and technology education, and their current classroom practice that may influence the teaching of a technology curriculum, teachers' professional development needs, and the impact of professional development on their technology teaching. The social context in which cognition activity takes place is an integral part of that activity not just its surrounding context. Vygotsky (1986) claims that to understand an individual's psychological development, it is necessary to understand the system of social relations in which the individual lives and grows. In other words, by understanding the circumstances and the participants' construct of the situation, a valid interpretation of the cognitive activity can be made. Hence, we must consider the context in which technology education is delivered, in this case the Solomon Islands.

The Solomon Islands

The Solomon Islands is a small island nation in the Pacific region. The Solomon Islands gained formal independence from the British government in 1978. The main inhabitants of the Solomon Islands are Melanesians with a small number of Polynesians and Micronesians, and others. The Solomon Islands consists of six major islands and approximately 992 smaller islands, atolls and reefs, stretching over approximately 1,500 square kilometers, with a land area of some 28,369 square kilometers. The climate is maritime tropical with heavy rainfall, an average temperature of 27 degrees Celsius and periodic maritime cyclones (Honan & Harcombe, 1997). Honiara, the capital is located on the main island of Guadalcanal. The capital is the center for the government and commerce, as well the location of the international airport and main seaport. The main industries in the country are fish canning, agro-processing, timber, copra, palm oil and cocoa that form a major part of the country's economy (World Bank Report, 1993). The country's population is estimated to be about 409,000, and is increasing at a rate of 2.8% per annum, with some 47% of Solomon Islanders under the age of 15 years. These unusual demographics have resulted in an alarming unemployment rate, especially amongst youth (Kudu, 2001). There are nine main provinces in the Solomon Islands, and about 86 different languages across the country. Although Solomon pidgin is the main spoken language, English is official language (Honan & Harcombe, 1997).

Solomon Island Teachers' Perceptions of Technology and Technology Education

Solomon Island teachers' hold a variety of views of technology and technology education. *Technology* is seen as: an artifact; as a process of making something; and as the application of knowledge. Teachers who think of technology as an artifact talk about technology as either a foreign thing or a new thing, which did not exist in the country before. Teachers who see technology as a process of making something, talk about technology in terms of how things are made both in schools and in society generally. Teachers who see technology as an application of knowledge talk about technology in terms of science and see it as applying scientific knowledge to construct something industrial or involving construction.

Solomon Island teachers' views of *technology education* are related to their views of technology. So technology education is seen as hands-on activities or vocational training involving preparing students for a trade (e.g., in construction). Technology education also is perceived as educating people to become creative, and making new things or putting new ideas into making new products. Other teachers see little difference between technology and science, so technology education is seen as coming from science, and being about developing things that 'work'.

Technology Teaching in the Solomon Islands

Technology teaching in the Solomon Islands involves heavy use of prescribed teaching documents, is rather fragmented in terms of the relationship between teaching of theoretical material and practical lessons, and overall presents technology rather narrowly as a form of technical training. Solomon Island teaching documents are mainly based on detailed curriculum documents and textbooks produced to implement the curriculum. Teachers tend to use them in a very formulaic manner, as recipe books on which their teaching lessons can be based. Learning outcomes are seldom clearly stated, because they do not appear in textbooks. The teaching of particular procedures is the key emphasis in most lessons.

Technology theoretical lessons are taught separately from practical lessons, and this is influenced by two factors. First, by the way school timetables are organized, in which case teachers have no choice but to organize lessons according to the school timetable. Second, the background experience of teachers varies, with some keen on practical work, and others keen on theoretical work. Third, teachers often prefer to teach theoretical lessons in isolation from the practical lessons or tasks.

Most teachers' teaching approaches focus only on a narrow technical approach to technology. The teachers' practical tasks which give students hands-on experiences, are closed tasks confined to prescriptions outlined in textbooks or prescribed lessons. Students then undertake closed tasks, following procedures exactly as prescribed. Lessons are highly teacher-dominated meaning that almost all student tasks are directed by the teachers, and the teachers' instructions are precisely followed by students. The students have little input on how the task should or could be done. Similarly, most classroom tasks are taken directly from curriculum-based textbooks. Again making student learning very structured and tightly confined. Finally, the lessons taught, practical or theoretical, are all held in classrooms. Lessons seldom have either links to the community, or expose students to any community-based technological practices.

A key influential factor in the delivery of technology education in the Solomon Islands lies in the assessment practices, and even formative assessment, tends to be very one-sided. Even when teachers interact with students through questioning, they do not wait for students to respond. Students' responses are thus limited to the one 'right answer', as most questions asked by teachers are 'tell-me' type questions. Likewise, summative assessment practices are limited to written tests and assessment of students' skills based on practical tasks.

Professional Development for Technology and Technology Education

Based on the above views of technology and technology education, we developed a teacher professional development program based on ideas of teacher reflection, teacher support, and on-going professional development. The program focused on developing teachers' views of the nature of technology and learning in technology education, assisting teachers in planning for effective technology teaching and introducing assessment for effective technology learning. Encouragingly, this led to changes in teachers' perceptions of technology and technology education, their teaching documents, pedagogy, and assessment practices.

The teachers' changed their perceptions of technology, which broadened to view technology as the combined application of knowledge, skills and tools or resources, and technology as a process. Teachers who perceive technology as combined application of knowledge talk about technology as the use of knowledge, skills, and resources to solve practical problems *in real life situations*, in which society or individuals used knowledge, skills, resources and tools to solve problems or meet needs of societies or individuals. Likewise, technology as a process means the process by which society, communities and individuals identify problems, and seek to solve them to meet a specific need or solve a specific problem.

In a similar way, Solomon Island Teachers' views of technology broadened, with technology education being seen as developing technological knowledge, and technology education being more than a vocational path, but also about acquiring skills. Likewise their teaching practice changed; in particular they reduced their reliance on the use of textbooks, used new approaches to technology teaching, and used more innovative assessment practices. So there is now more focus on design that allows teachers to identify problem tasks for students to follow through and identify processes to find solutions. This focus on a design process approach also enables teachers to integrate their theoretical lessons with their practical lessons in a timely manner. Teaching has become more student-focused and less teacher-dominated. So students identify problems, and seek additional information from the community, visit various industries related to their assigned tasks. As one teacher, stated: "I sent my students to go down and visit the industries in the city, and seek from the knowledgeable local people in villages". Students are now more active in their learning as the focus of the teaching approach has shifted from teacher-centered to student-centered learning. Teachers' formative interactions are now more like a two-way conversation, as the teacher and students talk about the tasks undertaken. Likewise, summative assessment includes additional student work, such as student self assessment, student design journals, (design folio) and skills assessment of students' completed tasks.

CONCLUSIONS

Technology education in the Solomon Islands is traditionally perceived by many teachers in a very simplistic fashion consistent with international literature which highlighted that both primary and secondary school teachers often have a narrow perspective of technology and technology education (Jones & Carr, 1992; Rennie, 1987). Solomon Islands secondary school teachers' perceptions of technology predominantly focus on the physical elements of technology with technology education in turn focusing on making of artifacts or hands-on activities. Technology teachers in the Solomon Islands typically come from a technical education background, for which their primary tasks are in making things, so it is not surprising this perception remains today (Sade & Coll, 2003). However, on a positive note, recent work on enhancing teachers' perceptions of technology and technology education suggest that a well structured teacher professional development program can enable Solomon Islands teachers to become technologically literate; changing their views of technology and technology education to become more broad and to include the relationship between technology and society (Daker, 2005).

REFERENCES AND BIBLIOGRAPHY

Adams, J. (1993). *Flying buttresses, entropy and o-rings: The world of engineers*. Cambridge, MA: Harvard University Press.

Bell, B. (1993). *I know about LISP but how do I put it into practice? Final report of the Learning in Science Project (Teacher Development)*. Hamilton, New Zealand: Centre for Science and Mathematics Education Research, University of Waikato.

Bell, B. (2005). *Learning in science: The Waikato research*. London: Routledge-Falmer.

Bell, B., & Gilbert, J. (1996). *Teacher development: A model from science education*. London: Falmer.

Black, P.J. (1994). Technology in the school curriculum. *Paper 10 from Science, Mathematics and Technology (SMT) Education in OECD countries* (pp. 114-117). Paris: OECD.

Custer, R.L. (1995). Examining the dimensions of technology. *International Journal of Technology and Design Education, 5*(3), 219-244.

Cutcliffe, S.H. (1981). Technological literacy for non-technologists: In *Proceedings of the Technology Education Symposium II*. Menomonie, WI: University of Wisconsin-Stout.

Daker, J. (2005). Technology education as solo activity or socially constructed learning. *International Journal of Technology and Design Education, 15*, 78–89.

Dyrenfurth, M.J. (1984). Literacy for technological world. *Information Series No. 266*. Columbus, OH: The National Centre for Vocational Education.

Gardner, P. (1997). The roots of technology and science: A philosophical and historical view. *International Journal of Technology and Design Education, 7*, 13-20.

Gradwell, J.B. (1999). The immensity of technology ... and the role of the individual. *International Journal of Technology and Design Education, 9*, 241-267.

Hansen, R., & Froelich, M. (1994). Defining technology and technology education: A crisis, or cause for celebration? *International Journal of Technology and Design Education, 4*, 179-206.

Honan, M., & Harcombe, D. (1997). *Solomon Islands*. Victoria, Australia: Lonely Planet.

Jarvis, T., & Rennie, L. (1996). Perceptions about technology held by primary teachers in England. *Research in Science and Technology Education, 14*, 43-54.

Johnson, J.R. (1989). *Technology: A report of the Project 2061 Phase I Technology Panel*. Washington, DC: American Association for the Advancement of Science.

Jones., A. (2001). Theme issue: Developing research in technology education. *International Journal of Technology and Design Education, 31*, 3 - 14

Jones, A. (2003). Developing a national curriculum in technology for New Zealand. *International Journal of Technology and Design Education, 13*, 83-99.

Jones, A. (2005). The developing field of technology education in New Zealand: The last twenty years. *PATT Conference Proceedings*. PATT – 15; ITEA. Retrieved 10 August 2005, from: http://www.iteaconnect.org/PATT15/Jones.pdf

Jones, A., & Carr, M. (1992). Teachers' perceptions of technology education: Implications for curriculum innovation. *Research in Science Education, 22*, 230-239.

Jones, A., & Compton, V. (1998). Towards a model for teacher development in technology education: From research to practice. *International Journal of Technology and Design Education, 8*, 51-65.

Jones, A., & Moreland, J. (2004). Enhancing practicing primary teachers' pedagogical content knowledge in technology. *International Journal of Technology and Design Education, 14*, 121-140.

Kudu, D. (2001, July). *Population trends and their impacts on human resource development and planning*. Paper presented at the leaders seminar, Ministry of National Planning and Human Resources Development. Retrieved 10 July 2007 from hpp://www.peoplesfirst.net.sb/General /Population %20Seminer.htm

Lux, D.G. (1983, April). *Science and technology: A new alliance*. Paper presented at the American Industrial Arts Association Annual Conference. Milwaukee, WI.

Maykut, P., & Morehouse, R. (1994). *Beginning qualitative research*. London: Falmer.

McComick, R. (1993). The evolution of current practice in technology education. In R. McComick, P. Murphy & M. Harrison (Eds.), *Teaching and learning technology* (pp. 3-14). Workingham, UK: Addision-Wesley & The Open University.
Mitcham, C. (1994). *Thinking through technology*. Chicago, IL: University of Illinois Press.
Moreland, J.P. (2003). *Becoming effective technology teachers*. Unpublished PhD thesis, University of Waikato, Hamilton, New Zealand.
Naughton. J. (1992). What is technology? In R. MacCormick, C. Newey & J. Sparkes, (Eds.), *Technology for technology education* (pp. 3-10). London: Open University.
Northover, B.A. (1997). *Teacher development in biotechnology: Teachers' perceptions and practice.* Unpublished MEd thesis, University of Waikato, Hamilton, New Zealand.
Pacey, A. (1993). *The culture of technology*. Oxford, UK: Blackwell.
Petrina, S. (1998). Multi-disciplinary technology education. *International Journal of Technology and Design Education, 2*, 103-138.
Rennie, L. (1987). Teachers' and pupils' perceptions of technology and the implications for curriculum. *Research in Science and Technology Education, 2*, 121-133.
Sade, D., & Coll. R.K., (2003). Technology and technology education: Views of some Solomon Island primary teachers and curriculum development officers. *International Journal of Science and Mathematics Education, 1*, 87-114.
Symington, D.J. (1987). Technology in the primary school curriculum: Teachers' ideas. *Research in Science and Technology Education, 2*, 167-172.
Vygostsy, L. (1986). *Thought and language* (trans A. Kozulin.) Cambridge, MA: Massachusetts Institute of Technology Press.
Williams, P.J. (1992). Technology as a discipline. In *ACET Tenth Biennial National Conference Proceedings* (pp. 83-88). Gold Coast, Queensland, Australia.
World Bank Report (1993, May). Post-secondary education and the labour market: Issues and options. *Pacific Islands Series, 5*(2). Solomon Islands: Pacific Regional Post Secondary Education Study.

AFFILIATION

David Sade
Centre for Science & Technology Education Research
University of Waikato
New Zealand

TAYLOR, TALOGA AND ALI

IMPROVING PRIMARY SCIENCE EDUCATION IN FIJI BY USING A MULTIFACETED APPROACH

INTRODUCTION

The value of science education as a development strategy for the economic and technological modernization of developing nations is now widely acknowledged (Brown-Acquaye, 2001). Benavot (1992) has reported that the time spent on primary science correlates positively with economic growth rate for a variety of nations. At societal level, science education has the potential to improve living conditions through addressing local problems with respect to such basic needs as clean water, sound nutrition, and personal health (Lewin, 1993). Consequently, improved science education has been placed high on the agenda of tasks to be tackled in many developing countries (Kahn, 1990). Such counties have invested heavily in school science education since the 1960s, but by the beginning of the 1990s, concerns about instructional quality and student achievement were becoming acute (Lewin, 1993). Gray (1999) argues that the last few decades have seen a steady decline in the quality of science education in most developing countries.

This rather depressing view is reiterated by Lewin (1990) who comments that although far more children in developing countries study science than previously, research suggests that the great majority do not master more than a small proportion of the goals set for them. It seems that efforts to teach science in developing countries are often met with rote learning of strange concepts, mere copying, and a general lack of understanding on the part of local students (see, e.g., Hewson, 1988). Indeed the continued 'backwardness' of many developing countries in relation to science and technology is claimed by some authors to be the result of the quality of education, rather than the quantity given in these countries (e.g., Mbajiorgu & Iloputaife, 2001).

PRIMARY SCIENCE EDUCATION IN FIJI

Many of the generalizations outlined in the introduction can be applied to primary science education in Fiji, the context for this chapter; in particular, a strong culture of didactic teaching which encourages rote learning. Research on primary science education in Fiji is limited and somewhat dated, but what research there is presents a rather depressing picture. In the 1980s Muralidhar (1989) observed 289 lessons

Richard K. Coll, Neil Taylor (eds.), Science Education in Context: An International Examination of the Influence of Context on Science Curricula Development and Implementation, 55–65.

of the Fiji Basic Science Curriculum, and noted that students were rarely actively involved in what was intended to be an activity-based science program. Instead, teachers tended to employ a formalistic, teacher-centered style of science delivery in which they adhered rigidly to the prescribed text. As in other studies reported in the literature, lack of confidence was cited as a contributing factor: "Teachers' lack of confidence in handling certain activities sometimes resulted in their skipping an activity entirely or abandoning the activity halfway; students were simply asked to copy the expected results into their books" (p. 281).

Work by Taylor and Coll (1999) on Fiji pre-service primary teachers' self efficacy using the Science Teaching Efficacy Beliefs Instrument (STEBI) produced findings seemingly at odds with Muralidhar's assertions, as pre-service primary teachers scored higher on this instrument than similar cohorts of Western students. However, follow up interviews suggested that the Fiji participants attributed a very high status to highly prescriptive teachers' guides as a means of support for teachers. Thus, although self-efficacy appeared high, their competence was perhaps questionable, certainly in terms of their ability to provide worthwhile science learning experiences for their students.

The problem of extremely didactic science teaching in Fiji has been exacerbated by an intense regime of summative examinations. Students sit for these in Years 6, 8, 10, 12 and 13 of their schooling. Fiji is typical of the situation in many developing countries in which, according to Vulliamy (1988), the content and style of national examinations tend to be more important determinants of the content and process of teaching than syllabuses. Vulliamy (1988) goes on to point out that the vast majority of questions asked in school examinations in developing countries test factual recall rather than comprehension or application skills. This encourages rote learning of factual information that is promoted by a formalistic, didactic style of teaching. Such a situation prevails in Fiji where science examinations tend to concentrate on testing lower level cognitive skills such as recall (Taylor, 1991; Taylor & Macpherson, 1992a). This is a major contributing factor to a formalistic, teacher-centered style of science delivery, which is at odds with the apparent aims and objectives of the science courses developed for Fiji schools (Muralidhar, 1989; Taylor & Macpherson, 1992b).

The situation in Fiji is further complicated because of a marked difference in performance in science between the two main ethnic groups, ethnic Indians and indigenous Fijians, with Indians consistently performing better than Fijians in science examinations and tests (Fiji Government, 2000; Stewart, 1983; Tavola, 1992; Taylor, 1991). This is supported by data for summative examinations from the Ministry of Education (Macpherson & Taylor, 1994). Nabobo and Teasdale (1994) suggest that the formal school system in Fiji which, in general, places a strong emphasis on rote learning may be at odds with traditional Fijian ways of learning. Greenfield (1996) whose research in Hawai'i revealed that Hawaiian (Polynesian) students demonstrated low achievement and expressed the lowest educational aspirations related to science when compared to other ethnic groups, also points to school-culture incompatibility. She suggests that while traditional classrooms tend to stress student-adult interactions, extended time commitments, and rote learning, Hawaiian culture puts greater emphasis on peer relationships and interactions, periods of hard work interspersed with periods of leisure, and on-the-job 'learning by doing' - preferably with friends.

Work by Taylor (1997), Taylor and Lucas (2001) and Taylor, Lucas and Watters (1999) indicates that the same may be true for Fijian students. These authors reported that science taught within a constructivist learning environment, involving extensive use of group discussion based around problem-solving activities, significantly improved the performance of Fijian students without diminishing the performance of Indian students.

RECENT DEVELOPMENTS IN PRIMARY SCIENCE EDUCATION IN FIJI

Clearly, primary science education in Fiji has been facing significant problems. However, significant efforts are currently being made to improve this situation, with the Fiji Government (assisted by AusAID) adopting a multifaceted approach to change in primary science education as part of a strategic plan for development of education up until the year 2020 (Fiji Government, 2000). This approach has drawn heavily on a model developed by Goodrum, Hackling and Rennie (2001) (Figure 1) that links professional development experiences (pre- and in-service) to curriculum resources that exemplify best practices and participative inquiry where teachers can implement, analyze and reflect on their classroom practices including inquiry and assessment.

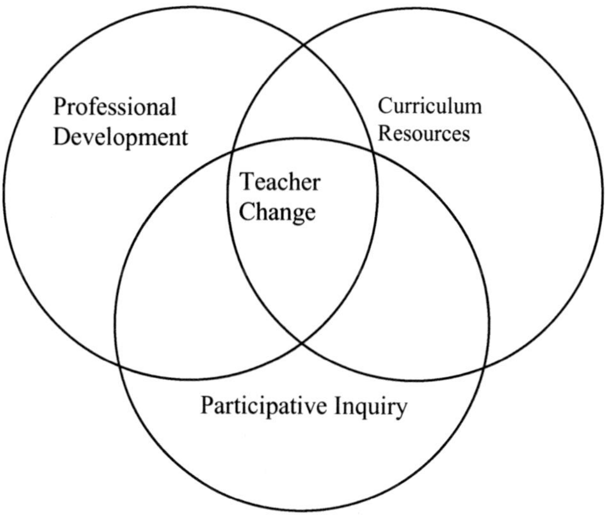

Figure 1. The role of professional development, curriculum resources and participative inquiry in teacher change (Goodrum, Hackling & Rennie, 2001)

This multifaceted or holistic approach to changing primary science education marks a significant move away from previous initiatives that were often piecemeal and had little impact or failed completely.

The remainder of this chapter examines one of the initiatives within the multifaceted approach in some detail, that of upgrading primary teacher training in science at Lautoka Teachers' College (LTC). It goes on to place this initiative in the context of the wider agenda of change that is taking place in primary science in Fiji.

UPGRADING PRIMARY TEACHER EDUCATION IN SCIENCE

The primary teacher education upgrade at LTC involved developing a new diploma program that would articulate with BEd degree offered by the University of the South Pacific in Fiji. It was also one part of the overall model for improving primary science education across the country as within the diploma a new science curriculum was developed. This comprised two compulsory core courses in science and a foundation science course for students who specialized in the humanities at secondary school.

Development of the New Science Curriculum

Prior to writing the new science curriculum, a subject Advisory Group was established to identify possible content. This Advisory Group comprised a variety of stakeholders including representatives from the Ministry of Education, the Curriculum Development Unit, the University of the South Pacific, primary school principals, practicing primary teachers and the college science lecturers. The group mapped out the aims, rationale and some general content, based on their experience of the problems encountered by primary teachers when teaching science. The curriculum was intended to develop both content knowledge and pedagogical skills, thus the initial sessions provided an explanation of the constructivist theory as the pedagogy underpinning course. This was to provide the pre-service teachers with some insight into why it is important to teach science in an active 'hands-on' manner. This approach was to be modeled by the lecturers in all of the workshop sessions. Having developed a general content outline, the college lecturers then worked with an overseas consultant to write the course material in detail. This process involved refining the content identified by the Advisory Group, developing a generic workshop format for the course, identifying learning outcomes, teaching activities to meet these and appropriate assessment strategies to assess the learning outcomes of each workshop.

It was decided to adopt a problem-solving approach within the workshop sessions, where pre-service teachers would research or be provided with background theory on a topic and would then have to apply this knowledge to explain simple problems. These problem-solving activities would be conducted within a small group format to allow for the social construction of knowledge through discussion. Research undertaken by Taylor (1997) at the same teachers' college had shown this approach to be effective in teaching science concepts to both Indian and Fijian students. Furthermore, such a strategy was likely to be quite different from anything most pre-service teachers had encountered during their schooling as science in Fiji schools is generally presented in a very didactic manner.

Figure 2 shows the general format used within the workshop sessions and content covered. A series of recent articles were identified from science teaching journals to complement each of the main topics covered in the science courses and compiled into a course reader. The pre-service teachers were expected to read articles pertaining to any topic in advance of that particular session.

The new curriculum was implemented at the start of 2005. Ongoing evaluation of the courses was conducted during the implementation and the data from this used to produce a report that was subsequently tabled at a final meeting of the Advisory Committee at which lecturers and students also commented on the course. What follows is a description of how the curriculum was evaluated and a discussion of the general findings from that.

THE EVALUATION

A number of data sources were used during the evaluation of the new science curriculum: reflective journals, focus group interviews, and individual interviews.

Reflective Journals

All of the pre-service teachers who completed the science course were given some instruction on the process of keeping a reflective journal. They were asked to make an entry at the end of each workshop noting what they felt they had learnt during the sessions and what pedagogies they had found particularly effective and why. At the end of the course the diaries were collected and analyzed by reading the entries for each session and identifying themes. The themes were largely based on the various teaching strategies used during the course. The lecturers also completed reflective diaries.

Focus Group Interviews

Three focus group interviews were conducted by the overseas consultant. In each case a convenience sample of pre-service teachers was used. A total of 12 first-year students were involved and the sample comprised equal representation of gender and ethnicity. During the interviews pre-service teachers were asked to compare the teaching in the new science courses with their school experiences of science.

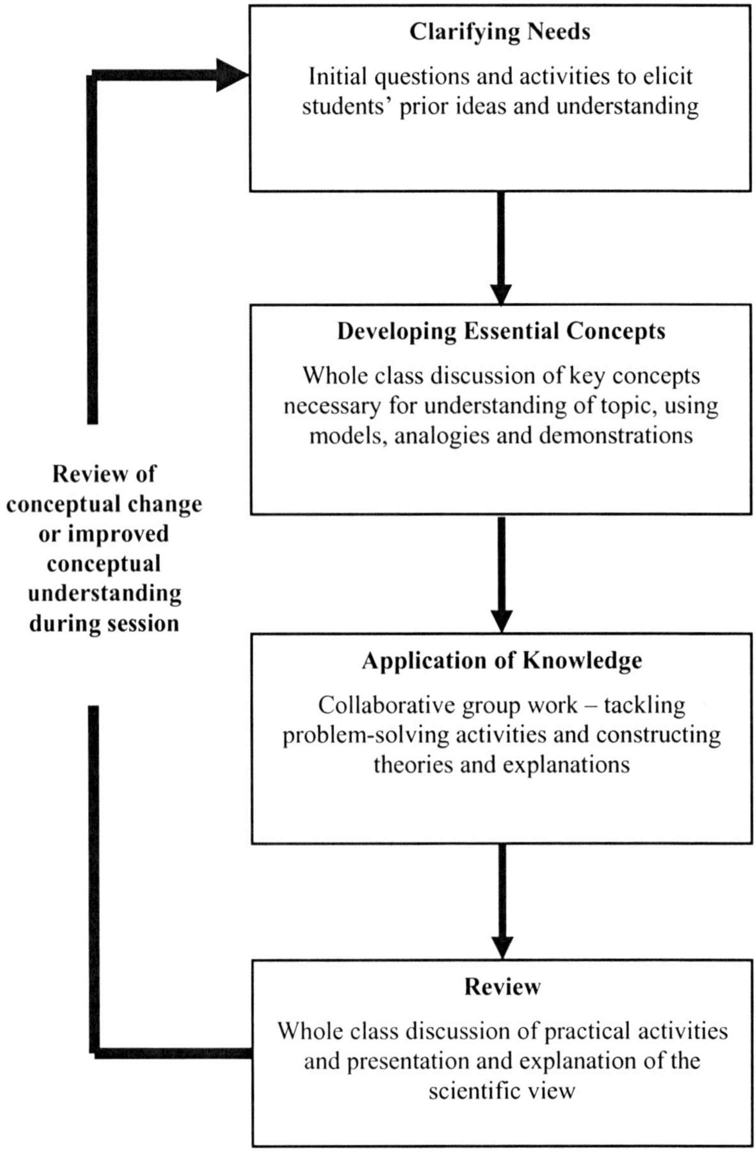

Figure 2. A flow chart of the sequence of a typical two hour science session within the new curriculum

They were also asked to comment on the overall learning environment in which the course had been presented and specific pedagogies employed. Finally, some probing of the pre-service teachers' science content knowledge took place as they were asked to provide scientific explanations of certain concepts that had been covered during the course

Individual Interviews

Individual interviews were conducted with each of the lecturers involved in delivering the new science courses. All interviews were audio-taped and transcribed and analyzed in a similar manner to the reflective journals.

FINDINGS FROM THE EVALUATION

Lecturers' Comments on the Course

A number of the lecturers' reflections and comments indicated that they felt at home using a constructivist approach with their pre-service teachers even though this type of learning environment, with a student-centered focus, is very unusual in Fiji. One lecturer commented: "Mostly I played the role of facilitator or advisor, providing opportunities for discussion clarifying points and raising questions which enable students to undertake further research and investigations."

This comment indicated that this lecturer had considerable insight as to how a constructivist learning environment might be established. Rather than directing the students' learning, a very common approach in Fiji, she provided opportunities for student-centered learning and then scaffolded that learning. Furthermore, the lecturers were aware of the importance of effective questioning and the need to elicit prior conceptions in a constructivist learning environment. Employing this approach seemed to have given this lecturer considerable insight into her students' prior conceptions in various areas of science and offered a basis on which to design activities and learning experiences to address these.

Pre-service Teachers' Comments on the Course

Comments from the pre-service teachers confirmed that the learning environment was constructivist in nature and that the lecturers acted as facilitators rather than being directive. The pre-service teachers also compared their experience of the new science very favorably with their school science that for many had featured highly didactic teaching. For some the contrast with school science had changed their view of science teaching and learning in a positive way.

Furthermore, most pre-service teachers could see the rationale for a constructivist approach to the teaching and learning of science and realized that this would result in better educational outcomes for their own students than the traditional transmissive or didactic approach. However, all of the pre-service teachers interviewed were aware of the influence the examination regime in Fiji had on teaching styles. The pre-service teachers claimed that they enjoyed the hands-on nature of the course, and some commented favorably on some of the specific pedagogies used within the constructivist environment: "We realized that group work developed team spirit and as we progressed we learned to work together and contributed equally by doing what each of us can do best". There was also evidence from the interviews that pre-service teachers had improved both their pedagogical and content knowledge of science. They could provide plausible explanations of difficult concepts such as friction and density, and they could offer good examples of pedagogy for delivering these.

DISCUSSION

In developing the new science curriculum course at LTC, a genuine attempt was made to employ constructivist principles of teaching and learning, and expose pre-service teachers to forms of pedagogy they were unlikely to have encountered during their school science career. Furthermore, the evaluation of the curriculum indicated that it was well received by pre-service teachers and appeared to have developed both their content and pedagogical knowledge in science.

However, as a stand-alone initiative it would probably have little impact on the teaching of primary science in Fiji. Given the present primary science curriculum and assessment regime in Fiji, it seems unlikely that these pre-service teachers would attempt to teach science within a constructivist learning environment on their own imitative. Such are the demands of various groups within society for examination success within a system largely testing lower level cognitive skills that many teachers revert to rote learning strategies that have a proven track record. 'Drilling' and 'cramming' are pedagogical strategies commonly used in primary science for this reason (Taylor, 1991; Taylor & Macpherson, 1992a). Consequently many previous curricular and professional development initiatives have delivered little change as these were largely stifled by the prevailing examination system that remained constant (Cook & Taylor, 1995; Taylor, Vlaardingerbroek & Coll, 2003).

This time, however, there has been considerable vision on the part of the Fiji Government who, instead of approaching educational reform in a piecemeal manner, has adopted a more holistic/multifaceted approach to change. This began with the Commission Report of 2000 that highlighted the following issues:

- the impact of the examinations on teaching encourages: didactic, teacher-centered approaches; teaching for rote learning rather than understanding; training to pass the test; shallow rather than deep learning;
- the need for students to learn to think, apply, reason;
- the need for greater flexibility in the curriculum, so teachers can contextualize to increase the relevance and meaning;
- the need to reduce the 'academic' nature of the curriculum; and
- life long skills such as decision-making, critical thinking, problem-solving not taught well in science lessons.

Furthermore, the report recognized that Years 7 and 8 of primary schooling in science were a major cause for concern and therefore a target area for reform:

> The majority of teachers in the target Years 7 and 8 do not have a formal science background and therefore have limited deep science knowledge and have had little experience with doing scientific investigation. This results in a lack confidence to move away from a structured, teacher-centered approach to more activity based, student-centered inquiry processes where students are exploring their own ideas. (Fiji Island Education Commission, p. 35)

The Commission Report of 2000 was followed by a 'Blueprint for Education in Fiji'. This is now being implemented and will involve an entire overhaul of the entire primary curriculum in Fiji in the next few years, with the reform of science education leading the way with a pilot project. Science at the primary school level has become a Key Learning Area and an outcomes approach has been adopted for the new curriculum. This mirrors the approach adopted in the new science curriculum at Lautoka Teachers' College and should help provide a consistent approach to the teaching of science.

Furthermore, a number of new resources in the form of Teacher and Pupil Books have been developed to meet the outcomes of the new curriculum. These resources have adopted a modular and integrated approach to the teaching of primary science. So, fro example, the module entitled 'Investigating Change' developed for Year 5 examines both physical and biological change. The activities within the module are firmly grounded in a constructivist view of teaching, and learning, and each module has an introductory explanation of constructivism and a rationale for adopting this approach. These new resources are crucial to achieving change in primary science in Fiji as the existing resources date back to 1976 and are highly prescriptive.

A system for progressively in-servicing existing primary teachers with the new resources is being developed and is likely to involve residential components at Lautoka Teachers' College to allow all teachers both pre- and in-service to be exposed to the same form of training in science education. Most importantly, there are also plans to change the assessment regime by reducing the emphasis on national summative examinations. There are plans to remove the Years 6 and 8 summative examinations and replace these with internal assessment carried out by classroom teachers. This should allow teachers the freedom to employ more appropriate models of pedagogy for developing deep understanding and should articulate well with the changes taking place in primary teacher education. In so doing, this overall initiative may improve the quality of science education as advocated by authors such as Mbajiorgu and Iloputaife (2001).

The impact of these combined changes remains to be seen, but if they are successful in changing the didactic culture of primary science in Fiji they may well be of particular significance to other countries in the Pacific region.

REFERENCES AND BIBLIOGRAPHY

Benavot, A. (1992). Curricular content, educational expansion, and economic growth. *Comparative Education Review, 36*(2), 150-174.

Brown-Acquaye, H.A. (2001). Each is necessary and none is redundant: The need for science education in developing countries. *Science Education*, *85*(1), 68-70.

Cook, A., & Taylor, N. (1995). The sustainability of aided curriculum development projects: The need for robust adaptive processes. *Pacific-Asian Education Journal*, *7*(1&2), 67-71.

Fiji Islands Education Commission. (2000). *Fiji 2020.* Suva, Fiji: Government Printery.

Goodrum, D., Hackling, M., & Rennie, L. (2001). *The status and quality of science education in Australian schools.* Canberra: Department of Education, Training and Youth Affairs.

Gray, B.V. (1999) Science education in the developing world: Issues and considerations. *Journal of Research in Science Teaching*, *36*(3), 261-268.

Greenfield, T. A. (1996). Gender, ethnicity, science achievement and attitudes. *Journal of Research in Science Teaching*, *33*(8), 901-933.

Hewson, M.G. (1988). The ecological context of knowledge: Implications for learning science in developing countries. *Journal of Curriculum Studies*, *20*(4), 317-326.

Kahn, M. (1990). Paradigm lost: The importance of practical work in school science from a developing country perspective. *Studies in Science Education*, *18*, 127-136.

Lewin, K. (1990). International perspectives on the development of science education: Food for thought. *Studies in Science Education*, *18*, 1-23.

Lewin. K. (1993). Planning policy on science education in developing countries. *International Journal of Science Education*, *15*(1), 1-15.

Macpherson, C., & Taylor, N. (1994). The national trial of the new standardised test of achievement in primary science for Fiji: A comparison of sub-groups performances within the sample population. *Journal of Science and Mathematics Education in South East Asia*, *17*(1), 8-14.

Mbajiorgu, N.M., & Iloputaife, E.C. (2001). Combating stereotypes of the scientist among pre-service science teachers in Nigeria. *Research in Science and Technological Education*, *19*, 55-68.

Muralidhar, S. (1989). *An exploratory study of a science curriculum in action*: *Basic science in Fiji.* Unpublished PhD thesis, Monash University, Melbourne, Australia.

Nabobo, U., & Teasdale, J. (1994). Teacher education for cultural identity in Fiji. *Directions*, *16*(2), 3-13.

Stewart, R.A.C. (1983). Fijian education: Its special demands. *Directions*, *10*(1), 12-26.

Tavola, H. (1992). Race is not the only issue in education in Fiji. *Directions*, *14*(2), 34-40.

Taylor, N. (1997). *Fiji pre-service primary teachers' understanding of physical science*: *A cultural perspective.* Unpublished PhD thesis, Queensland University of Technology, Brisbane, Australia.

Taylor, N. (1991). The Fiji Junior Certificate Basic Science Examination: Its implications for the teaching of science in Fiji. *Journal of Science and Mathematics Education in South East Asia*, *14*(2), 73-78.

Taylor, N., & Coll, R.K. (1999). Pre-service primary teachers' self efficacy in science teaching: The Fiji perspective. *Pacific-Asian Education Journal*, *11*(1), 32-45.

Taylor, N., & Lucas, K. (2001). Some learning outcomes from a science program for pre-service primary teachers undertaken in the context of a developing country. *Evaluation and Research in Education*, *15*(4), 228-249.

Taylor, N., Lucas, K., & Watters, J.J. (1999). Collaborative science activities and the social construction of understanding of physical science concepts by pre-service teachers in Fiji. *Research in Science Education*, *29*(4), 479- 500.

Taylor, N., & Macpherson, C.M. (1992a). Developing the Standardised Test of Achievement in Science for Fiji. *Journal of Science and Mathematics Education in South East Asia*, *15*(1), 19-30.

Taylor, N., & Macpherson, C.M. (1992b). Primary science in Fiji: Some reflections on activity-based lessons. *Journal of Science and Mathematics Education in South East Asia*, *15*(2), 40-45.

Taylor, N., Vlaardingerbroek, B., & Coll, R.K. (2003). Exploiting curriculum commonality in small island states: Some strategies for primary science curriculum development in the South Pacific. *International Journal of Science and Mathematics Education*, *1*(2), 157-174.

Vulliamy, G. (1988). Third world schools. In S. Briceno & D.C. Pit (Eds.), *New ideas in environmental education* (pp. 143-157). London: Croom Helm.

AFFILIATIONS

Neil Taylor
School of Education
University of New England
Australia

Kelera Taloga
Ministry of Education
Fiji

Sadaquat Ali
Department of Chemistry
University of the South Pacific
Fiji

CARIBBEAN

SADLER-MCKNIGHT AND RAINFORD

TOWARDS THE DEVELOPMENT OF A SCIENCE EDUCATION POLICY: AN ANALYSIS OF THE CURRICULUM INTERVENTIONS AND REFORMS OF SECONDARY SCIENCE EDUCATION IN POST-COLONIAL JAMAICA

INTRODUCTION

As a former British colony (1655-1962), the impact of this legacy on Jamaica's education system has been significant. The vast majority of the population is of African descent, sharing a history in which their ancestors served as slaves and were thus uneducated. With the abolition of slavery, the introduction of elementary schools was mainly for the purpose of maintaining the social structure - stratified according to upper, middle, and lower classes with the majority of the population falling into the lower social class. Education has always been viewed as a means of social mobility. Ironically, the many different types of schools that have at some time been a part of the Jamaican secondary education system, have served to short-circuit this anticipated social mobility as most of the nation's poor have little choice but to attend the schools which are perceived to be inferior to those attended by the dominant middle and upper classes. In some of these less prestigious schools, subjects such as science were not taught or barely introduced. In post-colonial Jamaica, the access to secondary education for members of the lower class has improved. Additionally, with the many different policies and curriculum reforms, all secondary students are required to do a common core of subjects which include one science subject. Inquiry-based methods are recommended for use in delivering science lessons. Although students from the lower class have significantly greater access to secondary schools, they are still under represented in the island's more prestigious secondary institutions where the quality of the education is perceived as being superior, and to which greater value is ascribed by employers, parents, and society as a whole.

Of particular concern is the persistently low level of performance in science among primary and secondary students in national and regional examinations. Various initiatives have not significantly influenced students' perceptions of science, interest in science or motivation to pursue careers in either science or science teaching. Further, the performance of students from the lower social class is significantly below that of their counterparts from the upper social classes in national and regional examinations.

Richard K. Coll, Neil Taylor (eds.), Science Education in Context: An International Examination of the Influence of Context on Science Curricula Development and Implementation, 69–81.

The failure of various curriculum reforms to impact significantly on teachers' methodology and students' interests and performance in science require an examination of some deeper issues associated with teaching such as the historical context from which our present secondary education system has emerged, an analysis of the reforms that have been introduced to improve science education as well as wider institutional and policy issues that impact on the curriculum delivery process.

HISTORICAL DEVELOPMENT OF SECONDARY SCHOOL EDUCATION IN JAMAICA

The establishment of a secondary education system in Jamaica took place through legislation in 1879. Schools under this system were funded through interest from trust funds or through fees, and were located mainly in urban areas catering to the needs of elite light-skinned upper class individuals. In a few instances, children of the working class who were very bright gained entry to these institutions on the basis of scholarships or as a token. The relatively few students who were able to penetrate the barrier to gain secondary education became the emerging middle class and, in this regard, education served the purpose of social mobility for the predominantly poor Negro population who were descendants of ex-slaves. In the majority of instances, lower and middle class blacks were catered for by government-run elementary schools and the teachers' college systems. Both systems of education were fashioned on that of the British, with very little focus on anything local. Teachers were either from Britain or were local whites or brown-skinned Jamaicans. Examinations that marked the end of schooling were set by agencies outside of the country. Significant changes to secondary education began with the establishment of the Ministry of Education in 1953 followed by independence in 1962. Sparked by the development of a national education policy which provided a framework for more substantial changes to the education system, the period was marked by an effort to expand the scope and access to education. During this period, greater priority was given to increasing the number of primary and junior high schools, and the construction of several schools was undertaken. Opportunities for post-primary education remained limited to a select few, mainly those that were able to obtain scholarships or persons with a strong financial background. Successive governments since independence have identified, among other things, issues of access, equity, productivity and quality as key factors to be addressed in order to improve the standard of education, and hence the socioeconomic status of the nation's citizens. These issues have become even more critical in this present globalized environment where national, but even more so, global competitiveness is at stake.

In some respects, the policies of various governments have served to widen rather than narrow the gaps in the quality of education offered to the rich and the poor, and hence served to stifle this upward mobility (Miller, 1990). It is also ironic that the many different types of schools that have at some time been a part of the

Jamaican education landscape have managed to keep one section of society on the margins, thereby curtailing the anticipated improvement in social status that education promises.

Research carried out by UNESCO in 1983 revealed that secondary institutions offered a variety of programs with little differentiation and limited achievement. Further, the report stated that there were two types of schools, the 'high school' and the rest, which suggests that up to that point the distinctions between the types of education experienced by the upper class was generally superior to that experienced by the lower class (Education Task Force Report, 2004).

By 1993, there were six different types of schools offering secondary education in Jamaica. These were traditional high schools (THS), technical, comprehensive schools (CHS), new secondary schools (NSS), all age schools (AAS) and vocational schools. These schools varied according to their perception by the wider society, the curriculum used, examinations taken, their mode of entry and the amount of subsidy they received from the Government (Rainford, 1997).

Since 1958, entry to secondary schools has been based on success in a high stakes examination at the end of Grade 6 - the Common Entrance Examination (CEE), and now the Grade 6 Achievement Test (GSAT). Success guaranteed access to the limited places in the island's most prestigious secondary institutions - the traditional high schools (THS). These schools offered classes from Grades 7 to 13 and were widely viewed as the route to a university education. The majority of students who were able to gain access to the relatively few THS were from the upper and middle classes. Success in the CEE resulted in approximately 25% of the cohort being selected for places in the THS where a range of traditional grammar school subjects were offered as compared to the more technical and vocational subjects offered in the New Secondary School (NSS). The THS students did Cambridge O- and A-level examinations at the end of Grades 11 and 13 respectively until 1978 when regional examinations set by the Caribbean Examinations Council were introduced. The Caribbean Secondary Education Certificate Examination (CSEC) replaced the Cambridge O-Level examination and in 2000, the Caribbean Advanced Proficiency Examination (CAPE), replaced the A-Level and continued to provide access to tertiary institutions, and hence higher education. The students in the NSS were expected to sit a locally administered SSC examination which was ascribed very little value by the society. Success in this examination would only permit access to a few professional organizations such as the police force, teaching and nursing and even then only students who achieved very high grades were considered as acceptable for matriculation into training in these areas.

The year 1993 marked a watershed for secondary education in Jamaica as major changes were implemented in an attempt to address problems of access, equity and productivity. For the first time a common curriculum was introduced for use in Grades 7-9 through the Reform of Secondary Education (ROSE) program. Students in non-traditional high schools had greater access to subjects such as science and resource and technology and social studies. The project was also intended to address institutional strengthening and the rationalization of examinations. The Grade 9 achievement test previously taken by Grade 9 students in the all-age school was to be replaced by a new examination - The Junior High School Certificate Examination to be taken at the end of Grade 9 by all schools under the reform. Based on performance in this examination, students from All Ages (AAs) schools would be placed in secondary schools to complete at Grades 10 and 11. Additionally, secondary schools were supposed to be built to provide additional places for the successful candidates. The traditional teacher-centered curriculum was replaced by a more student-centered one. Teachers were trained in the new teaching strategies during the summer by regionally located teams of teacher trainers who visited their classes during the year to support them in the curriculum implementation. Each school that introduced the curriculum received support from these teams for three years.

Successful implementation of the ROSE was, however, not achieved in many secondary schools perhaps because it was not mandated by the Government. Also, several of the traditional high schools were dissatisfied with the quality and standard of the content of some subjects in this program. With respect to science education, the ROSE curriculum required the teaching of integrated science in Grades 7 to 9. The content was considered adequate for a smooth transition to the discrete sciences in Grades 10 and 11. However, in many of the traditional high schools, the Grade 9 integrated science syllabus was replaced with the discrete science subjects (physics, chemistry and biology). This was done by individual schools to address their needs, and it consisted mostly extracts of some of the content from the syllabi for the CSEC examination. This was mainly due to the perception that the ROSE syllabus did not provide adequate foundation on which the Grade 10 syllabi for these subjects could be built. The NSS continued to use the ROSE curriculum to Grade 9.

While the ROSE program achieved the desired goal of a common curriculum available for use in all secondary schools, other components of the program, such as the institutional strengthening, was not as widespread as originally intended. As a follow on to this program, ROSE II was piloted in 2002 to address some of these concerns. This new program was to allocate more focus on the strengthening of institutional support as well greater equity and quality in the curriculum at the higher levels Grades 10 and 11.

Since 2005, the Government has taken a bold step and reclassified all secondary schools as High Schools. While changing the names of the institutions was carried out in a fairly short time, the general ethos, curriculum delivery and other institutional factors have been much slower to change. All students are placed in secondary schools based on their performance in the GSAT examinations but the more prestigious THS continue to get the top performing students. There continues to be disparity in the amount of resources available for funding the programs in these various institutions.

At present there are 245,124 students registered in public and private secondary schools (Table 1). The Government spends in excess of 30 billion Jamaican dollars on education (about 6.6% of the GDP & 9.2% of the Fiscal budget). The largest share of the education budget, approximately 38%, is allocated to secondary education. A very significant source of funding in most secondary schools is, however, through fund-raising initiatives of the parents, alumni and community support, well as through the cost-sharing program (i.e., parental contribution of school fees). It is, therefore, not surprising that the schools with more vibrant PTA and alumni and located in more affluent communities are able to offer a superior education. It was estimated in 2003/4 that the average operating cost per student was approximately $56,600 ($Ja). About 11,900 teachers are employed in the secondary education system, 80% of who are trained and 70% are females. The average pupil:teacher ratio reported in 2004 was 1:23.

Compared with neighbors Barbados and Trinidad, Jamaica's expenditure per student at all levels of the education system is significantly lower (see Table 2), and so too, is its literacy rate. The higher levels of economic development as well as students' performance in the CSEC examinations enjoyed by these other countries suggest that there is a correlation between educational expenditure and economic development.

Table 1. Enrolment in the Jamaican secondary level schools[1]

School Type	*Number of Schools*	*Male*	*Female*	*Total*
All Age (7-9)	348	8,535	3,924	12,459
Primary &Junior High (7-9/10)	89	14,929	8,976	23,905
Secondary High	139	8,531	96,712	182,243
Technical High	14	8,948	9,098	18,046
Agricultural High	2	174	122	296

[1] Source: Jamaica Ministry of Education, Youth and Culture Education Digest, 2003/4, Statistics Section, Planning and Development Division, 2004

Table 2. Expenditure per student for selected West Indian countries (US$) for year 2000[1]

Country	*Primary*	*Secondary*	*Tertiary*	*GDP/Capita*	*Literacy Rate (%)*
Barbados	1,871	2,432	5,634	14,553	97
Trinidad & Tobago	816	734	N/A	8,176	93.5
Jamaica	323	508	1,703	3,561	86.4

[1] Source: Task Force on Educational Reform: Jamaica a Transformed Education System: Final Report (2004)

It is ironic that the many different types of schools that have at some time been a part of the Jamaican education landscape have managed to keep one section of society on the margins, thereby curtailing the anticipated improvement in social status that education promises. Over the years, the various governments have identified issues of access, equity, productivity and quality as key factors to be addressed in order to improve the standard of education and hence the socioeconomic status of the nation's citizens. These issues have become even more critical in the present globalized environment where national, but even more so, global competitiveness is at stake. Science and technology are key components of this development.

INTRODUCTION OF SCIENCE IN THE CURRICULUM

Science in Elementary/Primary Schools

The date for the actual introduction of science in the curriculum is not well documented. In the elementary schools which catered to students from age 6 to 15 years, the curriculum focus in the early years was on the development of numeracy, literacy and vocational skills. Science was not a formal part of the curriculum and aspects of it were limited to domestic science and agricultural studies. In 1980, science was formally introduced in the primary curriculum. However, due to the lack of adequately prepared primary science teachers as well as the introduction of the Common Entrance Examinations, most schools did not seriously implement science as a subject. In more recent years, through the Primary Education Improvement Project, a National Assessment Program (NAP, 1990) was introduced as well as a Grade 6 Achievement Test where more emphasis has been placed on teaching science as a subject from Grades 4 to 6 and integrating it throughout the primary curriculum. The impact of these interventions has been and continues to be the subject of several research projects.

Science in Secondary Schools

The curriculum during the early years of Jamaica's secondary schooling was mainly liberal and classical in its orientation. However, natural science, elementary science and mathematics were taught as separate subjects. By 1914, natural and applied sciences were introduced in the secondary school curriculum. The syllabuses for these courses were identical to those of the London or Cambridge Universities. The range of subjects covered was wide and included chemistry, physics, biology, zoology, botany, agricultural science, human and social biology. In 1972, the governments of the Caribbean established a Caribbean Examinations Council (CXC) and, through this council, new regional curricula, syllabuses and examinations for secondary education in the Caribbean. The examinations CXC/CSEC developed replaced the previously taken General Certificate of

Education (GCE) London and Cambridge examinations. One of the major interventions for science was the introduction of school-based assessment (SBA). This school-based assessment (SBA) component has been a feature of the Council's CSEC examinations for biology chemistry and integrated science since 1985 and for physics since 1993. Introduction of the physics SBA was met with some resistance by the physics examination board (Bernard, 1997). For the science subjects, SBA took the form of practical work organized, implemented and marked by the classroom teacher over a two-year period.

The SBA is now deeply entrenched in the examination format of the CSEC. Initially, the students' SBA grades (after moderation by CXC examiners) were combined with grades obtained through a practical examination in order to arrive at an overall grade for experimental skills. Since 1998, the external practical examinations have been discontinued and the SBA grades now constitute 15% of the grades for experimental skills for chemistry physics and biology. Students' performance on a written paper contributes another 5%, to complete the 20% weighting for experimental skills. In spite of the variety of materials that are now available for teachers to use, the prospect of conducting SBA can be quite daunting. Apart from providing the opportunities for developing the requisite experimental skills, the teacher must present practical work in such a way that it forms a meaningful whole with the rest of the science being taught. Many teachers continue to find this component of the CSEC syllabus particularly challenging to implement.

Since 1998, the Caribbean Advanced Proficiency Examination (CAPE) has been offered by CXC in a range of subjects. The CAPE examination is the replacement for the GCE Advanced level and allows for easy matriculation of students into regional and extra regional universities as well as other professional courses. Overall, the regional collaborative approach used to develop these curricula and examinations has played a significant role in fostering awareness and understanding among Caribbean peoples of the important contribution that the Caribbean makes to the global economy. It has also contributed to the professional development of teachers in the region through training workshops and marking exercises that permit teachers from various countries to continuously network and contribute to improving the system. In other words, there is the development of an important and significant Caribbean school system.

STUDENTS' PERFORMANCE IN AND ATTITUDE TOWARDS SCIENCE

Despite the many initiatives taken to reform secondary education, for example, access to science for all students, rationalization of the secondary system - giving each school equal status, changes in methodology to include more inquiry-based approaches especially at the lower secondary levels, improvement of teacher qualifications through training and so on, performance in the science examinations has been mediocre. Many students continue to display little interest in science, and it is still perceived as being difficult and abstract with little relevance to their every-day lives. At the primary school level the lowest percentage scores in the

Grade 6 Achievement test (GSAT) have consistently been in mathematics and science, each averaging 48%. Over the past 20 years there has been significant improvement in passes (Grades 1-3) in biology in the CSEC examinations but chemistry and physics continue to average below 50% (Figure 1). In all three subjects, female students outperform their male counterparts. It is also interesting to note that performances in mathematics and English have also been poor.

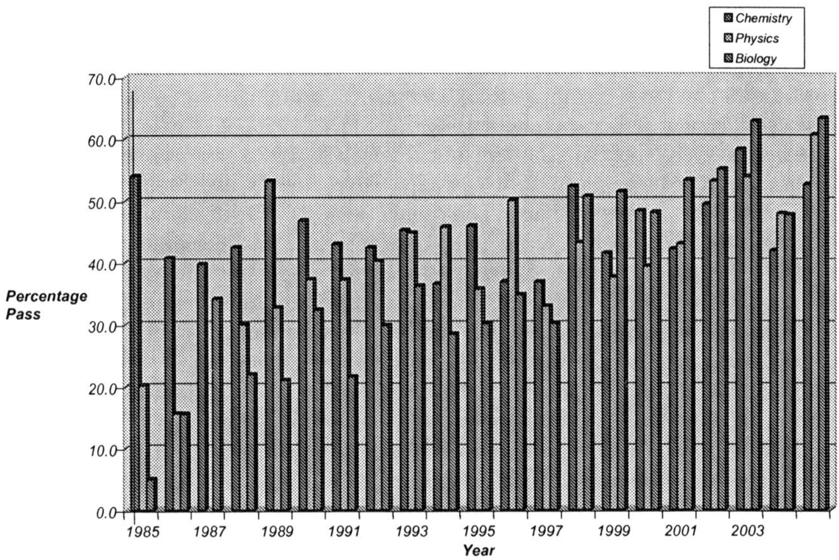

Figure 1. Percentage Passes (Grades 1-3) in CSEC Chemistry, Physics and Biology

The Medium Term Socioeconomic Policy Framework report for 2003-2007 laments the poor performances in sciences at the secondary level and indicates that given the centrality of success in science to national development, the Government of Jamaica should embark on measures to improve the state of science education and hence student overall performance. A joint project between the Government of Jamaica/OAS was centered at the College of Agriculture, Science and Education to provide a targeted 17 primary schools with improved equipment and resources for the teaching of science. In addition, the Joint Board of Teacher Education revised the primary school curricula to one in which science is integrated throughout the entire curricula. Several partnerships between the Ministry of Education and various stakeholders to encourage interest and promote participation through science competitions and clubs have also been developed and plans for improvement in science and technology continue to be the articulated in various policy documents by the National Commission on Science and Technology which is operated from the Office of the Prime Minister, and which will harmonize the aspects of the policy.

Table 3. Students performance in CSEC general proficiency level physics by school type 2003-5

School Type	*Total No. Sitting*	*Percentage Attaining Grades 1-3*							
		2003							
		Grade 1	%	Grade 2	%	Grade 3	%	Total	%
Secondary High	2,016	182	9	599	30	418	21	1,199	59
Upgraded High	284	0	0	41	14	37	13	78	27
Technical High	256	9	4	43	17	47	18	99	39
		2004							
		Grade 1	%	Grade 2	%	Grade 3	%	Total	%
Secondary High	1,641	155	9	407	25	351	21	913	56
Upgraded High	275	1	0	14	5	42	15	57	21
Technical High	315	4	1	42	13	52	17	98	31
		2005							
		Grade 1	%	Grade 2	%	Grade 3	%	Total	%
Secondary High	2,392	322	13	822	34	495	21	1,639	69
Upgraded High	360	2	1	24	7	48	13	74	21
Technical High	255	5	2	40	16	63	25	108	42

The uneven quality of publicly provided secondary education still remains a great concern. This results in examination performance of students in the newly upgraded high schools (US) in science being far lower than their counterparts in the traditional high schools (SH) as illustrated in Tables 3-5. The issue of quality is also further exacerbated by the large enrolment figures in some upgraded high schools and the resulting shift system being implemented in order to cope. In addition, the greatest proportion of trained teachers are employed in the traditional schools (see Table 6). The annual subject reports for chemistry, physics and biology consistently suggest that among the reasons for the poor performances in these subjects are language, comprehension and conceptual challenges. In keeping with these observations are the poor performances in Mathematics and English language. For example, analysis of CSEC results for 2003 revealed that only 4.7% of students in Grade 11 of the Upgraded High Schools attained passes in Mathematics and 9.2% for English Language compared with 38% and 55% respectively in Traditional High Schools.

DEVELOPMENT OF A SCIENCE EDUCATION POLICY

It is still difficult for science to penetrate our people's way of life and become a significant part of the societal and cultural fabric. It is also true that students view science as being "too difficult" or as an area for "bright people", or "white males". This reality, coupled with other persistent values, attitudes and socioeconomic relationships continue to deprive our young people of the motivation, drive and self esteem needed to inspire their deep interest in science (beyond the desire to enter the medical profession).

Table 4. Student performance in CSEC general proficiency level biology by school type 2003-2005

School Type	*Total No. Sitting*	*Percentage Attaining Grades 1-3 2003*							
		Grade 1	%	Grade 2	%	Grade 3	%	Total	%
Secondary High	2,903	156	5	653	22	1221	42	2,030	70
Upgraded High	698	3	0	37	5	223	32	263	38
Technical High	285	6	2	41	14	99	35	146	51
		2004							
		Grade 1	%	Grade 2	%	Grade 3	%	Total	%
Secondary High	2429	107	4	372	15	917	38	1396	57
Upgraded High	867	2	0	15	2	153	18	170	20
Technical High	347	3	1	42	12	126	36	171	49
		2005							
		Grade 1	%	Grade 2	%	Grade 3	%	Total	%
Secondary High	3,462	389	1 1	820	24	1,255	36	2,464	71
Upgraded High	842	3	0	53	6	223	26	279	33
Technical High	230	7	3	29	13	93	40	129	56

Table 5. Student performance in CSEC general proficiency level chemistry by school type 2003-2005

School Type	*Total No. Sitting*	*Percentage Attaining Grades 1-3*							
		2003							
		Grade 1	%	Grade 2	%	Grade 3	%	Total	%
Secondary High	2,184	66	3	369	17	977	45	1,412	65
Upgraded High	396	4	1	20	5	113	29	137	35
Technical High	276	1	0	24	9	88	32	113	41
		2004							
		Grade 1	%	Grade 2	%	Grade 3	%	Total	%
Secondary High	1,755	45	3	193	11	527	30	765	44
Upgraded High	411	0	0	3	1	48	12	51	12
Technical High	18	2	1	18	5	80	21	100	26
		2005							
		Grade 1	%	Grade 2	%	Grade 3	%	Total	%
Secondary High	2,689	55	2	408	15	1145	43	1,608	60
Upgraded High	566	0	0	12	2	128	23	140	25
Technical High	291	4	1	14	5	98	34	116	40

Students' performance in science examinations at the primary and secondary school levels continue to be below par, and at the tertiary level, various remedial interventions are becoming increasingly necessary in order to address some of these early deficiencies. Urgent attention must be, therefore, be given to a detailed understanding of the dynamics between the Jamaican culture, the sciences and their applications. There needs to be a greater focus on those who have made important contributions to Jamaican life and culture including noted scientists and engineers, and a focus how combined efforts of all ethnic groups of different cultures and races have contributed to development of what is now our common culture, and our attitude towards science and science related issues. Textbooks should therefore be written or revised to include examples, analogues, and metaphors illustrating the applications of various concepts in science to local foods, medicines, herbs and industries. The information gathered from such a study can be used to prepare science teachers who can cultivate and nurture scientific values, talents and creativity among our young people and influence the development of a national

Table 6: Distribution of science teachers by school type and level of training

Level Of Training	*Science Teachers*	*School Type*					*Total*
		AA	PJH	SH	US	TH	
Trained	Count	4	20	178	148	32	382
Graduate	% of Total	.3	1.4	12.9	10.7	2.3	27.6
Teacher	Count	14	75	129	308	46	572
Trained	% of Total	1.0	5.4	9.3	22.3	4.3	41.4
Pre-trained	Count	3	3	99	55	20	180
Graduate	% of Total	.2	.2	7.2	4.0	1.4	13.0
Pre-trained	Count	2	11	27	18	7	65
Teacher	% of Total	.1	.8	2.0	1.3	.5	4.7
Untrained	Count	4	14	50	97	19	184
Teacher	% of Total	.3	1.0	3.6	7.0	1.4	13.3
	Count	27	123	483	626	124	1,383
Total	% of Total	2.0	8.9	34.9	45.3	9.0	100

Key: AA – All Age schools; PJH – Primary and Junior High; SH Secondary High; US – Upgraded Secondary; TH – Technical High

science educational strategy/policy. The development of a national science education policy seems necessary to facilitate management and monitoring of the curriculum, long-term planning for science education and ensure sustainable development. Prior to the development of this policy, however, is the urgent need for detailed research focused on determining the following issues:

- The factors (positive and negative) which influence pupils at the secondary schools to learn and internalize scientific concepts;
- The ways in which aspects of the Jamaican culture, environmental and institutional factors impact on the learning and teaching of science;
- The prevailing teacher training strategies and the manner in which they affect the ability to foster creativity, competency, and intellectual growth amongst teachers in regards to their ability to inspire and guide the learning process in the Jamaican environment;
- The development of curriculum materials for the teaching of selected science concepts, based on the findings of (i) – (iii) above;
- Procedures by which the findings from (i)-(iii) can be used to influence programs for the training and professional development of teachers of science in secondary schools;

- A program for further design, development and building of prototype science teaching materials and equipment.
- Recognition of the demands of the local and global labor market for scientists at present and projections for the future;
- The exact cost of a sound science education and sources of funding for it. This data will enable specific meaningful budgetary allocations towards science education.

Of note is the agreement between Government and the Opposition Party for a National consensus on education, which is intended to improve the quality of education. A task force on education reform was commissioned by the Prime Minister with unanimous approval from the House of Representatives in 2004 to define a new direction for education in Jamaica. It remains to be seen whether these visions and projections will be realized.

REFERENCES AND BIBLIOGRAPHY

Bernard, M. (1997). Innovations in science teaching in Caribbean schools with special reference to school based assessment. *Journal of Education and Development in the Caribbean*, *1*(2), 131-151.

Lowe, H., Brown, Y., & Magnus K.E. (2000) *Discovering the future: The emergence, development and future of science and technology in Jamaica.* Kingston, Jamaica: University of the West Indies Press.

Miller, E. (1990) *Jamaican society and high schooling.* Institute of Economic and Social Research, Kingston, Jamaica: United Cooperative Printers.

Rainford, M. (1997). *An evaluation of grade seven children's performance on some of the Jamaican science ROSE components.* Unpublished MEd Thesis, University of the West Indies Mona, Kinston Jamaica.

Scientific Research Council. (1999) *Science and technology for human resource development.* Kingston, Jamaica: SRC.

Task Force on Educational Reform Jamaica. (2004). *A transformed education system: Final report.* Kingston, Jamaica: Ministry of Education, Youth and Culture.

UNESCO Jamaica (1983), *Development of secondary education.* Paris: UNESCO.

AFFILIATIONS

Novelette Sadler-McKnight
Department of Chemistry
The University of the West Indies, Mona
Jamaica

Marcia Rainford
Department of Educational Studies
The University of the West Indies, Mona
Jamaica

MIDDLE EAST
&
INDIA SUBCONTINENT

AMBUSAIDI AND ELZAIN

THE SCIENCE CURRICULUM IN OMANI SCHOOLS: PAST, PRESENT AND FUTURE

INTRODUCTION

The Sultanate of Oman located in the south east of the Arabian Peninsula, borders three Arab countries (United Arab Emirates, Saudi Arabia & Yemen) and Iran and Pakistan across the sea. It has a total land area of about 300,000 km^2 and a population of 2,350,000 according to the 2003 Census. Three quarters of the population are Omanis, the rest are mostly migrant workers (Ministry of National Economy, 2005). Some two thirds of Omani citizens are aged less than 24 years. Contemporary Omani society, like many others, is witnessing accelerating and long-lasting changes across all spheres of life including education. Education is seen in Oman as an important agent of change of Oman society.

The Oman economy was the subject of a national conference held in Muscat in 1995. This conference developed a vision of the Oman economy for 2020, and among other things, made an urgent call for reform of the education system to help the nation meet the requirements of the economy for the new century. In response to that conference, the Ministry of Education initiated a series of symposia[i] and conferences to develop the framework for a new education system. As a consequence of these symposia the Ministry of Education adopted 'basic education' and 'post basic education', as the alternative systems to the old 'general education' system. The development of the basic education and post basic education in Omani schools was a cooperative process involving international organizations such as UNICEF and UNESCO.

Basic education has been described as: "A unified ten-year education provided by the Sultanate for all children of school age. It meets their basic education needs in terms of knowledge, skills, attitudes and values enabling them to continue their education or training based on their interests, aptitudes, and dispositions and enabling them to face the challenges of their present circumstances and future developments, in the context of comprehensive social developments" (Ministry of Education [MoE], 2001a p. 1). The adoption of the basic education system was for several compelling reasons. These were:

- A need to reform education and improve its effectiveness, due to the challenges and requirements of the present and the future;
- The inability of the primary education system to meet basic educational needs of learners;

Richard K. Coll, Neil Taylor (eds.), Science Education in Context: An International Examination of the Influence of Context on Science Curricula Development and Implementation, 85–97.

- A need to integrate the early stages of education into a unified stage in order to minimize waste of resources;
- The importance of verbalism in early stages of education, and a lack of meaningful connection to application and practice;
- A need to respond to pertinent recommendations from national and international educational conferences, many of which advocated adoption of basic education by the end of the twentieth century;
- An emphasis in the 'Arab Strategy for the Development of Education' (Al Farsi, 2002) on the generalization of basic education, along with a reform of its content and structure, in order to achieve flexibility and diversity in education provision; and
- A call from the Future Vision Conference for Omani Economy (called Oman 2020) to develop Omani human resources of high caliber, with competencies and skills capable of keeping abreast of scientific and technological developments, able to professionally manage these changes, able to cope with continuing changes locally and worldwide, within a context of preserving Islamic values and Omani habits and traditions (Al-Farsi, 2002).

Clearly these aspects are interrelated and influence each other and this necessitated wide reform of all aspects of the Oman education system including curriculum content, assessment, teaching and learning strategies, the role of science teachers, administration, supervision, learning resources, and the role of parents.

THE OMAN EDUCATION SYSTEM: HISTORY AND STRUCTURE

The Oman education system initially consisted of three stages: primary education which involved six years of schooling, preparatory education which involved three years, and secondary education which involved three years. The new education system consists of basic education which involves 10 years of schooling, and post-basic education which involves two years. Thus primary, preparatory and the first grade of secondary education were in effect combined to form basic education.

Basic education is divided into two cycles based on the students' age group and characteristics of growth needs that are considered in the teaching and learning process of each cycle. The first cycle consists of Grades 1 to 4. The general features of this cycle are:

- The curriculum is divided into four areas:
    - Islamic Studies, Arabic and Social Studies
    - Science, Mathematics and information Technology
    - The English Language, and
    - Environmental Life Skills, Physical Education, Art and Music;
- Feminization of the teaching staff, to help the school represent an extension to the family, meaning children suffer less as a result of separation from their mothers; and

- Mixed-gender classes, in contrast to the old system where boys and girls schools were kept separate.

The second cycle of a six year period consists of Grades 5 to 10 with separate schools for boys and girls. In the old system science was taught as a single subject at primary and preparatory levels, and as individual subjects of biology, chemistry and physics at secondary schools. In basic education and post basic education, there is now a single integrated science curriculum from Grades 1 to 10, and separate subjects of biology, chemistry and physics at Grades 11 and 12.

GENERAL AIMS AND OBJECTIVES OF BASIC EDUCATION

The general aims of basic education as outlined by the Ministry of Education (2001b) are:

- Integration between theory and practice, thought and work, education and life;
- Comprehensiveness in developing all aspects of a whole personality;
- Acquisition of self-learning skills in the context of lifelong education;
- Inculcating values and practices necessary for mastery and excellence in teaching and learning; and
- Meeting the needs of human development in the context of comprehensive social development.

These aims were translated to more tangible, more specific, objectives to help practitioners (teachers, students, supervisors, parents) identify them and see how they might achieve them. These objectives include the development of life skills through communication; self-learning, scientific and critical thinking; ability to understand contemporary science and technology, as well the ability to adapt innovation; ability to deal rationally with problems of the present era: conservation and wise exploitation of the environment; and ability to internalize the values and ethics of mastery work, production, sensible use of leisure time and participation in civil life (MoE, 2001a).

DEVELOPMENT OF THE OMAN SCIENCE CURRICULUM

Science and technology are seen as an 'enzyme' that can activate positive changes in modern societies (Ambusaidi, 2005; BouJaoude, 2002) and Oman places strong emphasis on the teaching of science in basic education and post-basic education. This can be seen from the significant number of lessons allocated to science, the supply of science materials and equipment, and improvement in the quality of science textbooks. The reforms and development of various aspects of the science curriculum in Omani schools have focused on the goals of science teaching and the content of the curriculum.

Goals of Science Teaching

The goals of science education in schools vary depending on context. Barker (2004) lists 11 goals derived from international literature. Some of these goals, for example, acquiring scientific knowledge and understanding scientific methods, are common to different contexts and relate to educational goals. Braund and Reiss (2006) summarize the goal of science teaching in relation to societal aims identifying six goals: ensuring an adequate supply of future scientists, achieving scientific literacy, benefiting individuals, enhancing democracy, enhancing social justice and sociopolitical action, and developing critical thought. The goals of science education in Oman are similar to such goals, with additional goals of using science to strengthen the theological beliefs of Islam, and to point out the contribution of Arabs and Muslims to science. The latter goal is related to the historical context of science education in the Arab and Muslim world. Interestingly, there is virtually a one-to-one correspondence between the goals stated for general education in the old system and basic education in the new system. The learning goals are designed to support the development of attitudes, skills and knowledge in students. The goals as outlined in the Ministry of Education teacher guide are:

- Strengthening of students' Islamic beliefs through scientific observation of Allah's creation;
- Development of student interest in environment and the importance of living in harmony with nature;
- Obligations towards economical use of natural resources in Oman and the importance of international cooperation in preserving natural resources;
- Development of operational scientific thinking and skills as observation, classification, communication, prediction and experimentation;
- Problem-solving through scientific methodology. Students will learn how to apply a scientific approach in using data, knowledge and thinking skills to solve problems in other areas. In addition, students will develop their capabilities in applying individual and social values in decision-making;
- Gaining basic knowledge needed for students to play a role in contemporary technological world to which science has contributed significantly;
- Gaining the ability to obtain scientific information from books and journals in addition to utilization of computers and modern communication systems;
- Development of healthy habits and gaining knowledge about healthy nutrition, general health and safety;
- Gaining knowledge about scientific and technological professions; and
- Valuing the contributions of Arab and Muslim scientists (MoE, 2003a).

Ministry of Education documents specify learning objectives associated with the development of knowledge, skills and attitudes (MoE, 2003b). It is intended that students will gain knowledge in the areas of: life sciences, physical sciences, earth and space sciences, the nature of science, and science technology and society.

Learning skills are grouped into four broad areas (MoE, 2003b):

- *Initiating and planning* including skills of questioning, identifying problems, isolating variables and selecting one to investigate;
- *Collecting and presenting evidence* including skills in setting up an experiment or investigation, making observations, collecting and recording data;
- *Analyzing and interpreting data* including skills in examining observations or data, and presenting them in a way that can be interpreted appropriately, drawing conclusions, and evaluation of results; and
- *Communicating and teamwork.* Communication skills are essential when ideas are being developed, tested, interpreted, debated and agreed upon. Teamwork skills are important since development and application of science is a collaborative process.

The interaction processes among these skills are illustrated in Figure 1.

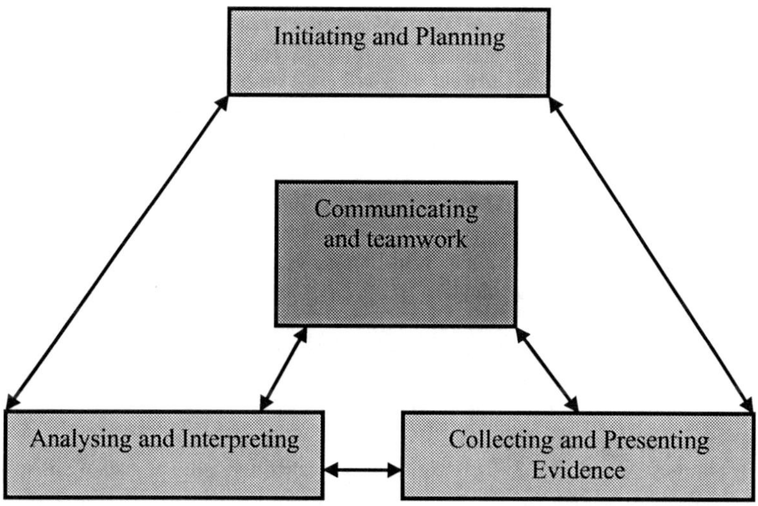

Figure 1. Interaction among skills specified in the new Oman science curriculum

In Grades 5 to 7 students are expected to attain qualitative skills for issues specified in the curriculum, whereas in Grades 8 to 10 students also are expected to advance their quantitative skills. The Ministry of Education (2003b) scoping document defines attitude as a quality referring to general aspects of behavior that are modeled for students by example, and reinforced by selective approval. Here it is expected that students will come to appreciate the role of science and its applications, the evolution of scientific ideas, and the nature of science. The document provides lists which detail the knowledge, skills and objectives for each topic in the curriculum.

General education in the old system presented no such detailed curriculum documentation, and instead there was a limited amount of information about skills and attitudes scatted throughout the curriculum.

Content of the Oman Science Curriculum

A significant driving force for reform of science education curricula has been a view that it is linked to technological advancement. However, equally important is the goal of producing a scientifically-literate society. This latter thinking in particular is important in that it represents a shift in emphasis from producing few scientists to science-for-all. A core objective of school science education in the view of many authors is thus to achieve scientific literacy (Trefil, 1996). It is hard to define scientific literacy precisely as it consists of a potpourri of facts, concepts, history, philosophy, and ideas, all connected to each other by strands of logic. A scientifically-literate person knows some basic facts about say the way the universe operates, and has some sense as to how scientists came to hold such knowledge, but does not necessarily know how to 'do' science. To illustrate the basic ideas of scientific literacy, Trefil (1996) explores the notion by consideration of understanding the universe. He likens scientific literacy in relation to the universe as being like a spiderweb: around the edge of the web are located physical phenomena like supernovae, beetles, radio transmissions, trees, and so on. Starting from any part of that web and asking questions like, 'What is this?', 'How does it work?', and 'Why is it the way that it is?', one begins to work one's way towards the center and eventually reaches the center, where one finds a small number of general, overriding principles. These then are the laws of nature, and Trefil lists some central ideas which a scientifically-literate person would understand. The list in the case of the universe includes: causality and order, energy and entropy, electricity and magnetism, the atom, chemical bonding and reactions, natural selection and evolution, cellular and biochemistry, the genetic code and molecular medicine, plate tectonics, the big bang, stars and galaxies, Mendelian genetics, properties of materials, Earth cycles, ecosystem, relativity, and quantum mechanics.

This chapter now explores such thinking and considers part of the Oman science curriculum, specifically the second cycle of basic education, to illustrate these principles, and contrast the old and new curricula. The preparatory school science content of the general education curriculum is built on knowledge from three areas: mankind and matter, mankind and energy, and mankind and environment (MoE, 2003c). The corresponding content in basic education is more specific, and covers the following topics:

- Living things and their interactions within an ecosystem;
- The structure, function and interaction of systems of the human body;

- The structure, properties, changes, and uses of natural or human-made substances;
- The forms and transformation of energy, and the need for people to use energy wisely;
- Cycles and changes through the study of local and global environments and the universe;
- Scientific inquiry, the application of scientific knowledge to technological developments, and the achievements of Arab scientists; and
- How science and technology affect, and are affected by, social and global issues (MoE, 2003b).

The Ministry of Education provides an official standard textbook for each grade, and the topics are introduced gradually for each area. It is interesting to compare the content and its presentation of textbooks used under the old science curriculum with those of the new. This is now done using physics as an example. Physics content of general education in preparatory schools is introduced sequentially. The textbook starts with a description of linear motion, distance, time and speed, followed by the introduction of force and its effect on motion without reference to acceleration. The origin of a number of forces and the resultant forces applied along a line are discussed next; a definition of pressure and its applications are also included in this chapter. Similarly, sound and its applications start with the introduction of sound in a simplified form, without dealing with the wave nature of sound apart from speed and frequency characteristics. The concept of heat is covered qualitatively, but thermal expansion is treated quantitatively. Light is presented in terms of geometric optics, but this again is treated as qualitative, and no equations are used in the description of images. The second semester textbook of the second year in preparatory schools is fully dedicated to physics. This covers energy and machines, where work is introduced and various machines, including the heat engine, are presented. Electricity is then introduced, including electrostatics and direct current, followed by astronomy. In the third year of preparatory education, electricity and magnetism, direct and alternating currents and their applications are introduced in more detail. This is followed by electromagnetic waves and their applications. Solar energy and heat energy are then treated quantitatively. Moving on, in the first year of secondary school of general education, physics is taught as a separate subject. The school textbook covers physics and measurement where basic and derived quantities and their units of measurements are presented. Coverage of the mechanical properties of matter occurs, including the properties of fluids. There are chapters on mechanical, thermal and electrical energies, and throughout the textbook topics are introduced formally, and are treated using a more mathematical approach.

The physics content of the basic education is, at a given level, much the same as the corresponding content of the old general education. However, the presentation of the material in the new school textbook is very different. Textbooks for basic education, for example, stress group work, and attempt to introduce physical concepts using conceptual and context-based approaches. Students carry out experiments in groups and compare their results with other groups. The Grade 5 textbook covers forces and machines, but includes topics on space and astronomy. Since motion is not covered at this level, its relation to force is not referred to, this being left to Grade 10. The Grade 10 presentation is more formal and is intended for the more mathematically-inclined students, although the topics are covered under a unit called 'Energy Flow in Technological Systems'. The unit also discusses thermal energy through the description of thermal machines and introduces the first and second laws of thermodynamics. Grade 6 physics covers different forms of energy, with emphasis on electrical energy and its applications. Grade 7 has little about physics concepts, but covers force, structures and machines, centered on static equilibrium. Thermal energy, with an emphasis on temperature, is covered in Grade 8 in addition to waves and their applications. Sound and electromagnetic waves are treated later. The textbook attempts to introduce the concept of waves through wave-riding (i.e., surfing), which is somewhat surprising, given that wave-riding is not familiar to many students at this age. Additionally, the textbook cites a number of Quranic verses to illustrate phenomena, but these are not related to wave motion. The textbook presents waves as traveling harmonic waves, which may leave the impression that only such disturbances are identified as waves. The geometric optics also comes under this unit and its presentation is far less satisfactory compared with the corresponding section in general education. The unit on the electricity and its technological applications in Grade 9 covers electricity more quantitatively, and is almost equivalent to the corresponding unit in the second year of preparatory school in general education.

In terms of Trefil's (1996) list, under the new curriculum a graduate of the second cycle of basic education would have been exposed to almost all physical concepts apart from relativity and quantum mechanics, although some applications of quantum mechanics like energy levels and chemical bonding are included in chemistry. The physics curriculum is completely conceptually and laboratory based, but learning experiences outside the classroom are not included. In terms of clearly relevant science based facilities, there is currently only one science museum in the country (The Child Museum), which includes a number of excellent demonstrations and illustrations of many physical phenomena. However, this is located in the capital Muscat, and this limits accessibility to the museum for students living outside the capital.

TEACHING AND LEARNING STRATEGIES

Another important aspect where significant development has happened for the science curriculum has been in the area of teaching and learning strategies. In response to changes in the nature and content of the science curriculum, teaching and learning strategies also were subject to reform. The Ministry of Education now expects teaching to be learner-centered rather than teacher-centered (MoE, 1998). The two main learner-centered teaching strategies used in science lessons are inquiry-based learning and cooperative learning. Inquiry-based learning is aligned to the nature of science, and students are expected to acquire skills such as scientific process skills (e.g., observation, prediction, interpretation and classification, measurement), and problem-solving skills (e.g., identification of the problem, planning and collecting evidence). The inquiry-based learning activities in most cases are conducted in groups of four or five. Cooperative learning is given high priority and is the main teaching strategy besides inquiry-based learning. In cooperative learning, students are expected to acquire many social skills in addition to acquiring scientific knowledge.

Student self-learning and learning experiences outside the classroom activities are two other components emphasized in science teaching in basic education. Here students are expected to take some responsibility for their own learning. Students are asked to do out-of-class activities, such as conducting small projects (e.g., collecting rocks from the surrounding environment), or searching for information via the Internet.

Comparison of such the teaching strategies used in basic education with those used in the old general education system points to significant movement from teacher-centered strategies towards more learner-centered teaching.

ASSESSMENT

Assessment of student learning also has undergone major development and has shifted from mainly summative assessment based on examinations to a combination of summative and formative assessment (Murphy, Ambusaidi & Beggs, 2006). The philosophy behind adopting formative assessment derives from a view that to encourage learning for understanding as opposed to rote learning, formative assessment is crucial (Harlen, 1998). The focus here is on making judgments of learning and creating descriptions which can be used to inform the next phase of learning. Formative assessment in Oman is a highly formalized procedure based on detailed documents prepared by the Ministry of Education.

In basic education, two main components are assessed: knowledge and understanding, and problem-solving skills. In the case of knowledge and understanding, three skills are assessed: recalling, describing, and carrying out simple numerical operations. For problem-solving, five skills are assessed: initiating and planning, collecting and presenting evidence, analyzing and interpreting, communicating and teamwork, and report writing. The proportion of assessment allocated to each component varies for the first cycle (Grades 1-4), the second cycle (Grades 5-10) and post-basic education (Grades 10-12) as shown in Table 1 (MoE, 2006a).

Table 1. Proportion of assessment (%) allocated to Knowledge and Understanding, and Problem-solving for each school grade under the new Omani science curriculum

Grade	*Knowledge & Understanding %*	*Problem-solving %*
3-4	50	50
5-9	70	30
10-12	80	20

To assess the problem-solving skills especially, Omani science teachers are now expected to use different types of assessment methods. These include short tests based on multiple choice questions, observation during group work activities, in and out of classroom activities like collecting samples from the surrounding environment and doing projects (i.e., individual and group tasks). However, in order to apply such tools, preparation is needed for science teachers, students, schools staff as well as parents. Biddulph (1998) points out that unless teachers have had first-hand experience of alternative forms of assessment it is not easy to implement them in the classroom. This is what the Ministry of Education has been forcefully doing since the reform started.

The science teachers are the key players in implementing the formative assessment inside the classroom. They use their personal knowledge of the students and their understanding of the context of the assessment and the curriculum targets to identify particular needs (Earl, 2006). The Omani Ministry of Education conducted several workshops to familiarize teachers with the requirements of implementing formative assessment inside the classroom. However, there are still some difficulties that affect implementing formative assessment properly such as the heavy load of teachers' duties, the large number of students per class (on average more than 35 students), and the insufficient training of teachers in how to implement and use formative assessment.

THE ROLE OF THE SCIENCE TEACHER IN OMAN UNDER THE NEW SCIENCE CURRICULUM

Attempting to reform science teaching without considering the role of science teachers is likely to be worthless. It is most unlikely we can achieve the aims of the new science curriculum, even with a good science curriculum, excellent teaching materials and equipment, and highly able students on the one hand, but weak science teachers on the other. Consequently, in order to achieve the aims and objectives of science teaching for basic and post basic education, the role of science teachers was reviewed and new roles were introduced. These are based on many of the preceding issues, such as the curriculum documents, teaching and learning strategies and the assessment of student learning. There had to be a shift from teacher-centered teaching to student-centered learning (i.e., science teachers becoming a facilitator rather than transmitter of knowledge), and from totally summative assessment to more formative assessment, although it is noted that this typically means more work for teachers in assessing student learning. Science teachers needed to use information and communication technologies (ICT), meaning ICT should be a normal part of teaching and learning, and they have to engage in science education research to improve the quality of their teaching and assessment processes.

To help teachers adapt to this dramatically different role, pre-service training and in-service training regimes were revised. During teacher training pre-service science teachers are taught about new approaches to teaching and learning, the use of ICT in teaching, and formative assessment. In-service professional development is provided by the Ministry of Education to help science teachers adapt to their new roles. The programs are in the form of lectures, workshops and training courses addressing the needs of the science teachers and supervisors.

THE FUTURE OF OMANI SCIENCE EDUCATION

Science education in the Omani schools has a bright future as a result of the on-going substantial support of the Ministry of Education. For example, the introduction of ICT in science teaching is given priority by the Ministry, and this has been accompanied by the establishment of a fully equipped resource center in each school with extensive software that science teachers can use to teach science. Similarly, electronics laboratories in which microcomputers are interfaced to other laboratory equipment, have also been established in all schools (MoE, 2006b). According to Krajick and Layman (1993), an electronics laboratory provides students with a myriad of learning opportunities; allowing them to ask and refine questions, make and test predictions, design and plan experiments, collect and analyze data, debate and communicate, draw conclusions and ask new questions. Ebenezer and Haggerty (1999) argue that, a particularly promising aspect of the use of electronics and computers in the classroom is that experimentation and experimental analysis occurs simultaneously, thus minimizing laborious data collection and instead providing more time for data analysis. The Ministry of Education is training science teachers and supervisors (inspectors) on how to conduct experiments using the electronic laboratory.

The second development related to the use of ICT in science classrooms in Oman is the adoption of the 'portable laboratory', supported financially by the Oman Petroleum Development Company. Each laboratory is in the form of a trolley that consists of various science kits. The trolleys have built-in sockets for electrical connections to audio and video systems. Depending on the class size, the portable laboratory, forms and integrated system consisting of audio and visual technical material and efficient remote control system to handle student responses, and electronic devices for taking measurements. The portable laboratory has many features that serve schools in the provision of basic and post-basic education, and is easily connected to existing electronic laboratories. The Oman Petroleum Development Company provides a number of laboratories to support schools in remote areas that lack traditional laboratories.

CONCLUSIONS

The introduction of the new Oman science curriculum involving basic and post-basic education represents a major development in the educational system in Oman. In basic education the position of science is elevated, and has received valuable support in the form of a better planned curriculum, better teacher training, better science laboratories, and better assessment approaches compared with the previous general education system. The full impact of these reforms is yet to be ascertained. The acid test will be when graduates of the new school science education system engage in tertiary education. Then a more concrete conclusion about the new system will be available.

NOTE

i For example, The Regional Symposium for Developing Post-Basic Education in the Arab World (Grades 11 & 12); The International Conference on the Reforms of the Secondary Education; Secondary Education for a Better Future; Trends, Challenges & Priorities; and The Secondary Education Development Symposium

REFERENCES AND BIBLIOGRAPHY

Al-Farsi, F. (2002, December). *Sultanate of Oman, a position statement: Results of the National Conference Education for a Better Future*. Paper presented at the International Conference on the Reform of Secondary Education. Muscat, Oman.

Ambusaidi, A. (2005). *The necessity competences needed for 11th and 12th grade students in the 21st century according to the scientific literacy requirements*. Paper presented at the Regional Symposium for Developing Education after the Basic Education in the Arab World (Grades 11 and 12). Muscat, Oman.

Barker M. (2004). *Key aims of science education in New Zealand schools in the 21st century: Messages from the international literature*. Retrieved 10 May 2007, from www.tki.org.nz/nzcurriculum

Braund M., & Reiss M. (2006). Validity and worth in science curriculum: Learning school science outside laboratory. *The Curriculum Journal, 17(3)*, 213-228

Biddulph, F. (1998). Peer and self-assessment in teacher education. In F. Biddulph & K Carr (Eds.), *SAMEpapers* (pp. 26-40). Hamilton, New Zealand: University of Waikato.

BouJaouds, S. (2002). Balance of scientific literacy themes in science curricula: The case of Lebanon, *International Journal of Science Education, 24(2)*, 139-156.

Earl, L. (2006). *Assessment as learning: Using classroom assessment to maximize student learning*. Retrieved 25 February 2006, from http:/www. ltag.education.tas.gov.au

Ebenezer, J., & Haggerty, S. (1999). *Becoming a secondary school science teacher*. Upper Saddle River, NJ: Merrill.

Harlen, W. (1998). Classroom assessment: A dimension and procedures. In F. Biddulph & K Carr (Eds.), *SAMEpapers* (pp. 75-87). Hamilton, New Zealand: University of Waikato.

Krajcik, J., & Layman, L. (1993). Microcomputer-based laboratories in science classrooms. *NARST News: Research Matters to the Science Teachers, 35*(1), 3-6.

Ministry of Education. (1998). *The general framework of science and math curriculum for basic education*. Muscat, Sultanate of Oman: MoE.

Ministry of Education. (2001a). *Basic education in the Sultanate of Oman: A guide to the first cycle Grades 1-4*. Muscat, Sultanate of Oman: MoE.

Ministry of Education. (2001b). *Basic education in the Sultanate of Oman: The theoretical framework*. Muscat, Sultanate of Oman: MoE.

Ministry of Education. (2003a), *Teacher's guide, third class, preparatory schools*. Muscat, Sultanate of Oman: MoE.

Ministry of Education. (2003b). *Science, scope and sequence: Grades 5-10, basic education, second cycle*. Muscat, Sultanate of Oman: MoE.
Ministry of Education. (2003c). *Teacher guide for preparatory schools*. Muscat, Sultanate of Oman.
Ministry of Education. (2006a). *Students' assessment document in biology, chemistry and physics for Grades 10-12*. Muscat, Sultanate of Oman: MoE.
Ministry of Education. (2006b). *Technology in science laboratories*. Muscat, Sultanate of Oman: MoE.
Ministry of National Economy. (2005). *Census results across the Sultanate*. Muscat, Sultanate of Oman: MoNE.
Murphy, C., Ambusaidi, A., & Beggs, J. (2006). Middle East meets West: Comparing children's attitudes to school science. *International Journal of Science Education*, *28(4)*, 405-422.
Trefil, J. (1996), Scientific literacy. In P. Gross, N. Levitt & M. Lewis (Eds.), *The flight from science and reason* (pp. 543-550). New York: Oxford University Press.

AFFILIATIONS

Abdullah Ambusaidi
College of Education
Sultan Qaboos University
Oman

Mohamed Elzain
College of Science
Sultan Qaboos University
Oman

RANADE

SCIENCE EDUCATION IN INDIA

INTRODUCTION

India has had a rich tradition of science and technology with significant achievements in astronomy, mathematics, medicine and engineering from ancient times. After attaining independence from British rule in 1947, the Government of India has consistently emphasized the importance of science as key to the development of the country. This sentiment was lucidly expressed by Jawaharlal Nehru, India's first prime minister, while delivering the convocation address of Allahabad University in 1946:

> It is science alone that can solve the problems of hunger and poverty, of insanitation and malnutrition, of illiteracy and obscurantism of superstition and deadening customs, of rigid traditions and blind beliefs, of vast resources going to waste of a rich country inhabited by starving millions. (India National Science Academy [INSA], 2001)

The Government of India places a high priority on imparting compulsory education to all children. Education for children in the age group of 6–14 years is a *fundamental right.*

In the post-independence era, the commitment of the Government to science education is evident from more than 40 years of planned science education through national-level commissions and committees. Much effort has gone into curriculum design. However, there is a large gap in execution due to various constraints as well as the sheer size of the problem.

HISTORY OF SCIENCE CURRICULUM DEVELOPMENT IN INDIA

The process of curriculum development in India lies between the two extremes of centralization and decentralization. Periodically, the central Government formulates the National Policy on Education, providing broad guidelines regarding content and process of education at different stages. These guidelines are further elaborated by the National Council of Educational Research and Training.

This centrally-prepared curriculum framework provides a broad overview of the school curriculum, including general objectives, subject-specific objectives, suggested schemes of study, and guidelines for the transaction of the curriculum and the evaluation of pupil outcomes. It is always suggestive rather than prescriptive, and it is not enforceable by law in the states. The states consider whether to *adopt* or *adapt* the Council syllabi and instructional materials. However,

Richard K. Coll, Neil Taylor (eds.), Science Education in Context: An International Examination of the Influence of Context on Science Curricula Development and Implementation, 99–114.

it is readily accepted by the states because of the Council's credibility and the participatory development approach it follows (National Council of Educational Research and Training [NCERT], 2007).

This chapter will focus on the history of science curriculum development from the mid-twentieth century, when India attained freedom from British rule, a major landmark in Indian history. Several national commissions and committees undertook the Herculean task of developing a perspective on the curriculum for the country with its vast diversity of regions, peoples and languages.

Indian Education Commission

The Indian Education Commission began work in 1964 on curriculum reform, with subsequent revisions in 1977, 1986, and 2005. The Commission of 1964, chaired by Dr. Kothari (1966), emphasized the importance of science education for India, seeing it as essential for the country to play a role in the ongoing science and technology revolution. It recommended development of a strong research community in universities; vigorous interaction between science, technology, and production; and centers of advanced study that would act as major suppliers of teachers and researchers. It also recommended uniform and free education through publicly-funded schools. This led to the introduction in 1975 of a uniform structure of school education (10+2+3), and a shift from science education to general education.

A National Curriculum Committee drew up guidelines for the new structure through a curriculum framework document (NCERT, 1975). Some of its recommendations were:

- Science and social sciences to be taught as a single subject, Environmental Studies, in primary schools;
- Science to be considered as one composite subject in upper primary and secondary stages; and
- All subjects, including science and mathematics, to be compulsory for all students up to Class 10.

A review committee chaired by Sri Ishwarbhai Patel in 1977 recommended that science at secondary stage be offered through two equivalent alternate courses. Course 'A' would teach physics, chemistry and biology as separate subjects, while Course 'B' was a composite course taught through a single textbook. However, the unpopularity of course B, due to the perception of its being a lower-level course, eventually led to it being dropped from the curriculum in 1984.

National Policy on Education 1986

The next important development was the National Policy on Education in 1986 which led to a framework document in 1988 (Department of Education [DoE], 1986; NCERT, 1988). It reiterated the teaching of science as a single subject rather than three separate disciplines, and gave special emphasis to equality in education for women, disadvantaged castes and tribes, and the handicapped. It suggested that science education should be redesigned to help the learner develop problem-solving and decision-making skills, and to relate science to health, agriculture, industry, and other aspects of daily life. Furthermore, it said the burden of the examination system should be reduced by using a semester system, and using grades instead of marks.

National Curriculum Framework 2005

The National Curriculum Framework (NCF) of 2005, is the latest policy document on education. It recommends an integrated approach to teaching science, technology and environment and an emphasis on learning by doing (as opposed to rote learning) for both mathematics and science. The policy makers also wanted computer hardware, software, and connectivity made available at the school level, but recognized this as a major infrastructural challenge (NCERT, 2005).

Trends in Curriculum Development

The Government of India treats science and mathematics as an essential component of general education for every child, during the first 10 years of schooling. Several major trends in science curriculum design have occurred in the past four decades.

There is now a uniform structure for education. In this structure, the first 10 years of schooling comprise eight years of elementary education, of which the first five years are of primary and next three years of upper primary education. This is followed by two years of secondary education. The students then undergo two years of higher secondary education to complete their schooling. This format is referred to as the '10+2 structure' of education (Rajput & Shrivastava, 2001). At the same time, there has been a shift from teaching science as separate disciplines to an integrated approach that today encompasses science, technology and environment. This was accompanied by recognition of the importance of computer education in today's global society, and an emphasis on learning by doing or activity-oriented learning, rather than lectures.

CURRENT CURRICULUM DESIGN ISSUES

The previous section described how curriculum renewal programs in science education have evolved to keep pace with global trends and changing needs of the society. However, this has not necessarily been reflected in the achievement of a high quality of science teaching in schools.

Contextual Flexibility

India is a large and diverse country. It has 18 major regional languages, four major religions with several sects, and its people occupy a wide range of socioeconomic status levels. There is a wide urban-rural spread as well, with about 30% of the population being in urban conurbations. This presents a challenge for developing a national curriculum, especially at the school level. The curriculum must accommodate this diversity in culture, language and student context in order to be effective.

Content Overload

Curriculum makers in India, as elsewhere, are faced with the issue of an explosive growth of information in science. They responded by adding more topics and by pushing subject matter down to lower levels. A general complaint is that science is being taught as a collection of unconnected facts. A National Advisory Committee was appointed by the Government to look into the frequent complaints of excessive burden on school children in 1993, and found that the perception of academic burden is tied to *incomprehension.* One of the root causes was identified as an inability to distinguish between *understanding* and *information* (NCERT, 2000). However, NCERT is reported to be revising the curriculum framework to resolve the complex problem of curriculum load (Rajput & Srivastava, 2001).

New Three-pronged Approach in Science Curriculum: Science, Technology and Society

The National Curriculum Framework developed by NCERT (2005) has incorporated a new 'three-pronged approach' in framing the science curriculum. Recognizing that a focus on learning pure science does not meet the country's needs, the committee recommends a treatment that integrates science, technology and society. Society refers to issues such as learning to live in peace, protecting the environment, and building a just and democratic society. The need to master technology and the ability to apply it appropriately are also given increased importance.

Introduction of Project-based Learning

Science in India has traditionally been taught from textbooks, mostly through lectures, with some laboratory demonstrations, and hardly any lessons in which students 'experiment' in the true sense of the word. The result has been rote learning with no application of the learning beyond the ability to answer examination questions. The real question is how to inculcate the spirit of scientific inquiry and scientific thinking on a large scale and in a long-lasting fashion. To this end, project-based learning is being tried out as a partial alternative to examination-oriented learning. Students will design and implement science projects that will be assessed and marks awarded to them will be counted to their total score for 'science'. This implies that the written science examinations will carry less weight than before.

Incidentally, a curious modification by the CBSE board may be mentioned in passing. Observing that the marks given for practical work by the school are often not an accurate assessment of the practical science skills expected to be gained by the student, most of those marks have been transferred to a newly created 'practical work' section in the written exam that will specifically test practical knowledge of the students through a written test paper!

Environmental Orientation

Understanding and dealing with environmental problems has become an urgent issue in India today as the country climbs the 'developmental ladder'. Global warming, pollution control, and water conservation are no longer theoretical topics; they have begun to impact everyone. This concern is being reflected in curriculum design today.

CURRENT SCHOOL ENVIRONMENT

It would be very challenging to accurately portray the school environment in the approximately one million schools in the country today (Seventh All India School Educational Survey, 2002). The diversity of this large country has an impact on the school environment. The state and central governments follow different curricula; textbooks need to be written in several languages, and there are differences along the rural-urban dimension as well as along caste lines. This is described below in greater detail.

Central and State Government

India is a republic of 28 states and several union territories. Each state government runs its own education boards, while the central Government runs the Central Board of Secondary Education. Curricula are set by individual boards although they tend to follow the central curriculum to a large extent. Nevertheless, the education provided by the Central Board of Secondary Education (CBSE) is perceived to be of higher quality (Wikipedia, 2007).

Textbooks in Regional Languages

School textbooks in a particular regional language are generally prepared by the corresponding textbook bureaus of the state where that language is spoken. Many states also publish textbooks in English or Hindustani in addition to the regional language. Preparation of a textbook is a 'tightrope walk' where a single book has to satisfy urban as well as rural students, and students of varying abilities. The quality of the textbook is constrained by the cost of the book that must be affordable to all students. Due to lack of resources in most schools, the textbook serves as the sole guide for teaching-learning.

Caste

The issue of caste, although thought to be diminishing due to Government policies, laws and affirmative action, still exists in the education system. Especially in rural India, children of lower caste have lower enrolment in schools, higher dropout rates, and attend schools with poorer standards. An article by Srinidhi, Nehru, Pandit, Chatterjee and Raghupathi (2005) presents an interesting perspective. The authors say that India has morphed into a three-tier education system: children of the rich and famous attend five-star English-medium schools, middle class children attend Government-aided English-medium schools, and those at the bottom rung attend municipal and Government schools, providing low quality education in vernacular languages.

Rural vs. Urban Differences

Rural literacy rates have climbed from 23% in 1961= to 60% in 2001, but they still lag behind urban literacy rates of 55% in 1961 and 80% in 2001 according to an analysis based on 2001 Indian Census data (National Literacy Mission, 2001).

In a nation-wide NCERT-approved survey initiated in April 2000 regarding the level of achievement at the end of Class 5 in environmental studies (science at the primary level), the performance of urban students, both boys and girls, was significantly better than their counterparts in rural areas.

Technological backwardness in rural areas, lack of well trained teachers, lack of facilities for the teaching of science and apathy on the part of students and parents towards formal education are some important causes of the dismal state of teaching-learning of science in rural areas. Rural students find the content of science inordinately taxing and passing in the subject is felt to be a daunting task.

Students in many schools are 'first-generation' school goers. There is no support from home to help these children study and learn outside the school environment. The school must take on the burden of fostering a taste for learning. Students from a rural background find it hard to understand the language used in science textbooks, even when the medium of instruction is the regional language, because of differences in dialects and since the terminology and language style used is often incomprehensible. Furthermore, they do not see the of relevance science in solving their day-to-day problems.

Recently, the education minister of the state for Maharashtra announced a plan to include A=agriculture as a subject from Class 8 so as to change the way agriculture as a profession is viewed by society ("Maharashtra plans", 2006, April 15).

Rise of Coaching Classes

In India science is generally perceived as a difficult subject to study. The focus on succeeding in examinations, coupled with lack of confidence in regular school teaching, has lead to the development of a parallel industry in urban and semi-urban areas. Popularly known as *coaching classes*, their objective is to increase the examination scoring skills of their customers, especially for Class 10 and 12, when students face board exams. They range in size from individuals handling a few dozen students to state-wide franchises. This has become a very successful industry, with most students enrolled in coaching classes despite their high fees. However, the focus on drill and practice in these classes discourages independent thinking and the spirit of inquiry in science.

BASIC PROBLEMS

As we have seen in the previous sections, all the India education commissions and committees have recognized the need for science and technology education for all children. A recent item in the news stated that the Government of India acknowledges the poor quality of science education in the nation (Nanda, 2006). For the nation to provide leadership in science, a substantial number of the brightest and most creative children should take up a career in science. Unfortunately, it has been found that the percentage of students taking up pure science has dropped from about 30% in 1950s to about 20% in 2000. Even worse, most bright students opt for careers in technology instead of science, depleting the talent available for pure science (INSA, 2001). A much smaller fraction of students progress to tertiary studies: the gross enrolment ratio at tertiary level in 1995 stands at 60 students per 1 million population, compared to 54,000 per million for USA. Another concern is the high dropout rate: only 14-18% of those enrolled in tertiary level science obtain a degree in the natural sciences; a peculiarity of the Indian educational system (INSA, 2001).

A survey was recently undertaken by the National Council of Applied Economic Research (NCAER) on the state of science education in India (Shukla, 2005). Most people surveyed saw science and technology as beneficial. Of the total, 60% of children of all classes were interested in further study in science, engineering, or medicine, but only 22% children in Classes 6 to 8 said they would like to study *pure science* at higher levels of education, while just 13.4% in Classes 11 to 12 felt the same way. Of the graduates who were unemployed at the time, 22.3% had studied science, and for unemployed postgraduates, 62.8% had a science background. More children in rural areas prefer higher education in the arts, in contrast to those in urban areas (25% vs. 11%).

It appears that one of the reasons for young people not taking up a career in pure science is the perceived lack of economic benefits. Parents also influence children's preference a professional career such as engineering or medicine where the economic rewards are perceived to be higher. A press release by the Government (Science Education, 2006) noted that about 1.6 million students finish their Bachelor of Science degree every year, but very few of pursue a career in science.

Examination-dominated System

Education in India is still driven largely by the desire of children and their parents to obtain a qualification (or certification) rather than for any intrinsic learning or development of thinking and reasoning skills. The primary cause of this is economic: entry into higher education requires high scores at earlier levels; higher degrees with high scores are prerequisite to good jobs. The methods adopted to evaluate performance in the field of science are quite inadequate. This is a major stumbling block to improving the quality of education in India. The subject is studied solely for the purpose of succeeding in the examinations. Any material outside the scope of examination is considered irrelevant and ignored by both staff and students (Rajput & Srivastava, 2001).

Poor Teaching Resources

Resources such as well-equipped laboratories, teaching aids, extra reading material in science or Internet access are not available in most schools in India, so that the only important resource is the textbook. A press note by the Government of India (Science Education, 2006) recognizes that science education in the nation suffers from the lack of reasonable experimental facilities; classroom demonstrations are very rare. According to an India Science Report by NCAER half the teachers interviewed said more computers/equipment were required for teaching science subjects since inadequate practical training in science was a serious issue (Shukla, 2005).

Inflexibility and Compartmentalization

The educational system is divided into major streams (science, commerce, and arts) from an early stage (Class 11). Once a non-science stream is selected, it becomes very difficult to get into the science stream later. Compartmentalization within different science streams is perceived as a problem and science is now being taught at the school level as a single subject instead of the earlier pattern of chemistry, physics and biology. However, teachers currently have their own limitations in teaching science. In-depth knowledge and insights are required to teach fundamental concepts in each discipline, but most science teachers possess this only in their chosen elective at college (e.g., physics, chemistry, or biology). Consequently, teachers cannot teach all units well (Ranade, 2006).

Poor Teaching Quality

There are about 5 million teachers and about 200 million schoolchildren in primary and secondary classes according to the Seventh All India School Education Survey (2002). Pre-service and in-service training of these teachers is a huge resource problem (Rajput & Srivastava, 2001). Regarding the poor quality of teaching, Dr. Arvind Kumar, the Director of the Homi Bhabha Centre for Science Education says, "It is unfortunate that the majority of schools across the country teach science in a boring and mechanical style allowing little room for original thinking and innovation. Most of the time of science students is spent on searching for 'correct' answers" (Yasmin, 2004, p. 25).

Large Classes

One of the effects of the lack of resources available to schools is the pressure to increase the number of students in a class. Class sizes range from 35 to 80, and since the teacher gets no assistance of any kind, it is difficult to use activity methods or allow experimentation by students. Large class sizes also preclude interactivities (Ranade, 2006). In the India Science Report survey by NCAER (Shukla, 2005), 40% of the children said the number of students in a class was too many for them to understand what was being taught.

Gender Inequality and Bias

In environmental studies (science at primary level), the achievement of boys is significantly better than that of girls, both in urban and rural areas according to a nation-wide NCERT approved survey, initiated in April 2000 (SSA, 2002). Gender bias on the part of parents and teachers, especially with regard to science subjects, is also a major setback to the education of girls in India. One reason attributed to lack of progress of girls in science and mathematics is low expectations set by parents and teachers for girls. Realizing the far-reaching effect of gender bias, the Government of India has taken major steps to address this situation. Gender equality has been declared as one of the 10 core elements in education (Sutar, 2006). The need for a reduction in gender bias is strongly emphasized in the 2005 National Curriculum Framework (NCERT, 2005). Textbook writers are made aware of this issue, and textbooks are specially reviewed before publication to identify and remove instances of gender bias.

Language Issues

Although prevalent wisdom says that students learn better through their mother tongue (especially in lower classes), it has been observed that the special Sanskrit-derived scientific terminology developed for science text books in regional languages proves to be almost as difficult as learning science in a foreign language.

Lack of basic language skills at the school level also hinders the understanding and expression of scientific concepts.

Lack of Activity-Based Teaching

Large classes, lack of resources, and lack of time lead to the teaching of science mainly through the lecture method. Science as exploration is not feasible in most schools.

Stimulating Inquiry and Creativity in Science

Although science curricula are well-designed, day-to-day teaching of science in school faces the problems of not rising above rote learning and of domination of the lecture method. Higher objectives of science teaching are rarely attempted.

Poor Academic Environment in Colleges

In the Report on University Science Education (Mukunda, 1994) the author observes that science education taught in affiliated colleges is of poor quality:

> In most universities and affiliated colleges one finds low educational standards and a poor academic environment. Colleges are generally under-equipped, overcrowded and poorly staffed. Not being directly involved in the framing of syllabi or in the evaluation process, it is perhaps not surprising that most teachers become unmotivated and are seen to be uninspiring. Questions from students are often discouraged, and experiments and demonstrations are few. Due to lack of experimental facilities, science is taught unimaginatively, and learnt by rote. Generally routine and unexciting topics are taught, basic concepts are poorly covered, and at the higher levels teaching remains divorced from research. Quality and excellence in teaching go unrecognized and unrewarded; the few good teachers there are work under trying conditions. All this has a snow-balling effect - the generally low standards of the output at the undergraduate level get reflected among the entrants to post-graduate education and research. (Section 2, subsection 5, sub-subsection g).

Secondly, talent in science is tied up in specialized science laboratories which have little contact with universities. Hence, students are not exposed to scientific thought.

Gap between Science Research and Teaching at Higher Levels

Despite having a large university system with a large student enrolment, the research output in India is very low and of poor to average quality. As Yasmin (2004, p.15) points out:

> According to Dr. S. Bhattacharya, the director of TIFR, Mumbai, one of the major causes of science education losing its sheen is the artificial division between science research and teaching, which obstructs automatic contemporisation of syllabuses. Indeed, most universities in India are just teaching shops with the faculty rarely engaged in any research work.

The sciences are practical and capital intensive subjects, but due to lack of funds in higher education and also outdated syllabuses, India is not making much headway in science education at higher levels.

SOME POSSIBLE SOLUTIONS

Solutions Implemented by the Government

Based on the findings and recommendations of various education commissions and committees, the Government of India and regional and local educational authorities has made efforts to improve the quality of science education. Some examples of efforts to strengthen science education follow.

- The National Science Talent Search scheme provides motivation for higher studies in basic sciences by awarding 1,000 scholarships each year at the end of Class 10 to talented students for study up to PhD level (National Talent Search, 2007);
- Sarva Shiksha Abhiyan (Education for All) is a national program being implemented to achieve universalization of Elementary Education. Special emphasis is being placed to educate girls and under-privileged/disadvantaged sectors of society. The National Programme for Education of Girls at Elementary Level (NPEGEL) was started in 2003;
- Homi Bhabha Centre for Science Education (HBCSE) is a National Centre of the Tata Institute of Fundamental Research in Mumbai, whose broad goals are to promote equity and excellence in science education from primary school to undergraduate college level, and encourage the growth of scientific literacy in the country It carries out a wide spectrum of interrelated activities, which may be viewed under three broad categories: Research and Development, Teacher Orientation and Science Popularization, and Olympiads and Other Students' Nurture Programmes;
- The central government's Department of Science and Technology designated 2004 as the 'Year of Scientific Awareness' and initiated several activities, including inter-school debates, exhibitions, and seminars, to attract the student community to science and research. The Kishore Vaigyanik Protsahan Yojna (KVPY) (scheme to encourage young scientists) is an ambitious scheme launched by this department in 1999 which awards generous scholarships ranging between Rs.2,000-5,000 per

month to Class 10 and 12 students who opt for basic sciences, engineering and medical education up to the master's level;

- Other science promotion initiatives of this department include the promotion of a Committee for Strengthening of Infrastructure in Science and Technology; Inter-University Centers; financing of computer centers in universities and computer laboratories for postgraduate courses; and the creation of University Science and Instrument Centers financed by the University Grants Commission (UGC) (Yasmin 2004, ¶19);
- The Delhi-based Indian National Science Academy (INSA), established in 1935 has also introduced several programs to popularize and improve science education at the school and university levels. It aims to increase awareness of the many careers available in science research and development. It provides quality reading material on science subjects to schools, colleges and universities; has formed a regular network with the print and electronic media to publish interesting science material on contemporary, relevant topics, and runs several science projects at the school and university levels. The academy also conducts competitions, such as the National Science Congress, for students and teachers in schools and colleges throughout the country;
- The Bangalore-based Indian Academy of Sciences has initiated various programs to enhance the appeal of science education. These include publishing the *Journal of Science Education*; awarding summer fellowships to students to work with fellows of the academy on research-oriented projects, conducting refresher courses for school and university science teachers, and organizing guest lectures and seminars;
- The University Grants Commission, the country's premier higher education funding agency, has announced the setting up of four National Institutes of Science (NIS) to address the growing concern for the continuing decline of science education standards at all level (Yasmin, 2004); and
- Navodaya Vidyalayas are special free-of-cost co-educational, residential schools started all over India to provide good quality modern education to promote and develop talented, bright and gifted children predominantly from rural areas who may otherwise be denied good educational opportunities (Navodaya Vidyalaya, 2007).

Other efforts, such as teacher training, are being made in education generally that may have a flow on effect to science.

Teacher Training

The National Council of Teacher Education (NCTE), National Council for Educational Research and Training (NCERT), District Institutes of Education and Training (DIET), and State Councils for Educational Research and Training (SCERT) are government agencies working to improve the quality of teacher education. The National Assessment and Accreditation Council (NAAC) has started the process of assessment and accreditation of all teacher education institutions under the jurisdiction of NCTE. It is making a special effort to train untrained in-service teachers. A six-month distance-learning program, Certificate in Primary Education, has been started by Indira Gandhi National Open University and NCTE. A Teacher Education Curriculum Framework has also been formulated (UNESCO, 2004).

Schemes Provided by the University Grants Commission

UGC has identified nine universities as 'Universities with Potential for Excellence', and allocated Rs 300 million to achieve and sustain a superior quality of education and research. They will be given freedom to change fee structure, research funding, admission criteria and recruitment of teachers to attain this goal. Similarly, 12 centers and 97 colleges are covered with an additional funding of about Rs 1500 million under 'potential for excellence' schemes by the UGC (Science Education, 2006). There are several other schemes as well. Further, an Indian Institute of Science for Education and Research (IISER) is being set up in Mohali in addition to those in Pune and Kolkata.

Some Innovative Programs

A number of innovative programs fro teaching science in schools in India have been developed recently.

School Complexes, Mobile Labs, Science Kits. The concept of school complexes as developed in Indian Education Commission Report, 1964-66 is based on the assumption that high and higher secondary Schools have better laboratory and library facilities, which can be utilized by the primary and middle schools (Kothari, 1966). Five or six primary and upper primary schools form a complex and obtain academic and administrative support from the nodal secondary/senior secondary school.

Non-government organizations such as Vidnyanvahini (in Pune) reach out to rural school children and offer them a means of learning science through laboratory experiments, using a 'mobile science laboratory' (Vidnyanvahini, 2006). Schools from rural or backward areas receive low-cost science kits that allow children to learn by doing simple experiments in physics, chemistry and biology; for example, the 'Eureka' low-cost science laboratory kits was developed by 15 scientists from IIT Madras and Institute of Mathematical Sciences (Sujatha, 2005).

Broadcast Technology for Science Education. A national program for technology enhanced learning has been started to develop technology-based programs that will supplement classroom teaching (UNESCO, 2004).

Private Initiatives

A number of private organizations also have been active in an attempt to enhance science and science teaching in India.

Hoshangabad Experiment. The Hoshangabad Science Teaching Programme (HSTP), started in 1972, focused on activity-based science teaching at the middle school stage; developing low-cost science kits affordable to rural schools, building a network of science teachers, effective teacher training, publishing a teacher resource journal, developing science teaching practices and workbooks tuned to the rural context. It gained popularity and was extended to several districts covering over a thousand schools (Hoshangabad Vigyan, 1977; HSTP Group, 2002). The important message from this experiment is that it showed that a spirit of scientific thinking can be inculcated in rural children having little external resources for science education, and that too at low cost. Unfortunately, the state government, after having supported it since 1978, shut it down in 2002, despite protests from various quarters (Menon, 2002; Ramachandran, 2002).

Exploratory. A science exploratory in Pune provides an environment for school children to discover and practice science in an open-ended way (BVB 2007). The exploratory conducts science talent searches, provides teachers' orientation, and publishes science-based books and self-learning multimedia materials. Around 2,000 students visit the exploratory every year.

Science Competitions. An education promotion firm, Aakar Knowledge Solutions, has organized Scientifica 2007, a science and technology competition for schoolchildren from Class 3 to Class 5, carrying total prize money of 1.5 million Rs (Science Beyond Textbooks, 2007). The competition will encourage the students to discover science beyond textbooks and examinations through a series of real-life problems.

Science Toys from Junk. Enterprising scientists like Dr. Arvind Gupta (Arvind Gupta Toys, 2007), have designed low-cost science teaching aids and science toys from low cost material - junk - and disseminate this knowledge to others, conducting workshops for science teachers and NGOs in education and science (Ray, 2004).

CONCLUSIONS

There are clearly a number of significant problems facing education and science education in India. However, the Government of India is making great efforts to bring education in general and science education in particular to its citizens. Several private initiatives are also working to promote science education. However, the sheer scale of the target population and the limited resources, combined with lack of economic incentives for science profession, means that we have a long road ahead before India is known as a nation that does great science and uses science to achieve uniform progress.

REFERENCES AND BIBLIOGRAPHY

Arvind Gupta Toys. (2007), Retrieved 12 February 2007, from http://arvindguptatoys.com/

Bhartiya Vidya Pratishthan. (2007). *Bhartiya Vidya Pratishthan Science Exploratory, Pune*, Retrieved 12 April 2007, from http://www.bvbpune.org/explor.html

Department of Education. (1986). *National policy on education 1986 (as modified in 1992).* Retrieved 17 September 2006, from http://education.nic.in/policy/npe86-mod92.pdf

Homi Bhabha Center of Science Education. (2007). *Home page.* Retrieved 1 November 2006, from http://www.hbcse.tifr.res.in/

Hoshangabad Science Teaching Programme Group. (2002). *Thirty years of Hoshangabad science teaching programme: 1972-2002*, HSTP Group, Ekalavya, Hoshangabad, India. Retrieved 10 December, 2006 from http://www.cisl.columbia.edu/grads/presi/EKLAVYA/HSTPNote.pdf

Hoshangabad Vigyan. (1977). *The Hoshangabad Vigyan: A unique adventure in rural science teaching.* Retrieved 1 November 2006, from http://ebooks.du.ac.in/edu-resources/Resources/books/readings/hstp.pdf

Indian National Science Academy. (2001). *Pursuit and promotion of science: The Indian experience.* New Delhi: INSA. Retrieved 6 April 6 2007, from http://www.iisc.ernet.in/insa/, http://www.insaindia.org/INSA-book.pdf.

Kothari D.S. (1966). *Report of the Education Commission 1964-66* (Education & National Development). Retrieved 12 December 2006, from http://education.nic.in/cd50years/g/T/W/toc.htm

Maharashtra plans to make Maths optional from Standard 7. (April 15, 2006). *The Hindu*, retrieved 10 May 2006, from http://www.hindu.com/thehindu/holnus/004200604152152.htm

Menon, M. (2002, October). Ignorance is bliss. *India together.* Retrieved 26 March 2007, from http://www.indiatogether.org/education/articles/hoshi.htm

Mukunda, N. (1994). *Report on university science education.* Indian Institute of Sciences, Bangalore. Retrieved 3 February 2007, from http://www.ias.ac.in/initiat/sci_ed/report.html

Nanda P. (2006, December 14). Government admits 'poor quality' of science education, *India eNews.* Retrieved 16 February 2007 from http://www.indiaenews.com/education/20061214/32438.htm

Navodaya Vidyalaya (2007). Retrieved 10 May 2007, from http://www.navodaya.nic.in/

National Council of Educational Research and Training. (1975). *The curriculum for the ten-year school.* New Delhi: NCERT.

National Council of Educational Research and Training. (1988). *National curriculum for elementary and secondary education. A framework.* New Delhi: NCERT.

National Council of Educational Research and Training. (2000). *India. Education policies and curriculum at the upper primary and secondary education levels. Globalization and living together: The challenges for educational content in Asia: Final Report 1999.* International Bureau of Education, UNESCO. Retrieved 26 December 2006, from http://www.ibe.unesco.org/publications/regworkshops/newdefinal.htm

National Council of Research and Training. (2002). *Seventh all India school educational survey.* Retrieved 29 December 2006, from http://gov.ua.nic.in/aises/

National Council of Educational Research and Training. (2005a). *National curriculum framework 2005.* Retrieved 26 December 2006, from http://www.ncert.nic.in/sites/publication/schoolcurriculum/NCFR%202005/contents2.htm

National Council of Educational Research and Training. (2005b). *Gender issues in education. National Curriculum Framework Review, 2005.* Retrieved 20 May 2007, from http://ncert.nic.in/sites/publication/schoolcurriculum/Position_Papers/Gender%20Issues%20in%20Education.pdf

National Council of Educational Research and Training. (2007). *Education policies and curriculum at the upper primary and secondary education levels.* Retrieved 2 May, 2007 from http://www.ibe.unesco.org/curriculum/Asia%20Networkpdf/ndrepin.pdf

National Literacy Mission (2001). *Literacy scenario in India.* Retrieved 18 May 2007, from http://www.nlm.nic.in/tables/scenario.htm

National Talent Search. (2007). *National talent search, National Council of Educational Research and Training, New Delhi*. Retrieved 15 May 2007, from http://www.ncert.nic.in/sites/nts/main_3.htm
Rajput, J.S., & Srivastava, V. P. (2001). Part II: Current trends and main concerns as regards science curriculum: India. In M. Poisson (Ed.), *International workshop on the reform in teaching of science and technology and primary and secondary level in Asia* (pp. 21-27). Beijing, 2000. Geneva: International Bureau of Education, UNESCO.
Ramachandran, R. (2002, September 13). *Axing a science teaching program*. Frontline, 19(18). Retrieved 26 March 2007, from http://www.hinduonnet.com/fline/fl1918/19180860.htm
Ranade, M.D. (2006). Development of CAI presentations for science teaching and overview of research findings. *International Journal of Science and Mathematics Education*, *4*, 763-789
Ray, K. (2004, June 3). Toying with junk. DH Education, *Deccan Herald*, Retrieved 26 March 2007, from http://www.deccanherald.com/deccanherald/Jun032004/edu4.asp
Science Beyond Textbooks. (2007, March 26). Science beyond textbooks, examinations, Times News Network. *Times of India*, Pune ed., p. 1.
The Press Information Bureau. (2006, December 12). *Science Education*. Government of India. Retrieved 10 April 2007, from http://pib.nic.in/release/rel_print_page.asp?relid=23139
Shukla, R. (2005). *India science report*: *Science education, human resources and public attitude towards science and technology*. National Council of Applied Economic Research, New Delhi. Retrieved 20 December 2006, from http://www.insaindia.org/India%20Science%20report-Main.pdf
Srinidhi, R., Nehru, A., Pandit, V., Chatterjee, G., & Raghupathi, H. (2005). India's unchecked textbooks racket. *India Together 1*6 February 2005. Civil Society Information Exchange Pvt. Ltd. Retrieved 26 December 2006, from http://www.indiatogether.org/2005/feb/edu-textbook.htm
Sarva Shiksha Abhiyan. (2002). *Executive summary*: *Learning achievement of students at the end of Class 5*. Department of Educational Measurement and Evaluation, Sarva Shiksha Abhiyan. Retrieved 26 December 2006, from http://ssa.nic.in/qualityedu/summ_exec.asp
Sujatha, R. (2005, August 10). Corporation schools get science kits. *The Hindu*. Retrieved 4 May 2007, from http://www.hindu.com/2005/08/10/stories/2005081008540100.htm
Sutar, B. (2006). *Position paper by National Focus Group on Teaching of Science*. National Council of Educational Research and Training, New Delhi. Retrieved 10 April 2007, from http://ncert.nic.in/sites/publication/schoolcurriculum/Position_Papers/TeachingofScience.pdf
UNESCO. (2004). *India*: *National Report on the Development of Education, 47th session of the International Conference on Education*, Geneva, 8-11 September 2004. Retrieved 17 March 2007, from www.ibe.unesco.org/International/ICE47/english/Natreps/reports/india.pdf
Vidnyanvahini. (2006, May 20) *Vidnyanvahini Pune*. Retrieved 4 May 2007, from http://members.aol.com/vvahini/index.html
Wikipedia. (2007). *Criticism of Indian education system, education in India*. Retrieved 18 May 2007, from http://en.wikipedia.org/wiki/Education_in_India
Yasmin, S. (2004). Science education on a slippery path. *India Together*. Retrieved 15 May 2007, from http://www.indiatogether.org/2004/oct/edu-science.htm

AFFILIATION

Mridula Ranade
Postgraduate Department of Education
S.N.D.T. Women's University
India

HALAI

CURRICULUM REFORM IN SCIENCE EDUCATION IN PAKISTAN

INTRODUCTION

Pakistan, like many countries in the region inherited its system of education from the British. Lord Macaulay (1835, Minutes on Education) aspired to make Indians into 'Brown Sahibs' who would mimic the British in dress, food and mannerisms, but who would lack the intellect to forge ahead on their own. Thus both the Hindus and Muslims of the sub-continent would become the 'workhorses' of the empire, and would do the bidding of the conquerors. As a consequence the system of education that the British installed produced not individuals who could think for themselves, but clerical staff who could deal with the bureaucracy of the Raj. Very little attention was paid to science education, and it was only included in the curriculum at the secondary level after the formation of Pakistan in 1947. This chapter provides a picture of science education in Pakistan, and examines how the gaps in science education were addressed with the help of huge investment in practical science teaching, but how this effort has not born fruit. Lastly, some ways in which science education might be enhanced and improved in Pakistan are proposed.

SCIENCE EDUCATION: POST PARTITION TO DATE

In 1947 the Indian subcontinent was divided into two countries, the Muslim majority provinces in the Northwest and Northeast of India was called (East and West) Pakistan and the rest became India. East Pakistan became Bangladesh and West Pakistan, became Pakistan. Since, its inception Pakistan has been disadvantaged as majority of universities and schools were in the territory that now comprise modern day India. However, through a huge effort Pakistan has enhanced its education sector, and there are now more than 100 institutions of higher education (Isani, 2002) and a literacy rate of more than 50% (World Bank, 2007). From the outset, science and technology were seen as a way to allow the young Muslim state to enter the twentieth century, and a concerted effort was made to improve the teaching and learning in science through the use of innovative strategies (Warwick & Reimers, 1995).

Until the 1950s science was taught only in post-secondary institutions, and very little science was taught at the primary and secondary school levels (Iqbal & Mamood, 2000). The topic nature study was introduced into primary classes in 1959 and, in principle, general science and mathematics were compulsory for

Richard K. Coll, Neil Taylor (eds.), Science Education in Context: An International Examination of the Influence of Context on Science Curricula Development and Implementation, 115–129.

Grades 1 to 8. But the implementation of science remained difficult, and the thrust of education in general focused more on the liberal arts.

More ambitious curriculum reforms, particularly in the area of active science learning, were initiated in the 1970s when practical work was first introduced in school curricula at Grades 9 to 12. In 1972 the new education policy of the Pakistan Government represented a massive shift from a general, to a more technical and scientific education, at the secondary and higher secondary school levels. However, appropriate long term planning was not undertaken. As a result, a severe shortage of science teachers at the time was solved by rather ad-hoc measures, such as inducting untrained teachers (or teachers with minimal qualifications in science) into the profession. This resulted in a dramatic drop in the standard of science teaching. In 1979 governmental education policy documents acknowledged this gap in educational planning, and observed that no system could be better than its teachers. Efforts were subsequently made with the help of loans from the Asian Development Bank to develop new science curricula. A new body called the Institute for the Promotion of Science Education and Training (IPSET) was established for in-service training of science teachers. Although these efforts helped to increase the profile of science in the school curriculum, science teaching remained rooted in the past for a number of reasons, which are discussed below.

In 2002 the Pakistan Higher Education Commission (HEC) was formed under the charismatic leadership of Dr. Attaur Rahman. Dr. Rahman was able to enhance the funding for higher education substantially, and a Task Force on Higher Education was formed. A large number of the Taskforce recommendations to improve the standard of education generally, and science education in particular, were accepted and are in the process of implementation as of 2007. However, none of these efforts questioned basic assumptions about practical work in science. It has been assumed, for example, that 'more of the same' will be effective in improving science learning, and hence the focus has been on improving practical work in the same way as practiced in the past, through provision of more laboratories and other resources. However, fundamental questions and issues about practical work have not been addressed, and these haunt science education in Pakistan. Some of the challenges facing practical work are specific to Pakistan; others are more general and impinge on the way practical work is envisaged all over the world.

Pakistan engaged in 'post-Sputnik' reform efforts in science education, and practical work in science was made compulsory for Grades 9 to 12. Every examination of science learning in schools includes a 'score' for practical work and this comprises 25% of the total marks. So, for example, a student assessed for physics at say Grade 10 would be allocated 75 out of 100 marks for a pencil and paper type test, and 25 marks from a hands-on practical examination in a laboratory. A large amount of funds were spent on teacher training, and for procuring science kits and other materials necessary for practical work. This was done with high hopes of increasing Pakistan's human resources in science and technology. However, the reforms have not produced the desired results. Practical work consists of activities that verify known scientific principles, or to illustrate scientific processes. The context of teaching and learning in the schools is such that even these activities are reduced to the level of rote memorization of the steps needed to complete 'the practical'. Hence, this form of practical work has not really helped Pakistani students develop understanding of scientific concepts or an understanding of 'doing science'.

CHALLENGES FACING SCIENCE EDUCATION IN PAKISTAN

The problems facing science education in Pakistan are two-pronged. One is the 'deficiency' problem that dominates many developing countries with limited resources. This results in consequential problems that include: (a) a shortage of science teachers, (b) poor training of science teachers, (c) poor quality of textbooks, (d) a system of examinations that encourages rote memorization, and (e) a lack of laboratories, equipment, and other resources needed to teach science. However, there is a deeper problem in Pakistan that I call the 'conceptualization problem'. The conceptualization of science education Pakistan requires us to ask questions such as, 'What kind of science education is needed for Pakistan?', 'What kind of practical work is needed for Pakistani students?' and 'What are the skills that we need to teach students through practical work in science?'. These have been left largely unanswered.

I first discuss briefly the deficiency challenges faced by Pakistan science teachers and educators as this relates to practical work, and then consider the conceptual difficulties and some ways I think we can address them.

Training of Science Teachers

Teaching science at different grade levels in Pakistan, as elsewhere requires different levels of specialization. The emphasis at earlier stages in Pakistan is on pedagogical skills of teachers, and their ability to convey concepts and ideas to students. Up to say Grade 5, after 10 years of schooling, primary certificate level training only is required. To teach Grades 6 to 8 requires a general certificate after 12 years of schooling. At higher levels, more advanced subject knowledge and technical expertise in the laboratory is required, in the form of BSc and BEd degree qualifications. Hence, primary level teachers with only 10-12 years of schooling are not able to do justice to the teaching of science. As in many other parts of the world, Pakistani primary teachers do not necessarily have an academic background in science, and yet are expected to teach science at the primary level. Not surprisingly, Pakistan primary school teachers often lack confidence about teaching science. Their teaching is at best textbook-based.

In addition, the training and practice teaching for Pakistan science teachers leaves much to be desired. Most of the pedagogy of science is taught via lectures and, rather ironically, pre-service teacher trainees are exhorted to use activity-based methods in science in these lectures. Hence, 'trained' teachers of science usually do not have much idea of how to teach through 'activity means', or about classroom management issues that often occur when more exciting activities or demonstrations are done in class. There has been a huge growth in the private education sector in Pakistan involving teacher training through private universities either in Pakistan or abroad. This has led to some improvement in science teaching in some private schools, but key core questions that affect science education and particularly practical work in Pakistan remain.

Shortage of Science Teachers

There is a significant shortage of science teachers in Pakistan. Statistics even for the largest and the most developed province of Pakistan show that at the elementary level one science teacher per school is the norm, while at the secondary level three science teachers is the norm (Social Policy & Development Center [SPDC], 2002-03). The situation is much worse in rural areas, and in rural Pakistan many primary schools have no science teachers, and most secondary school science teachers do not possess a degree in science. It is not unusual to find those teaching science who have never studied science in school. However, in large cities such as Karachi, it not difficult to find teachers who teach science at the secondary and middle school levels who have a BSc or MSc.

This shortage of science teacher then creates two kinds of problems, science teachers who are available are expected to teach a large number of classes, and cannot do justice to their teaching. Instead they tend to focus on covering the syllabi for the examinations. Second, science teachers, particularly at the college level, are expected to teach very large classes, often of more than 100 students, meaning it is not possible for the teacher to communicate well with students, let alone teach them well.

Quality of Science Textbooks

In many developing countries textbooks are often the only resource other than the teacher that the students can utilize in learning science. Hence, science textbook are even more important in science learning than in developed nations. "An examination of prescribed textbooks for secondary and higher secondary schools reveals that they:

- present science in a manner that is too abstract and formal;
- are deficient in content as well in presentation;
- are lacking in focus and precision that is essential in understanding and working with scientific phenomena; and
- are unsuitable in terms of developing concepts and creating and maintaining the interest of the students" (SPDC, 2002-03, p. 140).

In Pakistan, as elsewhere, textbooks lay a great deal of emphasis on definitions and facts rather than developing conceptual understanding of ideas. Nor do they try to make coherent the big ideas in science (American Association for the Advancement of Science [AAAS], 1993) so that students can, for example, see that light, magnetism and heat are all different forms of energy, all of which are governed by the same principles.

In Pakistan even after completing a science masters degrees the graduates who may become teachers may not develop conceptual understanding or and analytical capability in science, and are often unable to draw graphs, read and interpret data or otherwise analyze scientific data. Hence, poorly prepared graduates often become poor science teachers, who in turn teach inadequately in a cycle that reinforces a limited view of science.

Assessment of Science Learning

Secondary and higher secondary examinations in Pakistan are centralized, conducted by multiple boards of education, and set for this purpose in each province. The Federal Government of Pakistan also has a Board of Examination for schools under its gambit. The examinations occupy a central position in both students' and teachers' lives in both schools and colleges. Teaching approaches in the classroom, the assessment system in schools, the textbooks prescribed, and students' and teachers' attitudes towards teaching and learning are strongly influenced, if not governed, by these Board examinations (Halai, 2002). Schools and colleges are assessed, and assess themselves, on the basis of their students' performance in these exams. This would be acceptable if the examinations at least attempted to test conceptual understanding, reasoning ability or the ability to problem-solve. However, the question papers are structured in a very specific way, meaning those who teach to these examinations are rewarded. Only about 33% is required to gain an examination 'pass', meaning students need to know relatively little of the topics covered in the syllabus.

Competition for entry into each grade of study is very stiff, and this has given rise to the phenomenon of teaching Grades 9 and 10 science in Grade 8, and sometimes even in Grade 7, whether or not students are prepared for study at these levels. As an illustration, there is no requirement to use symbols and equations to denote chemical reactions in Grades 7 and 8. So the reaction between carbon and oxygen would be represented thus:

$$\text{Carbon} + \text{oxygen} \rightarrow \text{carbon dioxide}$$

However, many science teachers in the school expect the students to write it such an equation using symbols in the form,

$$C + O_2 \rightarrow CO_2$$

and then teach the balancing of equations; thus expecting Grade 7 and 8 students to memorize symbols and valences for elements and radicals, along with the concept that equations are expressions of molar quantities taught in Grade 9. Such teaching encourages rote memorization undertaken in order to keep up with teaching for

Grades 9 and 10. The examinations thus have become a test of memorization, and encourage even progressive schools to 'teach to the test'. Any efforts to bring about changes to teaching are resisted strongly both by parents and students because of the high stakes nature of the examinations.

Poor Implementation of the Science Curriculum

In Pakistan science is integrated up to Grade 8 after which the students are given the choice which stream to take, one of which is the science stream where the major branches, physics, chemistry, and biology, all taught as separate subjects. From a pedagogical standpoint, the argument in favor of integration is that that real life situations demand the application of scientific knowledge which is not 'compartmentalized' into physics, chemistry, or biology, but that exist as an integrated whole. In developing countries like Pakistan few students pursue higher education, and so a compulsory integrated science curriculum up to Grade 10 is deemed more appropriate. It is further argued that it is too early to stream at Grade 8, and to do so would deny students in non-science streams valuable learning not covered in general science. In Pakistan, streaming into science starts as early as the end of Grade 8, and student performance in science and mathematics at this grade thus acts as a 'gatekeeper' to further science study. Science 'graduates' are privileged, in that they can then reenter any other stream such as art or commerce, but the reverse is not allowed.

The implementation of an integrated science curriculum is not without problems. The main problem lies with the teachers, their capabilities and their resistance to change. Pakistan teachers find it difficult to hold mastery of all disciplines of science such as physics, chemistry, biology, geology, and astronomy, especially considering that they typically specialize in only one field. It is argued that if the different branches of science are to be taught with some depth in Grades 9 and 10, they must be handled by specialist teachers.

However, the weakest aspect of the science curriculum in Pakistan is probably in the implementation of the curriculum. The science curriculum documents overall are well written, but the resources necessary to put them into effect are simply not in place. Overall, the poor quality of the teachers is probably the major barrier to good science teaching in Pakistan.

Lack of Science Laboratories and Resources

A lack of availability of science laboratories for Grades 9-12 students is a cold reality in Pakistan, and much more needs to be done to ensure that the requisite materials are available to science teachers. However, it is also true that science teachers need to be empowered to obtain the required materials. It has been my experience that schools actually often have allocated funds which go unused, since no requests for materials are made by science teachers. Under-utilization of *allocated* funds is actually more of a problem in government schools than *lack* of funds. Additionally, the teaching of science in Grades 1 to 8 does not require much laboratory equipment. My work in teaching a science methods course for primary science teachers, suggests that using simple, readily available materials can lead to positive learning outcomes at this level (see Halai, 2002).

NEEDED REFORM IN PRACTICAL WORK IN SCIENCE

Practical work has the potential to deliver many benefits for students' learning of science (see, e.g., Nakleh, Polles & Malina, 2002) in allowing hands-on experience of natural phenomena, but there are also limitations that need to be considered. There are a number of assumptions about practical work that need to be very carefully considered when designing practical activities (Halai, 2002). It is, for example, assumed that 'teaching science' 'doing science', and 'teaching *about* science' are much the same thing. Through these three aspects overlap I would argue that they are distinct.

CHANGING CONCEPTIONS OF THE NATURE OF SCIENCE

There is reasonable consensus among science educators all over the world that teaching about the nature of science should be a part of science curricula (see, e.g., AAAS, 1993). While this thinking has been taken on board in the Pakistan science curriculum tacitly, no attempt has been made to define it more explicitly, or to actually include it in teaching of science. Most science teachers in Pakistan have been exposed to a very positivist conception of science. As a consequence, many local teachers do not understand the nature of science themselves, and assume, for example, that observations are independent of theory, and also assume that all students observing a given phenomenon will see the same thing as the teacher. A related assumption is that practical work 'speaks for itself', and that after practical work no discussion of outcomes is necessary.

Pakistan science teachers often assume that students *see* what the teacher does while looking at a science demonstration. All students have to do is to *see* the activities, and they will inevitably abstract the same principle or concept that the teacher wants to illustrate. Roth (1995, p. 3) expressed similar views when he started to teach science: "I began to teach science assuming that all students had to do was to look and recognize the truth about nature; I assumed that what I saw happening in a test tube should be seen equally by students". Roth soon changed his views with experience! The reality is, of course, that students often see things very differently from their teachers. Teachers bring their own theoretical framework with them, and look at the particular activity with a specific *lens* that guides their observations. Not sharing that lens (because of their own more limited experiences), students often see things very differently from the teacher. Studies in which students are expected to draw the internal structure of a plant cell after observing it under a microscope show clearly that students often focus on something entirely different to that of the teacher. Hodson (1998, p. 10) sums up the theory-dependence of observation very succinctly:

> The traditional school curriculum description of science says two things about observation. First, nothing enters the mind of the scientist except by way of the senses – that is, the mind is a *tabula rasa* on which the senses inscribe a true and faithful record of the world. Second, the validity and reliability of the observation statements are independent of the opinions and expectations of the observer and can be readily confirmed by other observers. Neither is true. In reality, we interpret the sense data that enter our consciousness in terms of our prior knowledge, beliefs, expectations and experiences.

Chalmers (1999) agrees that theory precedes observation, and says it is false to claim that science starts with observation. Even with highly guided activities, students do not see 'eye-to-eye' with their teachers. For example, in a science class, the teacher was using balloons to show that air has weight, but the students instead focused on the color of the balloons seeing this as an important variable in the experiment. Gunstone (1991) comments that 'looking at things' is an active process, in which students check out their perceptions *against their expectations*. In fact, students' personal theories may lead them to reject particular conclusions, by way of denying the legitimacy of observations.

This is a boarder problem than practical science Pakistan. In the 1980s as part of the *Learning in Science Project* (LISP), Tasker and Osborne (1985, p. 27) investigated difficulties students face when learning science in the classroom. They concluded that:

- There was a disparity between the *ideas* children brought to the lesson and the ideas the teacher assumed that they would bring to it;
- There was a disparity between the *scientific problem* the teacher would have liked the children to investigate and what they took to be the problem;
- There was a disparity between the *activity* proposed by the teacher and the activity undertaken by the children, despite considerable teacher intervention; and
- There was a disparity between the children's *conclusions*, and the conclusions proposed by the teacher.

Tasker and Osborne (1985) argue that the teacher's intentions cannot be directly transferred to students; rather the teacher has to *contrive* to see if the mental constructions of the students match what the teacher wanted to convey. Further, they state that appreciation of the perception the learner brings to the classroom (among other things) is essential if the gap between the teacher's intentions and the students' learning is to be reduced. This sort of thinking, about the importance of student prior knowledge, is what led to the development of constructivism and learner-centered education (Wheatley, 1991).

Meaning Making in Practical Work

Practical work in science, or 'using activities' has a long history in many parts of the world such as the United Kingdom (Wellington, 1998), but practical work was only introduced into Pakistan schools in the early 1960s (Iqbal & Mahmood, 2000). While practical work can motivate, illustrate, clarify and enhance understanding, it also has the ability to confuse. Driver (1994) attributes this problem to a simplistic and empirical view of science that *knowledge is based on observation*, and that scientific laws are acquired by a process of induction from the 'facts' of sensory data. The cliché 'I do and I understand' is often used to support the use of practical work in science teaching. If children are to understand science concepts more than that is needed. Driver comments that the slogan 'I do and I understand', instead is more likely to become, 'I do and I am even more confused'! Many science teachers have a rather naïve trust in the ability of practical activities to expose students to sensory data that will help them to construct scientific concepts. They may believe, for example, that observations give reliable information about the world, rather than being interpretations of phenomena based on one's own views and prior knowledge.

There has been an effort to reappraise the use of practical work in science teaching (Hodson, 1990; Wellington, 1998), and engage in a fair assessment of what practical work can and cannot achieve (Millar, 1998; Nakhleh et al., 2002). Science teachers think of activities as cognitively effective tools for teaching science because students can 'see' what is happening. But sometimes these tools became more like 'fun-work' rather than an activity that actually enhances understanding. Thus meaning making in science is often considered to be complete once the practical work ends. In fact the end of the activity is often the time when meaning making really starts. Driver (1983, p. 49) comments:

> It is common to see science lessons which end with the clearing up after the practical work is finished. The time for the important discussion of how the important experiences gained relate to the new ideas is missed. Activity by itself is not enough. It is the sense that is made of it that matters.

Gunstone (1991, p. 74) reiterates, "For practical work to have any serious effect on student theory reconstruction and linking of concepts in different ways, the students need to spend more time in interacting with ideas and less time interacting with apparatus".

Using Practical Science to Help Students See Science as More Than Received Knowledge

Let me now link some of these ideas from the literature to the situation in Pakistan. During a research project I was involved with I observed a science teacher explain the five major classifications of vertebrates as:

(a) Fish, (b) Birds, (c) Amphibians, (d) Reptiles and (e) Mammals

Over two class periods the teacher brought in some preserved samples of dog-fish, flounder, and a frog - all preserved in formalin. In each case she explained the dominant characteristic of each species. Then she explained that human beings are vertebrates, and she brought in four bones from the vertebrate column of a human skeleton. She finished her lessons by demonstrating the characteristics of all the five classifications. The treatment given to classification was as if nature *intended* the animals to be divided in such a way. Interestingly, this science teacher was well aware of the Swedish biologist Linnaeus and his achievement in devising the classification system for animals, as it is a part of the microbiology curriculum at the undergraduate level (which she had studied). However, in her instruction it appeared as if the classification system was received from a 'divine authority' rather than something devised by a human scientist. By pointing out the human effort that went into the classification process an important aim of teaching students something of the nature of science might have been achieved. Talking of the theory behind the classification of vertebrates, could then have added to a bigger picture of biological knowledge. Instead the students viewed these two lessons as separate examples of fish, amphibians, reptiles, birds and mammals.

This brings us to the nexus of knowledge and authority in Pakistan. In Pakistan society, as in many other countries of the world, knowledge, and hence scientific knowledge, is seen to reside with authority (Fuller & Snyder, 1991; Shumba, 1999). The concept of knowledge as truth, and 'truth is for all time', is a dominant way of thinking about science knowledge for many people (Krugly-Smolska, 1995). The idea of challenging authority is very difficult for Pakistani students in particular. Much of the knowledge in books is accepted *without question*. Questions in science such as, 'Why are animals classified in this manner?' are never raised. They are instead accepted as a fact, or as a truth. This is not only a problem in developing countries like Pakistan, but appears to be endemic even in developed countries. Lemke (1990, p. 175) goes so far as to comment that: "It is dangerous to society to have students leave school believing that science is a perfect means to absolute, objective truths, discovered by people of superhuman intelligence".

So Pakistan science teachers teach science in a way that adds to the 'mystique' (Lemke, 1990) or the 'mythology' of science (Smolicz & Nunan, 1975). They seldom make it clear that science is a human endeavor, and therefore, subject to human frailties. Questions such as, 'What are good reasons for believing that?' are very rarely encounter in Pakistan science classrooms. Acceptance of authority as a source of knowledge, and hence not challenging the knowledge provided by the teacher, is thus deeply embedded in the culture of Pakistan schooling (Aubusson & Watson, 1999).

Is Science a Superior, Value-free and Stable Way of Knowing?

In a similar way, science teachers that I have encountered in my role as a science teacher educator, generally see science as a body of knowledge based on empirical evidence. Science knowledge is then considered to be different from other kinds of knowledge because of its ability to yield empirical evidence. One can perform an experiment and 'prove' a claim in science, whereas one cannot do that in other forms of knowing such as religion. Pakistan science teachers thus lay a great deal of importance on experiments to 'discover the laws of science and develop scientific theory'. Generally, teachers see science as knowledge that is neutral and value-free, free from religious beliefs, the latter which falls within the realm of 'faith'.

Personally I do not view science as belonging to any religion or group, and like Pakistan's most famous physicist and only Nobel laureate Abdus Salam[i], I believe in maintaining a distance between science and religion (Hoodbhoy, 1998)[ii]. However, at the same time I acknowledge and value Sardar's vision, as quoted in Loo (1996, p. 288), of science as a human activity that is "subject to the moral philosophy of Islam", and envisage a *halal*[iii] environmentally-friendly science that promotes social justice and public interest. The key issue of the control of science by the community of scientists, or a higher moral authority, in my view needs rethinking. But the concept of the nature of science is essential for school science. Furthermore, there is a paucity of literature that discusses the nature of science from an Islamic perspective. Haidar (1999, p. 808), however, summarized the fundamental Islamic beliefs about the nature of science:

> The purpose of science is to discover God's wisdom in the universe; knowledge can be acquired by the scientific method as well as by other means, e.g., revelation; truth is not absolute, we see only what God permits us to see; and the only absolute truth is what God knows.

It is important to emphasize that these views may represent just a particular point of view, and that other denominations of Islam may have views that are somewhat different. However, in Pakistan, these views would certainly find a fair degree of acceptance. This is an important element to consider in response to the work of Lederman and Niess (1997), who question the ability of science teachers holding traditional views about science, to execute reform in science education. Does this mean that societies (not only Islamic societies) that espouse traditional views cannot reform science education? Or does it mean that science reforms need to adjust to belief systems in these parts of the world? Fuller (1997) explains the Islamic perspective about the nature of science, saying that Muslim thinkers question the sharp distinction that Western philosophers draw between *episteme* and *ethics*, and do not espouse a view of science that is value-free. Muslims view science as a knowledge that should be pursued for universal human good rather than for its own sake.

A WAY FORWARD

Teaching Science: A Balancing Act

Science teaching is characterized by a number of tensions. Tension between a desire to teach for understanding, and the need to complete the syllabus. Tension between a need to see that all students 'succeed' in the Board exams, and teaching for understanding. But teaching for success in examinations, and teaching for understanding ought not to be different, or have mutually exclusive goals. But in the less than ideal top-down, centralized examination system prevailing in Pakistan, it seems that skills needed to pass examinations are valued above understanding. Science holds the key to entry into fields such as medicine, engineering and computer science, which are highly valued in Pakistan. Hence, science marks are used as a 'weeders' to select students who can or cannot study science. These tensions are exacerbated in the teaching of science (and mathematics) as compared to subjects such as social studies and Islam.

Pakistan is a developing country, and has the ethos of a traditional Muslim society where conceptions of knowledge are very different from those prevalent in the modern and postmodern developed countries. In Pakistan, like many other developing nations, knowledge is seen to lie with authority and tradition (Shumba, 1999). Within such a framework students find it almost impossible to question and challenge their teachers – teachers are seen as the source of knowledge. In the same way, teachers themselves find it difficult to challenge authority, and are inclined to perform their duties in line with the dictates from 'above'. This contrasts markedly with modern views of teaching science, for example, inquiry-based learning, that requires students to construct their own knowledge and challenge and question traditional ways of knowing. Non-traditional modes of teaching also envisage a different role for a teacher, removing the teacher from the center of the classroom as the source of knowledge or the knower, and giving at least some power to the learner. The teacher's role then would become that of a facilitator, a part rather than the whole of the learning environment. This would clearly constitute a huge change in teachers', students', parents', and administrators' views of teachers and teaching and learning in Pakistan. I believe that this issue needs to be tackled much more seriously at the teacher education level than is being done at the present.

Knowledge of Subject Matter Is Not Enough

I suggest here that a teacher having strong content knowledge is an essential but not a sufficient condition, to ensure good teaching. Even if teachers have a good understanding of science, they may not know it in ways that help them to 'hear' their students, select appropriate activities and enhance student understanding. In Pakistan, a very large number of teachers, particularly those employed in private institutions, are not trained teachers. Additionally, the kind of teacher training available is so theoretical in nature that does not meet the needs of teachers. Thus there is something of a Catch-22 situation: teachers do not seek teacher education because the education and training does not meet their needs. School administrators do not insist on professional training for the same reason. As a consequence, teachers continue to use their own experiences as students as the basis for teaching. Thus the cycle continues. This cycle can be broken if teacher educators are seen to provide education that extends content knowledge, and also teaches how to present this knowledge in a pedagogically sound manner. One of the ways that might be done is to teach teachers using methods, activities and projects that they can use with their own students. This would expose teachers to new methods of learning, and having experienced these methods they might be more likely to use them in their own classes (Stofflett, 1994).

There is a basis to this suggestion in the literature. Research on science teachers' pedagogical content knowledge suggests that understanding of a subject matter and experience in teaching contributes to the development of pedagogical content knowledge (van Driel, Verloop & Vos, 1998). Adams and Krockover (1997) report beginning science teachers' pedagogical content knowledge depends both on their experience as a learner *and* as a teacher. Hence, a system of in-school professional development for teachers that suits the particular context, culture and needs of the school could go a long way to helping teachers who have not been exposed to professional education.

There is increasing evidence in the literature (e.g., Hodson, 1996, Nakhleh et al., 2002; Wellington, 2000) that practical work as currently conducted in both developed and developing countries does not necessarily help teachers achieve curricular goals. Open-ended investigations around relevant issues, project work, discussions, using reading and writing to learn science (Glynn & Muth, 1994) can enhance science learning - sometimes more than structured hands-on science activities. In Pakistan, teachers use demonstrations as their preferred model for bringing practical activities into the science room. A slight modification in the manner in which demonstrations are conducted, for example, incorporating activities like prediction-observation-explanation (Gunstone, 1991) may increase the effectiveness of demonstrations. However, the greatest challenge lies not in developing curricula and new ideas, but in supporting teachers to implement them and to further refine them to suit their particular needs and the needs of their students.

ACKNOWLEDGEMENT

This paper has been developed from the work done as part of my doctoral dissertation, and my work as a teacher educator at the Aga Khan University.

NOTES

[i] Abdus Salam received the Nobel Prize in 1979 with Steven Weinberg and Sheldon Glashow for developing the unified theory of elementary particles.
[ii] However, there are other Islamic philosophers, such as Jaafar Sheikh Idris (1987), who believe in the concept of Islamic Science.
[iii] That which is legitimate or accepted in the Quran.

REFERENCES AND BIBLIOGRAPHY

Asian Development Bank. (1988). *Impact evaluation study of the secondary science education projects. Nepal, Bangladesh & Pakistan*. ADB.

Aubusson, P.J., & Watson, K. (1999). Issues and problems related to science curriculum implementation in Pakistan: Perceptions of three Pakistani curriculum managers. *Science Education*, *83*, 603-620.

Driver, R. (1994). The fallacy of induction in science teaching. In R. Levinson (Ed.), *Teaching science* (pp. 41-48). London: Routledge

Driver, R. (1983). *The pupil as scientist?* Milton Keynes, UK: Open University Press.

Chalmers, A.F. (1999). *What is this thing called science?* (3rd ed.). Buckingham, UK: Open University Press.

Fuller, B., & Snyder, C.W. (1991). Vocal teachers, silent pupils? Life in a Botswana classroom. *Comparative Education Review*, *35*, 274-293.

Fuller, S. (1997). Western science from the outside in: The view from Islam and Japan. In S. Fuller (Ed.), *Science* (pp. 106-136). Buckingham, UK: Open University Press.

Glynn, S.M., & Muth, K.D. (1994). Reading and writing to learn science: Achieving scientific literacy. *Journal of Research in Science Teaching*, *31*(9), 1057-1073.

Gunstone, R.F. (1991). Reconstructing theory from practical experience. In B. Woolnough (Ed.), *Practical science: The role and reality of practical work in school science* (pp. 67-77). Milton Keynes, UK: Open University Press.

Haidar, A.H. (1999). Emirates pre-service and in-service teachers' views about the nature of science. *International Journal of Science Education*, *21*(8), 807-822.

Halai, N. (2002). *Munazza's story: Understanding science teaching and conceptions of the nature of science in Pakistan through a life history study*. Unpublished PhD dissertation, University of Toronto.

Hodson, D. (1998). *Teaching and learning science: Towards a personalized approach*. Philadelphia, PA: Open University Press.

Hodson, D. (1990). A critical look at practical work in school science. *School Science Review*, *70*(256), 33-64.

Hodson, D. (1996). Laboratory work as scientific method: Three decades of confusion and distortion. *Journal of Curriculum Studies*, *28*(2), 115-135.

Hoodbhoy, P. (1998). *Salam, science and secularism*. Retrieved 10 April 2001, from http://www.chowk.com/UniversityAve/hoodbhoy_jan0598.html.

Iqbal, H.M., & Mahmood, N. (2000). Science teacher education in Pakistan. In S.K. Abell (Ed.), *Science teacher education: An international perspective* (pp. 75-92). Boston: Kluwer.

Krugly-Smolska, E. (1995). Cultural influences in science education. *International Journal of Science Education*, *17*(1), 45-58.

Lemke, J.L. (1990). *Talking science: Language, learning, and values*. Norwood, NJ: Ablex.

Malinowski, B. (1948). *Magic, science and religion*. Boston: Beacon.

Millar, R. (1998). Rhetoric and reality: What practical work in science education is *really* for. In J. Wellington (Ed.), *Practical work in science: Which way now?* (pp. 16-32). London: Routledge.

Nakhleh, M.B., Polles, J., & Malina, E. (2002). Learning chemistry in a laboratory environment. In J.K. Gilbert, O. De Jong, D.F. Treagust & J.H. Van Driel (Eds.), *Chemical education: Towards research-based practice* (pp. 69-94). Dordrecht: Kluwer

Lederman, N.G., & Niess, M.L. (1997). Editorial: The nature of science: Naturally? *School Science and Mathematics*, *97*(1), 1-2.

Loo, S.P. (1996). The four horsemen of Islamic science: A critical analysis. *International Journal of Science Education*, *18*(3), 285-294.

Social Policy & Development Centre. (2002-03). *State of science education. Social development in Pakistan – Annual Review 2002–03*. SPDC.

Shumba, O. (1999). Relationship between secondary science teachers' orientation to traditional culture and beliefs concerning science instructional ideology. *Journal of Research in Science Teaching*, *36*(3), 333-355.

Smolicz, J.J., & Nunan, E.E. (1975). The philosophical and sociological foundations of science education: The demythologizing of school science. *Studies in Science Education*, *2*, 101-143.

Roth, W.M. (1995). *Authentic school science: Knowing and learning in open-inquiry science laboratories*. Boston: Kluwer.
Tasker, R., & Osborne, R. (1985). Science teaching and science learning. In R. Osborne & P. Freyberg (Eds.), *Learning in science: The implications of children's science* (pp. 15-27). Auckland, New Zealand: Heinemann.
Warwick, D.P., & Reimers, F. (1995). *Hope or despair? Learning in Pakistan's primary schools*. Wesport, CN: Praeger/Greenwood.
Wellington, J. (2000). *Teaching and learning secondary science: Contemporary issues and practical approaches*. London: Routledge.
Wellington, J. (1998). Practical work in science: Time for a reappraisal. In J. Wellington (Ed.), *Practical work in school science* (pp. 3-15). London: Routledge.
Wheatley, G.H. (1991). Constructivist perspectives on science and mathematics learning. *Science Education*, *75*(1), 9-21.
World Bank. (2007). *Pakistan: Data, projects and research*. Retrieved 5 September 2007, from http://www.worldbank.org.pk/WBSITE/EXTERNAL/COUNTRIES/SOUTHASIAEXT/PAKISTANEXTN/0,,menuPK:293057~pagePK:141159~piPK:141110~theSitePK:293052,00.html

AFFILIATION

Nelofer Halai
Institute for Educational Development
Aga Khan University
Pakistan

SELVARUBY, O'SULLIVAN AND WATTS

SCHOOL-BASED ASSESSMENT IN SRI LANKA: ENSURING VALID PROCESSES FOR ASSESSMENT-FOR-LEARNING IN PHYSICS

INTRODUCTION

> When the chef tastes the sauce, it is formative assessment. When the customer tastes, it is summative. (Anon.)

Assessment of student achievement is changing, largely because today's students face a world that will demand new knowledge and abilities. In the global economy of the twenty-first century, students will need to understand what are called the 'basics' of education, but will also acquire the ability to analyze, make inferences, and think critically. Helping students develop these skills requires changes in most forms of instruction and assessment at school and classroom level, as well as in approaches to large scale, high-stakes assessment at national level. Many countries are showing an increased interest in the formative dimension of assessment in the classroom (Airasian, 1991; Tunstall & Gipps, 1996). In general, such interest reflects a wish to teach for, and assess, high levels of conceptual knowledge, creative thinking, problem-solving, and effective communication skills (Pellegrino, Baxter & Glaser, 1999). To achieve this, we see a need to combine forms of assessment, blending formative and summative systems. While summative assessment provides some 'final form' picture of learners' achievements, formative systems provide continuous feedback on student learning and pays specific attention to the learning needs of students.

Over the last decade the Sri Lankan education system has sought to introduce a school-based assessment (SBA) program in schools. In this chapter we explore the teaching and learning of school science, and physics in particular, through some of the issues in assessment, and ways in which the problems of validating SBA have been developed.

BACKGROUND OF SRI LANKAN EDUCATION ADMINISTRATION

The nine provinces of Sri Lanka have 300 educational zones for the purpose of educational management of schools. Each zone has educational administrative officers, education development officers, subject specialists and master teachers (numbers depend on student population) to monitor and guide educational progress. Two important institutions, the National Institute of Education (NIE), for curriculum development and professional training of teachers and master teachers,

Richard K. Coll, Neil Taylor (eds.), Science Education in Context: An International Examination of the Influence of Context on Science Curricula Development and Implementation, 131–141.

and the Department of Examinations, National Evaluation and Testing Services (DOE/NETS), for assessment and evaluation, function under the Ministry of Education, Sri Lanka.

In Sri Lanka there are three different types of schools: 1AB, 1C, and Type II that have Grade 11 classes (senior secondary level). These schools differ in their physical and human resources to teach particular subjects, in this case science. Some have more resources, modern facilities, and qualified teachers than others. However, all schools follow the same curriculum and prepare their students to sit for the GCE (O-Level) Examination at the end of Grade 11.

The Sri Lankan national examination system has many strengths, and also some frailties. Information about students' progress, their strengths and areas of weakness, in appropriate areas of learning was not collected systematically by the examination system. With improvements to the quality of the examinations and better reporting of results, the national examination system has become known for supporting good teaching practices and effective learning.

The term *School-based Assessment* (SBA) has been used to refer to formative assessment on all subjects for the secondary grades in schools (Ministry of Education [MoE], 1998). The SBA is administered to students by teachers, on an on-going basis as a normal part of the teaching and learning process, in order to gather information about their progress. The information gathered enables teachers to modify instruction; to identify strengths and weaknesses in an individual's and group's performance, to provide feedback to each student, to make judgments about each student's progress, and to report for certification purposes. In particular, the teacher's role involves providing appropriate learning opportunities. The program began in 1999 in Grades 6 and 9 and was extended gradually to all grades. There are about 500,000 students attending about 10,000 senior secondary schools who have undergone this assessment program from 2001 at Grade 10.

The Asian Development Bank, Secondary Education Development Project report justified how formative SBA and summative GCE (O-L) examinations work together in a well-balanced assessment program noting that:

> Through SBA, a student's achievement of all the outcomes specified by the national curriculum can be measured, ensuring that even those competencies that cannot be measured easily on a paper and pen examination, i.e. the GCE examinations, will be given proper attention in each child's schooling. (Asian Development Bank, 1997b, p. 46)

Within Sri Lanka, DOE/NETS has overall responsibility for the implementation of SBA at senior secondary level (Grades 10-13). The results of the school-based assessment in Grades 10 and 11 are aggregated to build an internal summative result of each subject for reporting and certification purposes at the end of Grade 11. GCE (O-Level) examinations will be conducted at the end of the senior secondary level I (Grade 11) as an external summative assessment by DOE/NETS. Both results will be reported on student transcripts.

Given this, there is a need for secure validation of SBA because it serves both formative and certification purposes in the assessment system of Sri Lanka. Reviewing the class-based assessment programs implemented in selected countries, Papua New Guinea, Seychelles, Tasmania, Kenya, Hong Kong, Scotland, Netherlands, New Zealand and India, Perera (2000) identifies the following problems encountered:

- Inadequacies among teachers, students and parents in the conceptualization of class-based assessment;
- Doubts about comparability and credibility of the assessment procedures;
- Inadequacies in structural administrative support; and
- Lack of experience on the part of teachers and an increase in their workload.

Nor has Sri Lanka been without its critics. In addition to doubts about comparability and credibility of assessment procedures and methods, questions about the validity of assessment methods have been raised by several critics (see, e.g., De Silva, 2000) at junior secondary level. Towards this end, DOE/NETS has

- Conducted extensive training programs for education officers at provincial and zonal level, and for school principals in the implementation of SBA;
- Instructed principals of schools and directors of education of provinces and zones to establish School Boards, Zonal Boards and Provincial Boards of Assessment, to monitor the progress of SBA in their schools. These boards are expected to be responsible for training and continuing support necessary for teachers in making their judgments, and for monitoring of teacher judgments at their levels. This allows rectification of shortcomings, validation, and pre-processing of school-based assessment data on a term-by-term basis, first at school level and then at zonal and provincial level, before forwarding them to DOE/NETS for final tabulation and reporting; and
- Developed systematic validation processes for SBA, which we describe below. These have been necessary, not least to overcome problems experienced during previous attempts to introduce continuous assessment in the country, in 1972 and again in 1986.

The Notion of Validity

Early international research findings on the validity of testing have been based on the traditional conception of validity, where validity is seen as a single issue - Is a particular form of assessment valid or not? This has been the case in student experiences with traditional, normative forms of assessment, where students take little responsibility for their own learning. However, more recently, with the increased use of alternative methods of assessment, the definition of validity has been expanded to accommodate new techniques (Cunningham, 1998). Measurement theorists have proposed validation frameworks that are more consistent with both current theoretical understandings of validity, as a complex issue, and with the nature and potential uses of new forms of assessment. For example, in 1989 Messick presented a unified but faceted framework for validity by means of the two-by-two matrix, constructed by distinguishing two interconnected facets of the unitary validity concept. Such a model presents a useful framework for structuring arguments about validity. However, Messick provides little guidance about *how* the validation should be conducted. That is, there is still a need for a comprehensive framework for validating assessment and testing - but also constructing one that can be operationalized, in our case for the physics component of school science.

The research we report here develops such an operationalized framework based on Weir's (2005) language framework. Weir proposed a sociocognitive validation framework for language testing which, he argues, can form the basis of *any* test development and validation project. To examine the validity of a test, he says, requires both explicit theory and technique to guide the validation approach, and thus a validation framework to operationalize validity in its various manifestations (Weir, 2005). Our adaptation of his work we call the Selvaruby-Weir model (Selvaruby, 2006) for school-based assessment and national testing. Here, each element of validity is defined in relation to school physics (see Figure 1).

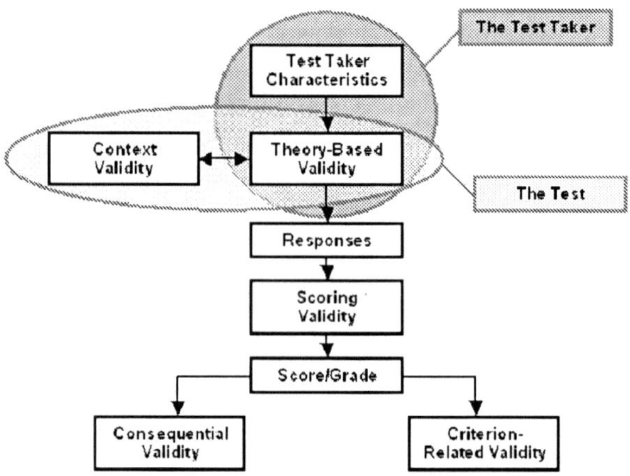

Figure 1. An outline of the Sociocognitive Test Validation Framework (Weir 2005)

The Model

Our model has six aspects of validity:

- Student (test taker) characteristics: How are the physical/physiological, psychological and experiential characteristics of students catered for by this assessment?

- Context validity: Are the characteristics of the assessment task(s) and its (their) administration fair to the candidates who are taking them?
- Theory-based validity: Are the cognitive processes required to complete the tasks appropriate?
- Scoring validity: How far can one depend on the scores/performance on the assessment?
- Consequential validity: What effects and impacts do the assessment have on its various stakeholders? and
- Criterion-related validity/reliability: What external evidence is there outside of the assessment scores themselves that it is doing a good job?

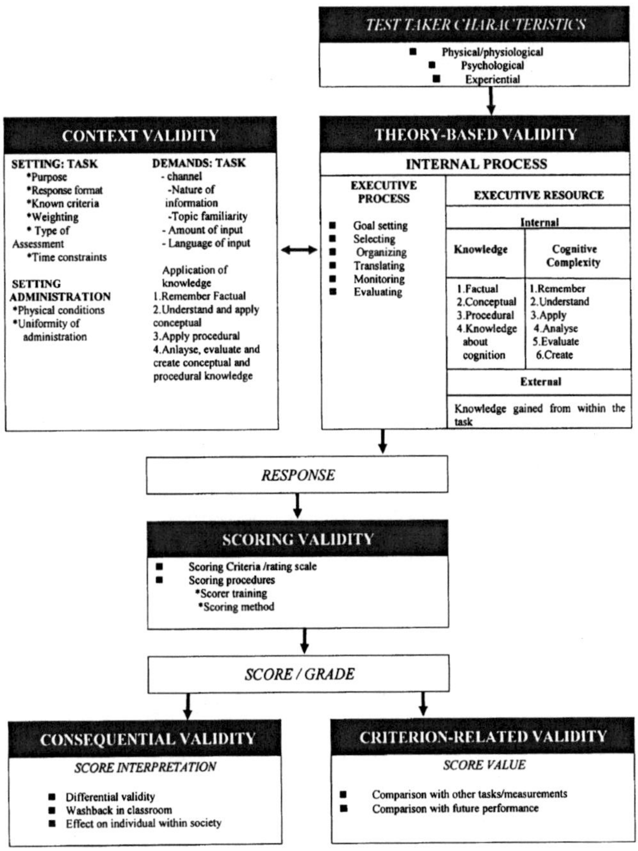

Figure 2: A sociocognitive framework for validating the school-based assessment in physics: The Selvaruby-Weir Model (Selvaruby, 2006)

These aspects are linked together as in Figure 2.

An Example Component of the Framework

In this chapter we explore just one of the six elements of this framework, that of theory-based validity. In this way we hope to illuminate the discussions and debates through students' data in the learning and teaching of physics.

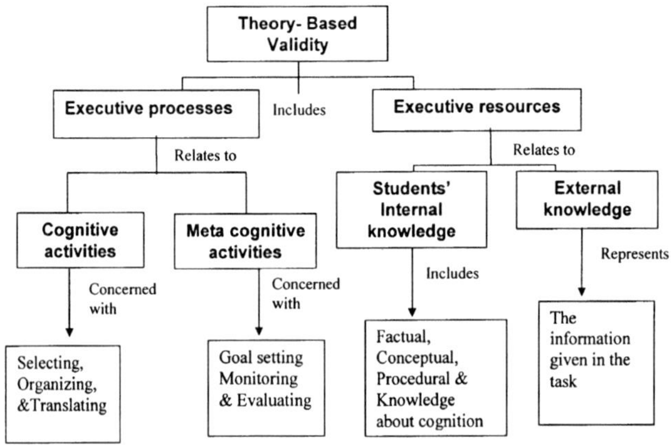

Figure 3. Elements of the theory-based validity in greater detail

In the context of the model, *executive resources* are those that are available to the students (test-takers) and include *content knowledge* of physics and *knowledge about cognition* of the student and the *external knowledge* given in the task. The student applies his/her *declarative knowledge (factual knowledge* and *conceptual knowledge*) and *procedural knowledge* of physics content in terms of relevant cognitive complexity, and *knowledge about cognition* to attend the given task. *Executive processes,* on the other hand, involve the cognitive and metacognitive activities of the student when he or she performs the test. The student has to attend to the input by accessing his/her executive resources, and internally processing this knowledge and information. So, for assessment and testing in physics, *executive processes* are identified as *goal-setting, selecting, organizing, translating, monitoring* and *evaluating,* to produce a response that is appropriate within the time and context given for the task. The executive processes act as the mediator between the context established by the way the test/task has been formulated and the executive resources available to the individual candidate.

THE STUDY

Teachers in Sri Lanka have adapted to the use of tests and examination for summative purposes, and since 1998, awareness of SBA is continuing to grow. That said, as we noted earlier, there are differences in physical facilities such as laboratories, science equipment, and professionally trained science teachers among schools that have an effect on student characteristics.

Our research used quantitative and qualitative methods (e.g., questionnaires, interviews, document analysis and participant observation) to generate data on each element of the framework, for both the SBA and the GCE O-Level exam in the physics components of science. For the study we selected 1AB schools that have science laboratories, trained teachers, high levels of student achievement and are also implementing SBA with a positive attitude, to conduct a questionnaire survey with students. All other participants were selected as representative samples: science teachers from 1AB, 1C, and Type II schools, experts and administrators from all districts.

Two questionnaires were constructed to focus on the two forms of assessment, the SBA and the GCE (O/L) exam. Questionnaire items were grouped according to the components of the framework under each section. Each section of the questionnaire was designed to cover the respondent's view of each element of the components of the validity.

Discrepancies between Student, and Experts and Teacher Opinions

So, is SBA valid? The broad answer is an 'overall, yes', but to arrive at that answer means considering various discrepancies in the data we generated. For example, if we return to theory-based validity, as set out above, we found discrepancies between students' opinion and experts and teachers' opinions about the executive resources of theory-based validity component.

At the core of the issue lie patterns of teaching and learning. Many teachers in Sri Lanka teach to the syllabus and most students learn by rote memorization. This is a situation that is well known and well documented in many parts of the world. In fact, it would be an unusual set of circumstances where this does *not* tend to happen. If we take one area of physics, for example, Newtonian mechanics, then didactic approaches to teaching, and 'surface' approaches to learning are endemic. There is evidence in the Sri Lankan context to substantiate this. For instance, Hestenes, Wells and Swackhamer (1992) developed the *Force Concept Inventory* (FCI) to assess the effectiveness of mechanics courses. The FCI systematically probes student abilities to make discriminations with respect to six fundamental aspects of the Newtonian force concept. The FCI was trialed on more than 12,000 students and 100 physics courses in high schools, colleges and universities through the United States. The results of this study suggest that students hold naïve beliefs about mechanics, which are incompatible with Newtonian concepts in most respects. Then, in 2000, Selvaruby reported that students rely heavily on memorizing Newton's laws and applying these to problems in an unthinking way, rather than understanding the basic conceptual aspects of the laws and their applications in real life situations.

While this may be expected of summative assessment, that is, that teachers and students 'cram' for the exam, it ought to be less expected during everyday teaching and learning, given that - as we pointed out at the beginning of this chapter - teachers *should be* providing 'appropriate learning opportunities' for this kind of mechanics. However, a closer inspection showed that teachers were not, in fact, doing this. A collected sample of assessment tasks allowed us to gauge the demands of SBA classroom tasks. Tasks were checked for content representativeness, scoring criteria and the demands of the tasks on scientific knowledge in terms of cognitive complexity. The findings are summarized in Figure 4.

Unit number	Unit 10 – 10 /16 objectives		Unit 11 - 4 /15 objectives	
Assessment task	Task 1	Task 2	Task 3	Task 4
Type of assessment	Written test	Practical assessment	Practical assessment	Practical assessment
Number of objectives covered	2	1	1	1
Suitability of the task for objectives	Suitable	Suitable	Suitable	Suitable
Number and appropriateness of criteria	No criteria	5 criteria Not appropriate and no specific description	5 criteria Not appropriate and no specific description	5 criteria Not appropriate and no specific description
The level of knowledge demanded by the assessment task	Remember factual knowledge and apply conceptual knowledge at lower level	Apply conceptual knowledge	Remember factual knowledge	Remember factual knowledge and apply conceptual knowledge at lower level

Figure 4. Content representativeness of a sample of SBA assessment tasks

As the final rows indicate, the emphasis - even of the classroom activities - was on rote memory and recall. Our second example comes from the teaching and learning of Ohm's Law in simple circuit electricity. The curriculum's unit objectives stipulate that the student at this level should be able to:

- Define and describe the concepts and units related to electricity (e.g., direct current, alternative current, electric potential difference, resistance) - *remember and understand factual and conceptual knowledge*;
- Analyze and describe how current varies with voltage in a conductor - *understand and analyze conceptual and procedural knowledge*;

- State Ohm's Law qualitatively and quantitatively - *understand conceptual knowledge*;
- Connect two or more resistance in series and parallel and find and compare the combined resistance - *understand conceptual knowledge and analyze procedural knowledge*;
- Analyze in quantitative terms simple circuit problems involving Ohm's Law - *analyze procedural knowledge*; and
- Recall and use electric circuit symbols - *remember factual knowledge apply procedural knowledge.*

In practice, the formative assessment task used to assess the pupils' understanding of this unit was a written classroom test. The written test required the students to:

- State Ohm's Law;
- Find current (I), resistance (R) and potential difference (V) in a simple circuit;
- Find the combined resistance of three resistors in series and parallel; and
- Recall some electrical symbols.

The assessment demands the pupils' ability to just apply equations of Ohm's Law and combined resistance in series and parallel, as multiple-choice items of the GCE (O-L) test at lower level of cognitive complexity.

In this context, more suitable assessment tasks would be practical assessments which demand that students understand, apply and analyze conceptual and procedural knowledge in terms of unit objectives that cannot be assessed by the GCE (O-L) examination items. For example, the practical assessment activities should assess students' knowledge in *classifying (understanding)/connecting* the resistances in series and parallel and *analyzing in quantitative terms* to find the combined resistance, and current in the circuit, analyzing the relationship between current and voltage and *deriving the Ohm's Law instead of stating Ohm's Law* as well as *implementing the procedures* given in the practical assessment worksheet. The assessment task should not only assess student knowledge in terms of cognitive complexity, but also permit a diagnosis of the student's underlying difficulties and also give an opportunity to learn.

FURTHER COMMENTS

So, we return to the question: Is SBA valid? The interesting feature here is that the pupils thought 'yes,' while the teachers and experts thought 'no.' To appreciate what this means, approximately 60% of the students responded through the questionnaire that they have clear knowledge of the organization and structure of physics subject matter in the curriculum and also have knowledge of their own cognition. On the other hand, mean and mode comparison of the questionnaire items relating to theory-based validity showed that experts and teachers were both insistent that students do *not* have clear knowledge of the organization and structure of the physics subject matter in the curriculum, nor knowledge of their own cognition. That is, students feel the assessment is valid because they have sufficient overview of the subject matter to allow them to make such judgments. The experts feel that the assessment lacks validity because - in their view - students do not have sufficient overview to make those judgments.

During interviews with the experts, they reinforced their opinions that the test encourages rote and superficial learning, which does not lead to improvement in students' knowledge in terms of cognitive complexity relevant to the course objectives. Hence the teaching-learning process does not provide a basis for developing higher order cognitive and metacognitive processes, and school-based assessment does not always provide a basis for developing higher order cognitive and metacognitive processes and also content knowledge in terms of relevant cognitive complexity.

So, based upon their experiences, the students felt they were well equipped to tackle tests and examinations; they had all the knowledge, skills, overview needed. However, based upon their experiences - both of the systems of teaching and learning, and of assessment - the experts thought otherwise, that students did not have the requisite skills to analyze, to make inferences, to think critically.

There are several conclusions to be drawn from all of this. First, physics education in Sri Lanka lacks 'constructive alignment'. This is an expression used by Biggs (1992) to suggest that curriculum objectives or intended learning outcomes (ILOs) should be aligned with methods of teaching and learning and that these, in turn, should be aligned with modes of assessment. That is, if you want students to think critically then you must teach to encourage learning in thinking critically and you must then assess critical thought in action. This has not been the case in Sri Lanka where, despite ambitious curriculum objectives, neither the teaching nor the assessment has been then aligned. In practice, the test developers do not have essential knowledge to design such test items. The O-level papers demonstrably tap factual knowledge and simple conceptual knowledge in relation to physics and require application of knowledge at low levels of cognitive complexity. As a result, teachers and students give priority to memorizing facts and preparing for tests, which does not lead to improved student knowledge in terms of cognitive complexity relevant to the course objectives. Hence the teaching-learning process does not provide a basis for developing higher order cognitive and metacognitive processes.

The underlying problems for lack of validity of SBA are inadequacy of teacher training in subject matter and assessment practices, lack of resources for physics instruction and the large number of students in the classroom. As a result of insufficient training, teachers are unable to plan and conduct assessment which assess the knowledge in terms cognitive complexity relevant to the curriculum objectives, and are unable to use assessment to monitor and improve students' learning and to examine and improve instruction practices. Hence students do not have an opportunity to improve their strategies and content knowledge through the assessment process. Due to lack of resources for science instruction in schools, no equity in assessment of performance, the reliability of assessment performance is questionable among stakeholders. Furthermore, due to the large number of students in a classroom, there are no opportunities for individual feedback and self-assessment, which help to identify the students' own weaknesses and improve their learning. Consequently, SBA performance was not a good indicator of students knowledge and skills in science and although not a good predictor for the GCE (O-L) science examination performance.

REFERENCES AND BIBLIOGRAPHY

Airasian, P.W. (1991). *Classroom assessment*. New York: McGraw-Hill.

Cunningam, G.K. (1998). *Constructing and interpreting tests*. London: Falmer.

Hestenes, D., Wells, M., & Swackhamer, G. (1992). Force concept inventory. *The Physics Teacher*, *30*, 141-158.

Ministry of Education and Higher Education. (1998). *Circular 1998/42*. Colombo, Sri Lanka.

Perera, L. (2000). *Different models of classroom based assessment in selected countries: Developments in Education*. Colombo, Sri Lanka: Faculty of Education, University of Colombo.

Pellegrino, J.W., Baxter, G.P., & Glaser, R. (1999). Addressing the 'two disciplines' problem: Linking theories of cognition and learning with assessment and instructional practice. *Review of Research in Education*, *24*, 307-353.

Selvaruby, P.S. (2000). *The differences in the knowledge representation about Newton's laws of motion between two groups of physics students*. Unpublished Master of Science Education thesis, University of Peradeniya, Sri Lanka.

Selvaruby, P.S. (2006). *The validity of the school-based assessment performance as against the national examination achievement in the physics component of science*. Unpublished PhD Thesis, Roehampton University, Surrey, UK.

Silva, D. (2000). *School-based assessment*. Marga Institute, Kohuwala, Sri Lanka.

Secondary Education Development Project. (1997). *School-based assessment program*. Asian Development Bank, Canadian International.

Tunstall, P., & Gipps, C. (1996). How does your teacher help you to make your work better? Childrens' understanding of formative assessment. *The Curriculum Journal*, *7*(2), 185–203.

Weir, C.J. (2005). *Language testing and validity evidence*. London: Palgrave.

AFFILIATIONS

Princy Selvaruby
Department of Examinations
Sri Lanka

Barry O'Sullivan
School of Arts
Roehampton University
United Kingdom

Mike Watts
School of Sport and Education
Brunel University
United Kingdom

EUROPE

TSAPARLIS

THE RIVALRY AMONG THE SEPARATE SCIENCE SUBJECTS FOR DOMINANCE IN SECONDARY EDUCATION: THE CASE OF GREECE AND BEYOND

INTRODUCTION

Historically, the various science subjects have developed as independent and almost entirely different disciplines. Physics was initially natural philosophy, while chemistry was preceded by ancient Greek *chumeutics* (the art of metallurgy) and by alchemy. On the other hand, biology started as descriptive natural history, classifying plants and animals in a systematic manner. Copernicus' mathematical theory (1543) about the rotation of celestial spheres was the forerunner of the *scientific revolution* - a theory that was accepted 50 years later with Galileo's and Kepler's contributions.

The seventeenth century was the era of formation of theories and the methods of physical science, which thus became independent of philosophy. Newton is the most well known, and perhaps the greatest physicist of his time. The discovery of his laws of classical mechanics (1687) constitutes one of the most important achievements of human civilization. On the other hand, at the end of the eighteenth century the chemical revolution occurred with the fall of the phlogiston theory, and the proposal by Lavoisier of a new theory on combustion on the basis of oxygen (1777). Biology underwent its own revolutions in the mid-nineteenth century: the proposal by Mendel of the laws of heredity (1865) initiated the science of genetics; and the formulation by Darwin of the theory of evolution by natural selection (*The Origin of Species*, 1859, and *The Descent of Man*, 1871).

From Newton's time it was assumed that chemistry could be explained on the basis of the laws of physics (to be reduced to physics: *reductionism*). However, it was necessary to wait until the second half of the nineteenth century for the substantial explanation of chemical phenomena in terms of physical theories with the creation of the science of physical chemistry. Finally, the development of chemistry and of biology in the twentieth century led to the new science of biochemistry that made possible the explanation of biological phenomena through chemistry.

The above historical circumstances had, with the passing of time, their impact on school programs of the various science subjects, which were treated, and in many ways are still treated today, as separate and nearly independent courses. One of the features, however, of modern science is that it aims at producing a unified and concise knowledge system, and while scientists are working toward such unification, students demonstrate quite different behavior. Research on students' ideas about science suggests that students' 'scientific' knowledge is weakly structured and frequently fragmentary, and research studies about the integration of

Richard K. Coll, Neil Taylor (eds.), Science Education in Context: An International Examination of the Influence of Context on Science Curricula Development and Implementation, 145–159.

scientific knowledge are limited (Taber, 2001). The term *conceptual integration* is used to describe a knowledge structure that will allow strong connections among the different areas.

In conclusion, it is widely accepted today that the various disciplines of science (physics, chemistry, biology, geology) are no longer independent of each other, as can be clearly seen by branches such as physical chemistry, molecular biology, biophysics, biochemistry, geophysics, and geochemistry. These interconnections have led internationally to many integrated science courses for the primary, lower secondary, and partly for upper secondary levels. At the same time, the teaching of the separate subjects continues to dominate in many countries, especially as we go up educational levels.

THE TEACHING OF SCIENCES IN PRIMARY AND SECONDARY EDUCATION

In primary education, the norm is for integrated science to be taught. It comes under different names (e.g., Nature and Science in Denmark). In some cases, or in the initial years of schooling (e.g., up to Grade 4 in Greece), science is part of a larger subject, such as Social and Environmental Studies in Ireland, or Environmental Studies in Scotland.

In lower secondary education, the situation varies. In many cases, an integrated subject exists, for example, in England, Scotland (age 12-14 years), Ireland, Israel and some federal states of Germany. In other cases, there is a grouping of physics with chemistry as occurs in Denmark, and of biology with geology as occurs in France. A final case is the teaching of separate subjects, as in Scotland (age 14-16 years), Greece, and federal states of Germany.

In upper secondary education, where specialization becomes a necessity, at least in the final one or two years, there are science courses as part of general education. Here again the situation varies, with some countries offering a general science course, others offering combinations of courses (e.g., physics with chemistry, and biology with geology), while other countries (e.g., Germany, Greece, Israel, Scotland) offer separate subjects. In addition, for students who opt for science, engineering, medicine and related subjects, there are additional more advanced science courses, that as a rule are offered as separate subjects. Students have a choice here, and can choose all three of physics, chemistry, and biology, any combination, or any one of them; for instance, in Portugal biennial combined double courses (physics with chemistry and biology with geology) are offered, but a single subject has to be selected in Grade 12.

In primary education, the teachers are typically general educationists, but there are cases (e.g., Portugal), where specialized teachers take over in the final years (Grades 5 & 6). In secondary education, the teachers are, as a rule, specialized science graduates, who often teach not only the subject of their specialty, but also all science subjects. On the other hand, this segmentation may be the cause of a rivalry among the subjects for dominance in the curriculum and in school. Such a rivalry is a feature in the Greek secondary education system.

Appendix 1 provides a brief description of science teaching at secondary education in a number of European countries.

THE EDUCATIONAL SYSTEM OF GREECE

Greece is a small European Union country, with a population of just over 10 million. Despite a brilliant ancient history of excellence in intellectual matters, the educational system of modern Greece, including the teaching of the sciences, has many problems, especially in secondary education - not least of which is the relatively poor ratings of the country in the PISA evaluation. Regarding the various disciplines within the curriculum, Greek language and literature (with large emphasis on Ancient/Classical Greek) and history courses play a dominant part, with mathematics also having great importance. Science follows, considered also important, but less than mathematics.

Primary education (*demotikon scholeion*) has six grades, and science is taught in these as an integrated subject. In the first four grades, science is part of a more general social and environmental studies subject. In Grades 6 and 7, integrated science forms a separate subject (Investigating & Discovering).

Lower secondary education (*gymnasion*) has three grades (Grades 7, 8 & 9), and science is taught as separate subjects in all three grades. In Grade 7, biology is taught for two 45-minute periods per week. In addition, there is a geography course (taught also for two periods per week). In Grade 8 we have physics and geography (each taught for two periods), and chemistry (one period). Finally in Grade 9, we have physics and biology (each taught for two periods), and chemistry (one period). We see that there is an inconsistency here with respect to chemistry, for which the time allocated is half that allocated to each of physics, biology, and geography. There is no rationale for this difference, but rivalries among the various specialists and politics play a decisive role here. Suffice it to say that some years ago, the *Pedagogic Institute* (a state organization under the Greek Ministry of Education) proposed the addition of an extra period of chemistry at both Grades 8 and 9. This period was to be taken from home economics that was and is taught for two periods per week at these two grade levels. There was a threat of closing down on the part of the *Charokopeion University of Athens*, where teachers of home economics are educated, and this forced the Ministry to hold back, and retain the status quo.

Another issue is that physical sciences (physics & chemistry) are absent from the first year of the lower-secondary level (Grade 7), while biology is present. According to the previous program of studies, Grade 7 biology was descriptive in nature (*natural history*), dealing with plants, animals and humans. But the current curriculum concentrates on the study of cells and organisms, and the knowledge of basic physics and chemistry is a prerequisite. For instance, how can students study photosynthesis, if they lack knowledge about gases (like oxygen & carbon dioxide), radiation, energy, and sugars?

Upper secondary education (*lykeion*) has three grades (Grades 10, 11 & 12), and here again science is taught as separate subjects. In *lykeion* we distinguish between subjects for general education, and advanced courses for specialized streams of studies. As general subjects, we have physics in all three grades (taught for three, two and one periods per week respectively), chemistry in Grades 10 and 11 (each taught for two periods a week), and biology in Grades 11 and 12 (each taught for one period a week).

All students follow the same courses up to the end of Grade 10. Starting at Grade 11, students have to follow one of three *streams*: The 'Positive Stream', the 'Theoretical Stream', or the 'Technological Stream'. In Grade 12, each stream has four specialized courses, which are tested nationally, and used for entrance to tertiary education. The Positive Stream is for students who want to study science, engineering, agricultural studies and related applied subjects, or medicine and related subjects. The Theoretical Stream is for students who want to study literature, law, humanities, and so on. Finally, the Technological Stream leads to the same studies as the Positive Stream, except for medical subjects.

THE RIVALRY IN GREEK UPPER SECONDARY EDUCATION

In recent years, there has been a strengthening of the Technological Stream at the expense of the Positive Stream in Greek upper secondary education, with students who want to enter health-related institutions mainly taking the Positive Stream. The reason is that the Technological Stream replaces chemistry and biology with two 'softer' courses (one on computer science and another on administration), while mathematics and physics common to the Positive and the Technological Streams. As a result, students can enter a department of chemistry or chemical engineering (and many others for which a background in chemistry is necessary, e.g., agricultural studies, civil engineering, mechanical engineering), without having chemistry as a nationally-tested subject used for university entrance.

This has caused an outcry on the part of the *Association of Greek Chemists* for the 'debasement' of chemistry in Greek upper secondary education. Despite heavy protests, the situation remains the same. To make things worse, biology (with physics, mathematics & history) is one of four optional general education subjects taught at Grade 12, and tested nationally in order for students to gain entry into tertiary education (students have to select two of the four). This has been another cause of inconvenience for chemists, who feel that chemistry is given an inferior status not only with respect to physics, but also to biology. There is a promise that chemistry should and would be reinstated in its proper place in upper secondary education, but this is related to politics and might have to wait for quite some time.

Another cause of imbalance is that Grade 11 biology starts its study from advanced biochemistry topics, such as sugars, amino-acids, proteins, DNA and RNA, at the same time as chemistry talks about hydrocarbons. Last but not least, is the problem that even in general education, physics and chemistry (but not biology) follow a formalistic curriculum, being dissociated from everyday life, technology and society (the context). This makes these subjects seem boring, irrelevant, of little use, and hence unpopular for many students.

THE NEED FOR, AND VALUE OF, INTEGRATED SCIENCE

According to the conclusions of the Varna International Conference on the teaching of integrated science in Bulgaria (CIES, 1968), integrated teaching contributes to students' general education by emphasizing the unity of science as well as its role in contemporary society. Gadsen, Becht and Dawson (1979) analyzed over 100 projects for integrated science teaching in various countries, and examined the motivations and the objectives of these projects. The main motivation was a conviction that the understanding of the separate subjects requires contributions from the other subjects.

At this point, we must be careful to distinguish between integration and coordination: although integration is more desirable (especially for lower ages), this is not always possible. There are, of course, topics with a truly cross-disciplinary character, such as matter and its states, water, air, environment. But there are many other topics for which coordination is equally preferable. Apart from content and concepts, the methodology of science is, to a great extent, interdisciplinary in nature.

Chemistry and physics are closely related disciplines. The integration of physics and chemistry not only can be accomplished successfully, but this would be beneficial for both subjects. From the outset, chemistry relies on basic concepts from physics, such as mass, density, weight, atmospheric pressure, temperature, heat, energy. On the other hand, physics also benefits from an exposure to basic chemistry and physical chemistry topics, such as substances and mixtures, elements and compounds, metals and non-metals. It is not then surprising that a physics course may precede a chemistry one, as in Serbia, where biology is taught first in Grade 5 (age 11 years), physics follows in Grade 6, and chemistry comes third in Grade 7 (Sisovic & Bojovic, 2000). Sisovic and Bojovic provide maps for physics and chemistry concepts, in which they draw the intra- and inter-disciplinary connections of these concepts. It is worthwhile to note that teaching chemistry with physics has been the theme of a special issue of the journal *Chemistry Education Research and Practice* (Taber, 2003).

At higher levels, chemistry is further dependent on physics. Thus, according to research (Harris, 1983), students who complete high school chemistry and physics achieve considerably higher in first-year college general chemistry than those who lack prior physics preparation. For this reason physics is deemed an important factor for success in college chemistry. Finally, mention should be made of a physical science discovery course for elementary school teachers, consisting of 24 exercises, designed to lead students towards an understanding of 24 selected topics in chemistry and physics (Jasien, 1995). The topics were based on the guidelines of the Science Framework for California Public Schools (1990).

Although the practice in most countries seems to be the integration of all science subjects, biology presents certain difficulties, as biologists themselves admit: "The concepts of biology are composite and reflect the complexity of the organization and function of the living organisms … In addition, very often the concepts of biology presuppose or include understanding of physics and chemistry" (Zogza,

1998, p. 121). In addition "in biology, it is not feasible to lead the students by means of simple experiments to the discovery of a law. This can be done in physics or chemistry; for biology, this is difficult even for organized research laboratories" (Pedagogic Institute, 1998, p. 190). Experiments in biology are more complicated, and so a large part of biology does not easily lend itself to a constructivist-based approach to teaching and learning, at least at this level. On the other hand, a coordination or combination of biology with say organic chemistry would make sense in that biology study would be supported by the necessary chemical and biochemical background.

AN INTEGRATED PHYSICAL SCIENCE (PHYSICS AND CHEMISTRY) PROGRAMME FOR GRADE 7

As mentioned above, in Greece, physical science (physics & chemistry) is absent from the first year of the lower-secondary level (Grade 7), while it is present in the final two years of primary school (i.e., Grades 5 & 6). This absence is surprising, if we take into account, on the one hand, the necessity for our modern technological society of the relevant knowledge, and on the other hand, the important contribution of physical science in the cognitive development of students in the Piagetian sense. For this reason, we have proposed an integrated program of physics and chemistry for a two-period per week course at Grade 7 (Tsaparlis & Kampourakis, 2000; Toomey & Garafalo, 2003). The basic aims of the program are:

- Repetition and revision of basic physical science taught in primary school, with some extensions and a more scientific approach;
- Preparation for the separate physics and chemistry courses that follow in Grades 8 and 9; and
- Application of theory and practice that arises from research work in science education. In particular, the employment of constructivist teaching methodology will contribute to the linking by the students of chemistry with physics.

The repetition and revision of basic concepts and knowledge that were introduced in primary school is necessary because, for various reasons, students do not maintain much of this knowledge, as was shown, at least for Greek students, in a relevant study (Kampourakis, Georgousi & Tsaparlis, 2001).

The program has been approved as an experimental program by the Greek Pedagogic Institute and went on trial during the school years 1998-99 and 1999-2000 in 11 schools, involving 24 teachers, and about 1,500 students. A textbook, suitable for the trial, has already been prepared, and includes experiments, theory, simple knowledge, and more demanding questions (Tsaparlis & Kampourakis, 2003). The content and the logic of the program are shown in Appendix 2.

There is a fundamental problem with the Greek education system here. A large number of experimental projects, such as that above, have been tried and evaluated, and then everything stops. No further application occurs, and they are not taken into account in curriculum reforms. What then is the use of such exercises? Any projects that have been judged operational and successful should be given a place in the school program (and why not in the curricula?), in formats similar to the ones used during their experimental trial.

CONTEXT-BASED APPROACHES TO SCIENCE TEACHING AND THE EUROPEAN PARSEL PROJECT

A European Commission report, Europe Needs More Scientists (2004), points out that "there is a need for more research into mechanisms for development of innovative science curriculum materials and associated teacher professional development initiatives" (p. 185). In addition:

> A complementary and more promising networking approach (to informing young people and the public in general) is based on the idea that extended dialogue and direct contact between citizens and scientist, schools and research organizations, is necessary in order to promote a scientific culture in society and to help citizens to acquire a better understanding of the role of science and technology in society. (p. 185)

Needs such as the above have led to context-based approaches to science teaching that have become, or are becoming, important in a number of developed countries (e.g., *ChemCom: Chemistry in the Community*, and *Chemistry in Context* in the USA, the *Salter's Approach* in the UK, *Physik im Kontext*, *Chemie im Kontext* and *Biologie im Kontext* in Germany, *Industrial Chemistry* in Israel, and *Chemistry in Practice* in the Netherlands). Some of the above innovative approaches have been adopted in a number of other countries.

Currently, a consortium of researchers from eight European nations has set up the *Popularity and Relevance of Science Education for Scientific Literacy* (PARSEL) project, which is funded by the European Union (PARSEL, 2007). It aims at raising the popularity and relevance of science teaching and also enhancing students' scientific and technological literacy through identification of suitable teaching-learning materials based on relevant context-based educational approaches (Gräber & Holbrook, 2007). These approaches, identifiable within the teaching-learning materials, are expected to focus on socioscientific issues, promote the acquisition of a range of personal and social skills and competencies (including cognitive skills associated with investigatory scientific problem solving and socioscientific decision making), and put forward an image of the nature of science relevant for the twenty-first century.

The PARSEL project will attempt to create a network community from those working in the field to develop relevant and interesting teaching-learning materials; assemble notions of 'best practice' from the diversity of considerations by partners; develop a model which encompasses philosophical consideration, learning theories and teaching approaches; modify existing exemplars, as appropriate, to reflect the model; and make such teaching-learning materials available in a range of European languages.

Returning to the Greek case, we must point out that context-based approaches are virtually unknown in Greece, where more traditional routes are followed. Both the educational policy-makers and teachers, being unfamiliar with such approaches, are almost always hostile to them. It is worthwhile mentioning here that a new program of study for chemistry in upper secondary school that was proposed in 1998 by a committee set up by the Greek Pedagogic Institute was turned down after the intervention of non-institutional, but politically influential, agencies. The program was for all students in Grades 10 and 11, and introduced chemistry through the separate study of the three states of matter (the States-Of-Matter Approach (SOMA), see Tsaparlis, 2000). For Grade 10, there were three major units in the program: air and gases; salt, salts, and solids; and water and liquids. Then in Grade 11, the program moved into the applications of chemistry: plastics and polymers; medicines; foodstuffs; and energy. Inorganic and organic chemistry were partially integrated. The overall approach was intermediate between a formal (academist) and a Science-Technology-Environment-Society (STES) (practicalist) one. In the place of the proposed program, a traditional program was enforced.

IMPLICATIONS FOR CURRICULA AND FOR SCIENCE EDUCATION

There are two approaches to correcting the problematic situation. One is to change or adapt the curricula; the other is to change the mood of all those involved in educational policy-making, curricula development, and the teachers themselves.

The following proposals are made with the Greek educational system in mind, but they may be relevant for other countries with similar educational issues. Note that we do not consider it appropriate to propose radical changes here; instead, our proposal constitutes a mild adaptation of a system from one where there is domination of the separate science subjects into one where there will be some integration and some coordination.

Regarding the curricula, let us start by taking into account that integration of science curricula is favored by the move toward interdisciplinary teaching, a move that is a strong and the main feature in the current educational reform in Greece (and in many other countries). Independent of this move, it might be appropriate to have an integrated or coordinated physical science course at the start of secondary education (usually in Grade 7). This may provide chemistry with the necessary physics tools for its development. At the same time, it could supply physics with basic chemistry concepts (substance, element, compound, mixture, material) that will facilitate beginning physics. Following this, in Grade 8, the two disciplines of physics and chemistry could have their isolated development. Finally, in the last year of lower secondary school, physics can remain on its own, while chemistry could be combined with biology.

Moving to upper secondary school, and not considering specialized advanced streams of study, Grade 10 can have physics and chemistry as general-education separate subjects to lay the foundations for a more advanced study. In Grade 11 and still with general education, physics could stay alone, while chemistry could be combined with biology. In this grade all three disciplines should pay more attention to applications, and less to theory - but biology still needs basic theoretical tools here. Finally, in the last grade, Grade 12, physics and chemistry could be combined into a context-based, applications-led approach, with emphasis on an elementary 'tour' over modern theories and applications (quantum mechanics, relativity theory, spectroscopy & lasers, modern materials & nanoscience, astrophysics & astrochemistry, etc.) On the other hand, biology now deserves a course of its own, in which emphasis could be laid on health/medical issues.

But changing the curricula is not sufficient. We need to change all those involved in educational policy and the production of curricula, but more importantly, the teachers themselves. The case of Greece is characteristic of a country where teachers are subject specialists (physicists, chemists, biologists, geologists), without having received special educational training. Despite this, in the state education system, they are called to teach all science subjects. Experience shows that the secondary teachers are generally enthusiastic and competent when teaching within their own specialism, but not necessarily outside their specialism. According to Aikenhead (2003), choices about integration rest with teachers' ideologies and their willingness to innovate: "One major challenge for chemistry and physics teachers is to rethink and reformulate their professional identities away from being loyal and accountable to their discipline towards another identity that celebrates views of relevance other than 'wish-they-knew' science" (p. 125).

In Greece, physics students receive very limited training in chemistry, and no training at all in biology and geology. On the other hand, chemistry students face their physics courses with reservation, as irrelevant and theoretical; in addition, they treat physical chemistry courses as abstract and difficult, with an over-emphasis on mathematics, and little connection with other branches of chemistry and with applications. There is then a need for an educational orientation of those students with an interest in becoming secondary science teachers. Not only should they be provided during their undergraduate career with general educational training as well as with special training in the didactics of their main subject, but also they must select a secondary subject. For instance, chemists who would like to be prepared for teachers should certainly receive training in physics and in physics education in addition to chemistry. Chemists should also be given a sound background in biology and a good biochemical education.

It is known that in different countries different approaches are followed for the preparation of secondary science teachers. These include the separate training of such teachers in special education departments and faculties. Greek science departments are yet not generally in favor of providing their students the opportunity to be prepared as science teachers. The situation is ripe, however, for the creation, in faculties of science and mathematics, of *education departments* that while not awarding their own first degrees, will support the departments (mathematics, physics, chemistry, biology, geology, information science) in preparing students to take up the teaching profession. These departments could be staffed both with general and special educationists, and would carry out their own graduate studies.

A final note has to do with the various scientists' unions or associations. In Greece (as in other countries) physicists, chemists and biologists are members of their respective association. In addition, all secondary teachers have their own teachers' union. It is necessary that the separate scientific associations should collaborate in setting up a common science-teacher co-association that will bring together professionally all science teachers. This co-association could also publish a teacher science journal or magazine in the language of the country that will contribute in bringing together the various specific subject teachers.

Last but not least, let us recall that the separation of the scientific disciplines is mainly of historical origin, and that science has a common unifying core. The various science teachers in Greece and elsewhere should have more in common and should not remain separated.

APPENDIX 1: BRIEF DESCRIPTION OF SCIENCE TEACHING IN SECONDARY EDUCATION IN EUROPEAN COUNTRIES

Denmark. At lower secondary level (Grades 6-9) the following subjects are taught: physics/chemistry (one subject), biology, and geography.

England. In lower secondary education, in most schools, the subjects are taught together under science until about age 16 years (the GCSE) and then the separate subjects take over. In a few of the most academically selective schools, the various subjects are taught separately. In a few years, it is likely that there will be greater emphasis on the separate disciplines.

France. At lower secondary school (*College*), the sciences of life and of earth (*Sciences de la Vie et de la Terre* (SVT), biology and geology), and the physical sciences (*Sciences Physiques* (SP), physics and chemistry) constitute two separate combined subjects. The teaching of SVT starts at the first year of college (*Sixième*), while SP starts at the second year (*Cinquième*). At upper secondary school (*Lycee)*, all students receive a common instruction in the first year (*Seconde*), where SVT and SP are separate subjects. The teaching of SVT and of SP continues at a deeper level in the scientific stream (*Filière S*) during the second year (*Première*) and the final third year (*Terminale*). The students of the other streams (*Filières*) also receive science instruction, but to a lesser extent.

Germany. The school system in Germany is a federal system: every one of the 16 federal states has its own syllabi and structure of school types. In some states, sciences are taught as an integrated subject up to year 7 or 9. In other states, there are separate science subjects right from the beginning.

Ireland. In lower secondary level there is a three year course - science - which is an integrated course of biology, physics and chemistry (but taught in separate strands). At the higher secondary level there are five science courses, over two years and offered at two levels: biology, chemistry, physics, agricultural science, and physics with chemistry.

Israel. In lower secondary education, science is taught as an integrated subject. In the upper secondary education, science is taught as separate subjects. Usually, the students in Grade 10 study biology, chemistry and physics. In Grade 11 and 12 students choose one, two or three scientific subjects.

Scotland. The Scottish Curriculum is defined for the first nine years of education as one document called 5-14. Here science is described in the context of environmental studies. In the first two years of secondary it is taught as Science. At age 14 years, pupils can choose to take one, two or three sciences from age 14-16 years. These are taught as separate disciplines. At age 16 years, pupils can choose to take one, two or three sciences from ages 16-17 years. These are taught as separate disciplines. At age 17 years, pupils can choose to take one, two or three sciences from ages 17-18 years. These are taught as separate disciplines.

Turkey. In lower elementary (primary) education (Grades 1-5, age 7-11 year) science is taught as an integrated subject in Grade 5. In upper elementary education (Grades 6-8, age 12-14 years) integrated science is taught in all grades. In secondary education (Grades 9-12, age 15-18 years) chemistry, biology and physics are taught as separate subjects in all grades.

APPENDIX 2: THE CONTENT AND LOGIC OF THE PHYSICAL-SCIENCE PROGRAM FOR THE SEVENTH GRADE

The program consists of eight units, which are composed of 36 lessons. The eight units are (a) matter and soil, (b) motion and force, (c) temperature and heat, (d) water, (e) air and atmospheric pressure, (f) chemical reactions, (g) energy, and (h) electricity. Table 1 has the titles of the 36 lessons. Figure 1 shows the distinction of the topics and the concepts of the program into purely physical, purely chemical, as well as into interdisciplinary ones. The main intra- and inter-disciplinary connections between various topics/concepts are also drawn. In addition, topics that are needed in both disciplines are shown in italics. Note that each unit is grouped separately, and that there is a top-down hierarchy, in accordance with the order of the topics in the program (see Table 1).

The concepts of *energy* (especially chemical energy) and of *interaction* are very difficult for young students (Duit, 1986; 1994). It is possible of course to do a lot of physics and chemistry without these concepts, that is, it is possible to leave energy entirely out of this course. We have opted not to do that; although energy formally appears late in the program, the concept of energy is introduced and used

Table 1. The thirty-six lessons of the program

0. Physics, chemistry and science.

UNIT A. Matter and soil
1. States of matter - Soil.
2. Substances and mixtures.
3. Separation of mixtures.
4. Measuring length.
5. Measuring surface and volume.
6. Mass.
7. Density.

UNIT B. Motion and force
8. Introduction to motion.
9. Forces.
10. Displacement, velocity, and force are vectors.
11. Weight, gravity, and gravity field.
12. Relation between weight and mass.
13. Pressure exerted by solids.

UNIT C. Temperature and heat
14. Temperature and heat.
15. Thermal equilibrium.
16. Expansion and contraction of solids on heating.
17. Thermal expansion and contraction of liquids.

UNIT D. Water
18. Water.
19. Solutions.
20. Crystallization, re-crystallization, and crystals.
21. Hydrostatic pressure.
22. Flow of liquids and communicating vessels.

UNIT E. Air
23. Air and the atmosphere.
24. Atmospheric pressure.
25. Behavior of gases on heating, cooling, compressing and decompressing.

UNIT F. Chemical reactions
26. The concept of chemical reaction.
27. Chemical reactions in aqueous solutions.
28. Compounds and elements.
29. Getting to know more about elements.

UNIT G. Energy
30. Combustion and fuels.
31. Energy.
32. Forms of energy - Energy transfer and conversions.

UNIT H. Electricity
33. Electrocharging and electric charges.
34. Electrical conductors and insulators.
35. Electric current.

from the introductory lesson. [Two other central concepts in the course are (i) the concept of *substance* that is introduced in a macroscopic manner, without reference to the structure of matter; and (ii) the concept of *chemical reaction* that is introduced following the method of De Vos and Verdonk (1985a, 1985b).] Here is how the introductory (0) lesson starts (Tsaparlis & Kampourakis, 2003, pp. 4-5):

Scientists, and especially physicists and chemists, basically are interested in two concepts of the universe: matter and energy .... All materials that are close to us or far away consist of matter; all materials contain energy. A large piece of iron contains more iron than a small piece of iron. This we express by saying that the larger piece has larger *mass*. Mass is a basic property of matter .... Another property of matter is that it occupies some space, that it has *volume*. A third property of matter is that it contains energy that we call *chemical energy*. For instance, food, fuels, batteries and dry cells contain chemical energy. Actually, everything contains energy, for instance wood and stone; even our bodies contain chemical energy that has been and is taken from food.

Take a stone from the ground, raise it, and then let it fall. In this action, first you did something to the stone: you took it and raised it; this we call an *interaction* between yourself and the stone. We *accept* that during this interaction, an amount of energy moved (was *transferred*) from you to the stone, where it was deposited in some way (but as we will understand later NOT as chemical energy). When you let the stone fall, the stone *interacted* with the earth: the earth attracted it. Again, we *accept* that there occurred an *interaction* between the stone and the earth during which the stone transferred to the earth the amount of energy that it previously had taken from you; part of this energy was transferred to the air during the fall, and the rest to the earth on touching it.

We *agree* then that during every change there is an interaction between materials that is always accompanied by a *transfer of energy*. Thus, when we eat, we '*interact with food*', taking energy from it. Unlike matter, energy has no mass, occupies no volume, yet it can be measured. The ability of matter to interact with other matter, giving or taking energy from it, is a fourth property of it.

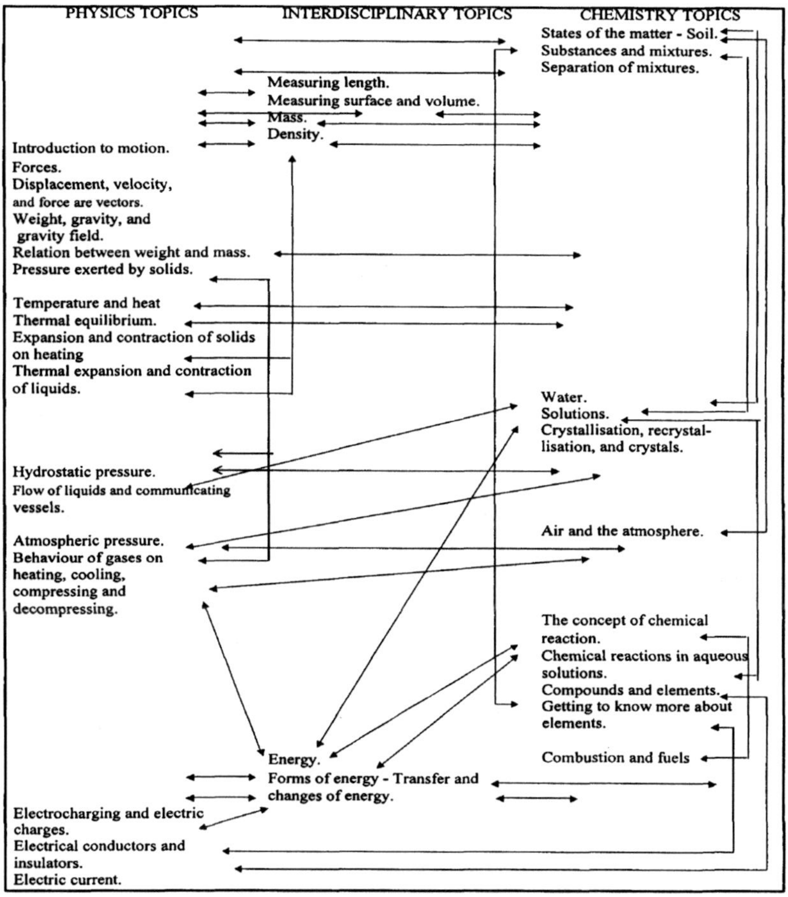

Figure 1. Distinction of the topics of the program into physical, chemical, and interdisciplinary and main intra- and interdisciplinary connections. For further explanations see text

REFERENCES AND BIBLIOGRAPHY

Aikenhead, G.S. (2003). Chemistry and physics instruction: Integration, ideologies, and choices. *Chemistry Education Research and Practice*, *4*, 115-130.

Conseil International des Unions Scientifiques (CIES). (1968). Congrès sur l'intégration des ensegnements scientifiques. Drujba (Bulgarie), 11-19 Septembre 1968. CIES, Bulevanrd Pasteur 3, Paris XV.

De Vos, W., & Verdonk, A.H. (1985a). A new way to reactions Part 1. *Journal of Chemical Education*, *62*, 238-240.

De Vos, W., & Verdonk, A.H. (1985b). A new way to reactions Part 2. *Journal of Chemical Education*, *62*, 648-649.

Duit, R. (1986). In search of an energy concept. In R. Driver & R. Millar (Eds.), *Energy matters* (pp. 67-101). Leeds: University of Leeds.

Duit, R. (1994). Learning and teaching energy. In P.J. Fensham, R.F. Gunstone & R.T. White (Eds.), *The content of science: A constructivist approach to its teaching and learning* (pp. 185-200). London: Falmer.

Europe Needs More Scientists. (2004). EU blueprint for action. Retrieved 6 September 2007, from http://ec.europa.eu/research/press/2004/ pr0204en.cfm.

Gadsen, T., Becht, P., & Dawson, G. (1979). The design and content of integrated science courses. In *New trends in integrated science teaching* (vol. 5, p. 41). Paris: UNESCO.

Gräber, W. & Holbrook, J. (2007). The PARSEL project (Introduction to Symposium). *ESERA 2007 Conference*, Book of Abstracts (pp. 53-54). Malmö, Sweden: Malmö University.

Harris, S.P. (1983). Physics, an important factor in the success of general college chemistry students. *Journal of Chemical Education*, *60*, 739-740.

Jasien, P.G. (1995). A physical science discovery course for elementary school teachers. *Journal of Chemical Education*, *72*, 48-48.

Kampourakis, C., Georgousi, K., & Tsaparlis, G. (2001). Physical-science knowledge and patterns of achievement at the primary-secondary interface. *Chemistry Education Research and Practice*, *2*, 241-252.

Popularity and Relevance of Science Education for Scientific Literacy. (2007). Retrieved 6 September 2007, from http://www.parsel.uni-kiel.de/cms

Pedagogic Institute. (1998). *Unified upper secondary school*: *Unified framework for the program of studies* (in Greek). Athens, Greece: Pedagogic Institute - OEDB.

Science Framework for California Public Schools. (1990). Sacramento, CA: California Department of Education.

Sisovic, D., & Bojovic, S. (2000). On the use of concept maps at different stages of chemistry teaching. *Chemistry Education Research and Practice*, *1*, 135-144.

Taber, K.S. (2001). Building the structural concepts of chemistry: Some considerations from educational research. *Chemistry Education Research and Practice*, *2*, 123-158.

Taber, K.S. (2003). Facilitating science learning in the inter-disciplinary matrix – Some perspectives on teaching chemistry and physics (Guest editor's Introduction to theme issue *Teaching Chemistry and Physics*). *Chemistry Education Research and Practice*, *4*, 103-114.

Toomey, F., & Garafalo, A. (2003), Linking physics with chemistry: Opportunities in a constructivist classroom. *Chemistry Education Research and Practice*, *4*, 189-204.

Tsaparlis, G. (2000). The States-Of-Matter-Approach (SOMA) to introductory chemistry. *Chemistry Education Research and Practice*, *1*, 161-168.

Tsaparlis, G., & Kampourakis, C. (2000). An integrated physical-science (physics and chemistry) introduction for lower-secondary level (Grade 7). *Chemistry Education Research and Practice*, *1*, 281-294.

Tsaparlis, G., & Kampourakis, C. (2003). *Introduction to physical science (physics and chemistry)* (in Greek). Ioannina, Greece: Pedagogic Institute.

Zogza, V. (1998). Symposium on biology education (Introduction, in Greek). In P. Koumaras, P. Kariotoglou, V. Tselfes & D. Psillos (Eds), *Proceedings of 1st Pan-Hellenic Conference on Science Education and New Educational Technologies* (pp. 121-122). Thessaloniki, Greece: Christodoulides Publications.

AFFILIATION

Georgios Tsaparlis
Department of Chemistry
University of Ioannina
Greece

ÇALIK AND AYAS

A CRITICAL REVIEW OF THE DEVELOPMENT OF THE TURKISH SCIENCE CURRICULUM

INTRODUCTION

The Turkish education and science education curriculum has undergone numerous reforms in four major phases since the country was first established. The principal driving forces behind such reforms have been to improve scientific literacy and satisfy a need for qualified scientists - the latter seen as a driver of economic development. Regular changes of governments, each with a different vision for education, means such reforms were not well organized or well thought through. As a consequence, the reforms lack a coherent strategy, and thus represent tinkering with the educational system. It seems a key factor is a need for a curriculum that takes cognizance of *local* needs and issues - that draws upon local educational research and expertise, and yet is open to international innovations. As noted in other educational contexts, enacting curriculum reforms without reform of the assessment regime and unaccompanied by substantive teacher professional development seems doomed to failure.

When the Ottoman Empire collapsed after the First World War, a new republic named Turkey was established in 1923. Turkey, with a population of 70 million and covering an area of 780,576 km^2, represents a bridge between Europe and Asia. The population is predominantly Muslim (98%) but includes ethnic and religious minorities such as Jews and Christians. Turkey has seven geographical regions, and the population density, climatic conditions and socioeconomic structure differs across these regions (see Table 1). The economy is heavily dependent on tourism, and supplemented with income from agriculture and light industries. In terms of education, there are now nearly 35,000 primary schools catering for about 10.7 million students with approximately 400,000 teachers (see Table 1).

OVERVIEW OF EDUCATION IN TURKEY

The Turkish Educational System was centralized by the Act of Law of Unification of Instruction in 1924 that located all educational institutions under the control of the Ministry of National Education (NME), incorporating religious schools apart from minorities' schools (i.e., Armenians, Greeks & Jews, see Ayas, Çepni & Akdeniz, 1993). The educational system comprises three levels: basic, secondary and higher education. Basic education commences at age six years, and consists of

Richard K. Coll, Neil Taylor (eds.), Science Education in Context: An International Examination of the Influence of Context on Science Curricula Development and Implementation, 161–174.

eight years of free (in public schools) compulsory education (Ayas et al., 1993; Educational Statistics of Turkey, 2006). Basic level school students first encounter science in Grades 4 and 5 in a course entitled 'science and technology'. They next formally encounter science at Grades 6 through 8. It is worthwhile to note that the current Minister of Education has signaled plans to require teachers to teach other science related topics at Grades 4 and 5 rather than primary teachers. Secondary education consists of general, vocational and technical training, is of four years in duration, and like primary schooling, free in public schools (Announcement of Reconstruction of Secondary Science and Mathematics Curriculum, 2006; Erdoğan, 2006; Gümüşeli, 2006). To continue into higher education, students first have to pass the University Entrance Exam (ÖSS). However, the number of students who graduate from secondary schools vastly exceeds the current capacity of higher education institutes in the nation, meaning that few progress even to the bachelor degree level (Ayas et al., 1993). For example, of the 1.74 million students who sat for the University Entrance Exam in 2006, only 600,000 eventually enrolled in higher education (this latter figure also includes students who gained access in other ways than through formal examinations).

Table 1. Mean number of enrolment of students and teachers per school and classroom sizes of schools in each of the regions in Turkey (adapted from Educational Statistics of Turkey, 2006).

Region	*Schools*	*Teachers per school*	*Classroom size*
Marmara	1821	27	37
Mediterranean	333	28	26
Aegean	292	23	29
Black Sea	173	21	25
East Anatolia	173	30	35
Southeast Anatolia	270	29	37
Middle Anatolia	322	23	31

The General Aims of the Turkish Educational System

Reflection on worldwide socioeconomic development trends during the 1960s suggested the Turkish educational system needed reform in terms of its objectives, content, teaching methods, and structure, in order for the nation to prosper economically. The net result was a new law known as The Basic Law of National Education, which was duly enacted. The principles of this law set the essential goals of the reform strategy (Binay, 1982). The main features of the law are summarized as follows (translated by Binay, 1982):

- *Integrity of the Educational System*: Formal and informal education will be organized in such a way that the citizens will be trained in accordance with the needs of society, individual interest and ability, and the principles of equality of educational opportunity;

- *Continuity in Education*: Formal and informal education should continue throughout life;
- *Selection, Orientation and Guidance in Education*: The orientation of students will be carried out according to their interest and abilities throughout their education. In this process, a specific emphasis will be given to their understanding, interpretation, scientific thinking and application rather than knowledge based on memory;
- *Productivity in Education*: In order to achieve this, rational use of teachers, physical facilities, and financial resources will be carefully utilized;
- *Programs, Methods, and Materials in Education*: Educational programs at every level are to be so designed that students are prepared for life and the manpower needs of the country;
- *Education Everywhere* (at home, in the community, and at work); and
- *Education and Development*: Educational objectives will be so designed that they would help industrialization, technical development, planned settlement and urbanization, balanced regional development, healthy living, and family planning.

In addition to the above features of the law, there was to be considerable emphasis on education for employment, and education related to technological developments (i.e., links with industry).

Curriculum Developments up until 2000

Since the establishment of Turkey as a republic, there have been numerous efforts to develop a new educational system and new curricula (including science) in order to address the nation's manpower needs in a variety of fields. The science curriculum itself changed many times and was tinkered with in 1924, 1926, 1936, 1939, 1948, 1968, 1974, 1977, 1992, and more recently in 2000 and 2005. However, these changes did not represent a well organized development of the curriculum but rather they consisted of relatively minor changes to the school syllabus. Up until the present, science curricula were essentially adapted from those of Western countries. Remarkably, there have been no comprehensive evaluations of the curriculum, but the common view is that the science curricula have consistently failed to achieve targeted aims - to improve scientific literacy and satisfy a need for qualified science staff. Below we outline some of the most important curricula efforts up until 2000.

Science education reform in Turkish schools went through four major phases influenced significantly by curricular developments in Western nations. The first phase began in 1932 and continued up until the late 1940s. This had its origins in the USA and later Europe, and affected science education in primary and secondary schools. Essentially this consisted of translating science textbooks in English into Turkish. The main purpose of this was to help students acquire knowledge of the facts and principles of science, and to give impetus to the notion of individual development through social adaptation. However, since the supplementary materials for textbooks were not translated into Turkish, it was difficult for teachers to know about things such as the philosophy of education, the teaching processes, and any assessment techniques. Furthermore, there was little or no in-service training provided. Not surprisingly, this lack of support, both in terms of supporting teaching materials and training, negatively affected the awareness of innovative ideas which could result in development and implementation of a national curriculum. In the latter part of this phase (in 1937) the integration of subjects such as chemistry or physics into science was abandoned, meaning each became a separate discipline of science with a unique structure of topics from elementary concepts, to abstract or complex concepts.

The second phase also came from the West, and began in the 1950s. This was not very significant, and really only resulted in the development of a few science experiments, by American educators supported by the Ford Foundation. Two examples are the Istanbul Girl Lycee Experiment, and the Ankara Trial Lycee Experiment (Özinönü, 1976; Soylu, 1984). Within this phase, there was another initiative, through the cooperation between National Ministry of Education (NME) and the Ford Foundation, resulting in the establishment of a National Committee on Education to overview the entire educational system. The Committee had responsibility for examining the educational situation in Turkey; investigating how other countries dealt with science education problems; reporting their findings; and making recommendations to the NME. The Committee subsequently prepared a list of problems and issues to be considered in discussion with overseas experts. Committee members visited several countries including Japan, the USA, Britain, Germany, France, and Italy and found that each country had its own unique system based on the local social, economic and cultural structure. The Committee concluded that it was not sensible to copy the curriculum from another country (Report of the National Commission on Education, 1961), and noted that in comparison with the countries visited, Turkish schools provided little opportunity for science learning to be tailored to individual needs. Even as Turkey grappled to find a way to overcome the educational problems in the country at the time, the rest of the so-called developed world, was already at another stage of development - the post-1950s curricula developments.

In the third and arguably most important phase, Turkish science education was again affected by Western curricula developments - this time at the secondary school level (although later lower levels also were affected). In this phase, a number of now famous American curricula programs, such PSSC, CHEM Study, and BSCS, were introduced into secondary schools through the financial support of the Ford Foundation. After the military coup of 1980, foreign-based projects for the NME essentially disappeared, including the influential Ford Foundation support for the Scientific and Technical Research Council of Turkey. At about this time, the Science Education Development Commission was abolished and it became very difficult to continue with any projects after 1984 (Turgut, 1990). Interestingly, a national conference held in Ankara in 1984 came to the conclusion that such projects were not very successful anyway, mostly because the implementation process was lacking in so many aspects. Key problems identified at this important conference included a perception of poor implementation of projects, inadequate pre-service and in-service training, crowded classrooms, and lack of chemicals and equipment (Scientific Meeting of Turkish Education Society, 1984). However, probably the most regrettable thing was that the experiences gained from this implementation did not provide much insight into how we might design original curricula programs suited to the nation's context and needs.

The fourth phase of curricula development occurred after the abrupt demise of the Eastern block - particularly the breakdown of the USSR. These were dramatic times, and together with technological developments occurring at the time, resulted in three new attempts to reform the science curriculum in Turkey - in 1992, 2000 and 2005. The first of these in 1992 emphasized student-centered learning, but because of a lack of in-service and pre-service teacher education, little change to the traditional teacher-centered modes of instruction actually occurred. Probably the main reason for this was the fact that the supposed learner-centered curriculum actually consisted only of course subjects, aims, and goals, and did not incorporate sample teaching activities and strategies or, new assessment strategies (Demirbaş & Yağbasan, 2005). The Turkish government then obtained a loan from the World Bank with the aim of developing new curricula, building schools, and improving the infrastructure of existing schools. The loan also was to help develop teacher education methods courses as well as equipping teacher education institutions with new teaching technologies and laboratory equipment.

Next we consider science curriculum developments from 2000 to 2005. The development of the most recent curriculum had started with Grade 4 progressing up through the grades with the Grade 8 curriculum at the pilot testing stage.

FEATURES OF NEWLY RELEASED SCIENCE CURRICULUM

Philosophy of Recent Science Curricula

The most recent versions of the Turkish science curriculum seek to be an innovative science curriculum, to result in better student engagement, and to drive the use of better teaching methods and strategies compared with traditional teaching approaches. In particular, there is now a clearly articulated philosophy underpinning the science curriculum - specifically it is based on constructivist learning theory. This represents a substantial departure from early Turkish science curricula, which were based on more traditional behaviorist learning theories (Kavak, 2005). However, although the science curriculum launched in 2000 contains elements of constructivist learning theory, it is only in the 2005 version that constructivism is explained in detail and a cognitive learning approach is presented (Ayas et al., 2006; Report of Investigation and Evaluation Related To New Curricula, 2005; Result Report of Evaluation Meeting of Curricula between Grade 1 and Grade 5, 2005). Interestingly, the 2005 curriculum was launched before the 2000 curriculum was fully evaluated. As a consequence, the "newly structured science curriculum disregarded all attempts of development of science curriculum" (Result Report of Evaluation Meeting of Curricula between Grade 1 and Grade 5, 2005, p. 4), and hence was criticized by many science educators. The 2005 curriculum introduced as the 'first and the most extensive education reform' and sought to link students' theoretical knowledge domain with practical skills, and to recognize the interrelationships between science subjects, and science and other disciplines (Ayas et al., 2006; Report of Investigation and Evaluation Related to New Curricula, 2005; Result Report of Evaluation Meeting of Curricula between Grade 1 and Grade 5, 2005). A number of local educators have criticized the new curriculum saying its theoretical base is solely constructivism (Result Report of Evaluation Meeting of Curricula between Grade 1 and Grade 5, 2005).

Although constructivist learning theory has a central role in the 2005 curriculum, conceptual understanding, multiple intelligence theory, active learning, reflective thinking and cooperative learning, are all taken into account (Ayas et al., 2006; Report of Investigation and Evaluation Related to New Curricula, 2005).

Other points of the 2005 curriculum have also been criticized. For example, several concepts such as 'psychomotor science and technology skills' were included in the curriculum, but without clarification. The curriculum was introduced in various national advertisements, talking, for example, of "relinquishing the positivist Newton paradigm" (The Turkish Academy of Sciences (TUBA)'s General Views and Suggestions of New Curricula, 2005, p. 1), whatever that might mean. Finally, it has been noted that it may take as long as 10 to 15 years to evaluate the 'fruitfulness' of a new curriculum, and science knowledge may well change in that timeframe. Hence, the notion of 'stability of knowledge' is retained from the 2000 curriculum (Kavak, 2005; Turkish Academy of Sciences General Views and Suggestions of New Curricula, 2005).

Perhaps the major criticism of the 2005 science curriculum is that input from Turkish science educators was not taken into account. As a consequence, the 2005 science curriculum contains a number of inconsistent features. For example, international science issues and topics dominate, and the curriculum does not take into account the reality of the country's context and needs. This is particularly disappointing given the fact that there is widespread belief that past attempts to transfer Western-based curricula in Turkey have failed (Result Report of Evaluation Meeting of Curricula between Grade 1 and Grade 5, 2005). For this reason, the 2005 science curriculum already needs revision to take into account local contextual factors such as national philosophy, culture, and history.

The Content of the 2005 Science Curriculum

The 2005 science curriculum aims to enable students to become literate not only in science but also in technology. This is signaled by a change to the name from Science, to Science and Technology. In the curriculum, in contrast to previous documents, students are encouraged to explore knowledge, to link their theoretical knowledge with daily life experiences, to track scientific research methods to work out the problems encountered in each phase of their lives, and to follow innovative developments in science (Ayas et al., 2006; Report of Investigation & Evaluation Related to New Curricula, 2005).

The newly structured curriculum is based on a 'thematic approach' and consists of four subject-content learning fields: Alive and Life, Matter and Change, Physical Events, and Earth and Universe, with an emphasis on the interrelationships amongst these fields. The complexity of content knowledge, and skill and comprehension levels increases as a student goes through the grades. Earlier curricula comprised unit-based rather subject-based content and each unit was considered to be a separate entity with any interrelationship amongst units neglected (Ayas et al., 2006; Kavak, 2005; Report of Investigation and Evaluation Related to New Curricula, 2005). In contrast, the 2005 curriculum focuses on three learning fields: Scientific Process Skills (SPS), Science-Technology-Society-Environment (STSE), and Attitudes and Values (AV). The use of the term Learning Field by the curriculum developers may bring about confusion because the same term is used for both the subject-content and SPS-STSE-AV fields mentioned above. A further criticism is that there is some confusion between statements and implementation. For example, *objectives* under Attitudes and Values, were organized using Bloom's taxonomy (Report of Investigation & Evaluation Related to New Curricula, 2005), the use of the term objectives being a consequence of the philosophy of the 2005 curriculum (in contrast to 'aim' or 'goal' used in earlier curricula). Finally, some important topics, such as the atom, were eliminated at lower grades (specifically Grades 4 and 5), based on the notion that 'less knowledge is concise knowledge' (Ayas et al., 2006; Report of Investigation & Evaluation Related to New Curricula, 2005).

Learning-Teaching Process and Modification of the Role of the Teacher

None of the earlier curricula clearly articulated preferred learning and teaching processes - although some incorporated a few sample activities or teaching tasks (Report of Investigation & Evaluation Related to New Curricula, 2005). The 2005 curriculum explains the expected learning and teaching process and the role of the teacher at a deeper level. The curriculum presents more concrete examples including specific teaching activities. In each activity, how scientific process skills match with Science-Technology-Society-Environment is explained. But, remarkably, the same process is not employed for the Attitudes and Values learning field. It is plain from the curriculum that it is expected that the students should conduct the activities. In this process, the teacher should play the role of a guide or facilitator of student learning, in a clear expectation that learning will be more learner-centered in nature (Ayas et al., 2006; Report of Investigation & Evaluation Related to New Curricula, 2005).

Finally, the 2005 curriculum also includes more direction about how to implement the curriculum (in addition to those described above) and more instructional materials; for example, materials such as transparent film, VCD, CD, and Internet are now supplied to each school. Furthermore, students have two books, a course book and a work book, both supplied free of charge.

Measurement and Assessment

The 2005 curriculum also includes new ways of measuring and assessing student learning. As well as traditional measurement and assessment techniques, alternative approaches such as the use of portfolios, concept mapping, self-assessment, peer assessment, project, and so forth are proposed. Thus, the 2005 curriculum not only involves 'learning products' but also seeks to evaluate the 'learning processes' (Ayas et al., 2006; Report of Investigation & Evaluation Related to New Curricula, 2005). However, some teachers think these notions of assessment are not that clear - meaning they must devise their own assessment and measurement tools.

DISCUSSION

Research suggests that these science curricula, including the 2005 version thereof, developed by Western countries and adapted to the Turkish educational context, are such that the targeted aims have not been achieved (Ayas et al., 1993, 2006; Ünal et al., 2004). This is mainly because of differences in things such as culture and socioeconomic conditions in Turkey compared with its Western neighbors. Certainly, the 2005 curriculum is a better fit for the national needs and Turkish context, but it does not represent an original science curriculum effort, or a science curriculum developed for Turkey to meet the nation's special needs. For example, the notions of constructivist learning theory, a student-centered teaching approach, ideas about scientific and technology literacy, science process skills, and taking into account the differences in students' capabilities, all reflect Western thinking about science curriculum (especially the USA, Canada, and Ireland). In particular, some features of the 2005 curriculum are highly contextualized to Ireland and Canada, for example, the content of the units in the 2005 Turkish science curriculum are much the same as Ireland's, and the Science and Technology part of the 2005 Turkish science curriculum reflects Canadian emphasis on Science, Technology and Society (particularly under 'suggestions for students, teachers and parents').[1] The 2005 curriculum, however, provides more direction for teachers; it incorporates examples of student-centered activities, science process skills, and science, technology, society and environment in regard to goals. This helps teachers to facilitate student learning by providing useful guidance on how to implement the teaching activities recommended in the curriculum.

The main problem in developing the Turkish science curriculum has been that rather than improving an existing curriculum, developers simply devised a whole new curriculum and expect to achieve all targeted aims in a very short timeframe. International experience suggests that a new curriculum may take as long as 10 to 15 years before improvements in learning occur. However, in Turkey there is little patience with education reforms of this nature, meaning new curricula are developed almost continuously! Moreover, since Turkey is driven by a centralized educational system, curriculum development also operates via a centre-periphery model that is very top-down in nature. This creates many problems, since there are seven geographical regions with different cultures, socioeconomic conditions, quality of teachers, physical environments, and for which student prior learning experiences are different (Evaluation Form of New Curriculum Course Books in Grade 5, 2006; Ünal et al., 2004). Curriculum development is a substantial exercise that takes significant resources in terms of time and money. However, as noted above, this process in Turkey often happens over a very short timeframe (e.g., 1924, 1927, 1936 & 1939). Additionally, developing new curricula requires sound pilot-testing, but in Turkey this is normally done in schools for whom conditions are better than school conditions around the country generally (Ünal et al., 2004), reducing the effectiveness of such pilots as an evaluation tool.

It is evident that a key feature of the success of any new curriculum is the role of the teacher. Naturally, then, teachers must be informed about any new curriculum especially if the curriculum seeks to dramatically change traditional teaching ways, beliefs and attitudes about teaching, and so on (Evaluation Form of New Curriculum Course Books in Grade 5, 2006). The development of a Turkish science curriculum has typically lacked a process of professional development; although inspectors and master teachers have tried to help teachers understand innovations and developments (Result Report of Evaluation Meeting of Curricula Between Grade 1 & Grade 5, 2005; Turkish Academy of Sciences General Views and Suggestions of New Curricula, 2005; Ünal et al., 2004). This points to a clear need for substantial in-service professional development to convince teachers that the new curriculum is better than the previous one; to help them understand the theoretical background; and to provide them with examples to aid implementation. However, in Turkey there are a huge number of science teachers - far more than can be involved in in-service professional development. As a result, teachers are mostly left to their own devices when it comes to curriculum implementation, particularly in rural areas (Akdeniz, Yiğit & Kurt, 2002).

Finally, the pilot-testing of the curriculum, and the measurement and evaluation of its effectiveness and fruitfulness has typically been done by the curriculum developers rather than external experts or reviewers without a vested interest (Result Report of Evaluation Meeting of Curricula Between Grade 1 & Grade 5, 2005). As a result, the evaluation for each Turkish Science Curriculum is rather superficial in nature, and does not actually inform the development process (Report of Investigation & Evaluation Related to New Curricula, 2005; Ünal et al., 2004).

CONCLUSIONS

Reviewing the developments of the Turkish science curricula throughout its history suggests the aims have been to: (1) help students become scientifically literate, (2) develop student-centered learning, (3) improve the quality and educational level of

Turkish society, (4) keep up with technology and contemporary science, (5) help students to link theoretical knowledge with daily life experience, and (6) provide new techniques, methods and equipment for learning and teaching, and also the measurement and evaluation of learning.

However, as noted above, the duration of curriculum development is usually too short, meaning few issues can be discussed in depth by experts, science teachers at school, and Turkish society at large. The piloting process is not done appropriately, professional teacher training is not adequate, and when teachers encounter difficulties in implementing the curriculum in the classrooms they have no one to contact. This suggests there are significant communication problems between curriculum developers and key stakeholders such as teachers, particularly in-service teachers. The rushed nature of new curriculum developments also means there is little opportunity to reflect upon or take account of experiences gained during the implementation of prior curricula. It is conceivable that there is, in fact, no need to develop a new curriculum; the current curriculum may fulfill societal needs given some revision.

A final important mitigating factor in curriculum development in Turkey is the presence of a national summative examination, the Selection and Placement of Students in Secondary Schools (OKS), which is used to select pupils for enrolment in the nation's elite schools. Even if teachers wanted to implement new curricula, many students and their parents are unhappy about change, given the high stakes of such an assessment regime. Shymansky, Yore and Anderson (2000), suggest that there is a need to get students and parents involved in the learning process and to establish an interactive learning process involving the teacher, student and parents. However, such efforts were made in the implementation of the 2005 Turkish science curriculum, but the parents were either reluctant to participate in the process, or had little understanding of science and science education, meaning they felt they could not contribute to the process (Evaluation Form of New Curriculum Course Books in Grade 5, 2006). In summary, despite literally decades of reform and numerous versions of science curricula, the fact that the current science curriculum supposedly focuses on conceptual understanding and meaningful learning, teaching in Turkish science classes remains didactic and teacher-centered in nature, mostly because of domination of the external examination system. Thus the Turkish science curricula have not been developed taking into account local and national needs and the characteristics of the Turkish educational context. The net result is a rather depressing picture in which all of the Turkish science curricula have been developed via a centre-periphery model without taking into account the geographical, cultural, or socioeconomic differences within the country. As a consequence they have failed to achieve the targeted aims.[ii]

IMPLICATIONS

In the light of the above, it seems that a centralized system of science curriculum development has not worked in the case of Turkey, mostly because it fails to take into account local conditions and needs. This, then, suggests a less centralized model might be preferable. Such a model should also draw upon what is now a quite extensive body of local research about science teaching and learning locally (including doctoral and masters' theses - see Coştu, Ünal, Çalık & Ayas, 2004).

Whatever model is used to develop science curricula (or modify existing curricula - see above) it will always be necessary to convince teachers that changes will result in better teaching. This then points to a need for regular and substantial in-service teacher professional development. This should be managed by the Ministry of Education, but incorporate feedback from teachers who evaluate the relationship between the targeted and achieved aims, and honestly confront issues about their own competency (e.g., in science content knowledge) to develop meaningful in-service training (see Fung, 2000, for an example of how this helped in Hong-Kong). In a similar vein, it would be helpful if teachers could get support from their colleagues when trying to implement new curricula; this would help alleviate feelings of isolation, particularly in rural areas. This might take the form of a consultation service (perhaps Web-based), where a teacher can access information and share experiences and ideas with others.

It seems little will change unless the examination system is modified to be consistent with the purported aims of any new curriculum. The present examination that dictates student selection for entry to upper schooling focuses mainly on rote learning which does not match the supposed learning outcomes of the 2005 science curriculum. None of these issues will be easy or cheap to introduce. But there seems little point in continuing with a seemingly endless series of curriculum developments and implementations (all of which, of course, cost money and consume vast amounts of time) that are pretty much doomed to failure.

Finally, the main difficulty in curriculum development in Turkey is related to her political situation, in that each new government has tried to substantially change the existing curriculum based on some new vision rather than improving the existing curriculum (Ünal et al., 2004). This, as noted above, results in short term thinking and planning. What Turkey needs is a more long-term strategic plan for curriculum development and implementation; one that takes a much more long term view than past efforts. If the nation has a well thought out, long term, plan of say 20 or more years, successive governments can still focus on targeted national aims which will be more stable, yet still open to educational trends and innovations from around the world.

A planning model for curriculum development in Turkey based on the above arguments is presented in Figure 1. This has been formulated for science education, but is flexible enough to serve as a template for other subjects.

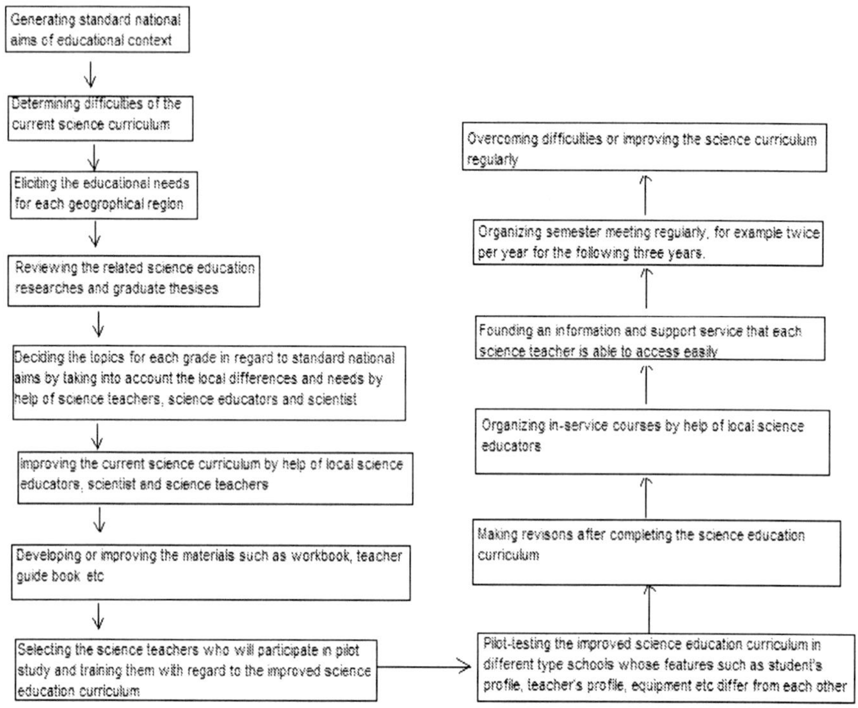

Figure 1. A model for developing curriculum in Turkey proposed by the authors

NOTES

[i] Indeed the name 'Science & Technology' seems to have been lifted directly from the equivalent Canadian documents (see Report of Investigation and Evaluation Related to New Curricula, 2005)
[ii] The 2005 curriculum is yet to be evaluated, but there is no reason to expect it will fare any better.

REFERENCES AND BIBLIOGRAPHY

Akdeniz, A.R, Yiğit, N., & Kurt, Ş. (2002). *Teachers' views of new science curriculum* [Yeni fen bilgisi programı ile ilgili öğretmenlerin görüşleri]. Paper presented at fifth National Science and Mathematics Education Congress, Ankara, Turkey.

American Board Publication Department. (1961). *Report of the National Commission on Education.* (1961). (M. Austin, trans.). Istanbul, Turkey: ABPD.

Announcement of Re-construction of Secondary Science and Mathematics Curriculum. (2006). Retrieved 15 October 2006, from http://ogm.meb.gov.tr/yap_b_ac.asp

Ayas, A., Çepni, S., & Akdeniz, A.R. (1993). Development of the Turkish secondary science curriculum. *Science Education, 77*(4), 433-440.

Ayas, A., Çepni, S., Akdeniz, A.R., Özmen, H., Yiğit, N., & Ayvacı, H.Ş. (2006). In S. Çepni (Ed.), *Science and technology teaching from theory to practice* [Kuramdan uygulamaya fen ve teknoloji öğretimi] (pp. 15-20). Pegema Yayıncılık, Ankara, Turkey.

Binay, H. (1982). *Turkey: Sociocultural information.* Strasbourg, France: Council for Cultural Cooperation.

Coştu, B., Ünal, S., Çalık, M., & Ayas, A. (2004). *A suggestion for the designing process of instructive activities that can be used in teaching chemistry* [Kimya öğretiminde kullanılabilecek öğretim etkinliklerinin geliştirilme sürecine ilişkin bir öneri]. Paper presented at the symposium of Reconstruction in Secondary Curricula. Ankara, Turkey.

Demirbaş, M., & Yağbasan, R. (2005). An analysis of science curriculum development attempts towards elementary schools for effective science teaching in Turkey and suggestions for problems encountered [Türkiye'de etkili fen öğretimi için ilköğretim kurumlarına yönelik olarak gerçekleştirilen program geliştirme çalışmalarının analizi ve karşılaşılan problemlere yönelik çözüm önerileri]. *Gazi University Journal of Kırşehir Faculty of Education, 6*(2), 53-67.

Educational Statistics of Turkey in 2005/2006. (2006). Retrieved 15 October 2006, from http://sgb.meb.gov.tr/daireler/istatistik/TURKIYE_EGITIM_ISTATISTIKLERI_2005_2006.pdf

Erdoğan, İ. (2006). *Four-year lycée education [Dört yıllık lise eğitimi].* Retrieved 15 October, 2006, from www.beykent.k12.tr/ilkedirne/makale/irfanerdo%F0an.htm

Evaluation Form of New Curriculum Course Books in Grade 5 [5. Sınıflar Yeni Ilköğretim Program Kitaplarının Değerlendirme Formu]. (2006). Retrieved 15 October, 2006, from http://duzce.meb.gov.tr/08_DUYURULAR

Fung, Y. (2000). A constructivist strategy for developing teachers for change: A Hong Kong experience. *Journal of In-service Education, 26*(1), 153-167.

Gümüşeli, A.İ. (2006). *Can we implement 'four-year lycée education' project?* [Dört yıllık lise projesi hemen uygulanabilir mi?]. Retrieved 15 October, 2006, from http://www.beykent.k12.tr/ilkedirne/makale/aliilker.htm

Kavak, Y. (2005, November). *Reflections in education VIII: After evaluation symposium of new elementary curricula* [Eğitimde yansımalar VIII: Yeni İlköğretim programlarını değerlendirme sempozyumunun ardından]. Paper presented at Evaluation Symposium of New Elementary Curricula, Erciyes University, Kayseri, Turkey.

Özinönü, A.K. (1976). *Innovation and changes in secondary science curriculum.* Faculty of Arts and Sciences Publication Number 30. Ankara, Turkey: Middle East Technical University.

Report of Investigation and Evaluation Related to New Curricula [Yeni Öğretim Programlarını Inceleme ve Değerlendirme Raporu]. (2005). Retrieved 15 October, 2006, from http://www.erg.sabanciuniv.edu

Result Report of Evaluation Meeting of Curricula between Grade 1 and Grade 5 [İlköğretim 1-5. Sınıflar Öğretim Programlarını Değerlendirme Toplantısı. (Eskişehir) Sonuç Bildirisi]. (2005). Retrieved on day month, year from ilkogretim-online.org.tr/vol5say1/index.htm URL-5

Scientific Meeting of Turkish Education Society (1984). *Science teaching and problems in secondary schools (Orta öğretim kurumlarında fen ögretimi ve sorunları)*, 12-13 June, Ankara, Turkey:TES.

Shymansky, J.A., Yore, L.D., & Anderson, J.O. (2000, April-May). *A study of changes in students' science attitudes, awareness and achievement across three years as a function of the level of implementation of interactive-constructivist teaching strategies promoted in a local systemic reform* effort. Paper presented at the annual meeting of the National Association for Research in Science Teaching, New Orleans.

Soylu, H. (1984, June). *New approaches in science teaching* [Fen öğretiminde yeni yaklaşımlar]. Paper presented at the Scientific Meeting of Turkish Education Society. Ankara, Turkey.

The Turkish Academy of Sciences *(TUBA)'s* General Views and Suggestions of New Curricula [Türkiye Bilimler Akademisi'nin Program ile Ilgili Genel Görüş Ve Önerileri]. (2005). Retrieved 5 July 2007, from http://www.tuba.gov.tr/files_tr/haberler/mufredat.doc

Turgut, M.F. (1990). Re-construction attempts of Turkish science and mathematics curricula [Türkiye'de fen ve matematik programlarini yenileme çalışmaları]. *Hacettepe University Journal of Education, 5*, 1-14.

Ünal, S., Coştu, B., & Karataş, F.Ö. (2004). A general look at the science curriculum development studies in Turkey. *Journal of Gazi Faculty of Education*, *24*(2), 183-202.

AFFILIATIONS

Muammer Çalık
Department of Primary Teacher Education
Karadeniz Technical University, Fatih Faculty of Education
Turkey

Alipaşa Ayas
Department of Secondary Science and Mathematics Education
Karadeniz Technical University, Fatih Faculty of Education
Turkey

AFRICA

MATSOGA

HANDLING SCHOOL SCIENCE CURRICULUM IN BOTSWANA: LOCAL CONTEXT REALITIES AND EXPERIENCES

INTRODUCTION

A careful observation of the history of education in Botswana indicates that since independence education has come a long way. It has moved from teaching under the trees, where learners wrote on slates, to modern classrooms with electronic media like computers, overhead projectors, and digital cameras. At independence, there were 239 primary schools, compared to 723 in 2006; there were nine secondary schools, which increased to 233; there were two teachers' training colleges and now there are six. What was then the University of Botswana, Lesotho and Swaziland, has now become the University of Botswana and the student population has grown from 170 students in 1966 to over 14,000 students in 2006.

This chapter aims to explore different context-specific factors that should be considered so as to establish the extent to which they influence and shape science curriculum development and implementation in Botswana primary and secondary schools. The chapter acknowledges that though the primary school curriculum may have its unique educational challenges, there are, nevertheless factors whose impact is felt across all educational levels. Similar challenges to those found in the primary schools exist in the secondary schools when the teaching and learning of science is examined. Before the actual factors are discussed, in order to put the country in perspective Botswana's basic physical, demographic and economic characteristics will be described. After the factors are unpacked, and discussed, conclusions and recommendations on the issue are made.

BACKGROUND

Botswana is generally a young (politically, economically, technologically, etc.) country. With a rapidly expanding economy there is a need to put in place trained personnel with skills that match modern technological developments prevalent in the country. In the education sector, specifically the teaching of science, this remains a huge challenge that calls for immediate attention from administrators.

Richard K. Coll, Neil Taylor (eds.), Science Education in Context: An International Examination of the Influence of Context on Science Curricula Development and Implementation, 177–186.

The need to have a more technology-literate society has been reported in several Botswana governmental policy documents, including the Revised National Policy on Education, Vision 2016, and the science syllabus for different cadres of education primary, junior and senior secondary curricula. Acknowledging and emphasizing the significance of science and its teaching, the Botswana Vision 2016 stipulates that:

> Botswana must recognize the rapid international developments in science and technology that are shaping the societies in the world. Botswana has no choice but to adopt high technology practices. The pace of international developments means that we will be forced to do so to compete in the export market. Science and technology must be emphasized through the education system. All children with appropriate aptitude, male and female, should be encouraged to study science. (Presidential Task Force, 1997, p. 40)

This illustrates the extent to which the teaching of science requires serious attention and a concerted effort from all stakeholders. The Botswana education system, just like in other African/Non-Western states, is challenged to produce a 'technically skilled workforce' that can be readily absorbed into the economy. The system needs to prepare individuals who are capable of participating actively in the so-called knowledge economy. To respond to this challenge, amongst many subjects offered in the school curriculum (primary, junior and senior secondary curriculum) is science. Whilst all other subjects are considered important and relevant for a country's development, it is the teaching of science and mathematics that are considered most likely to produce the desired scientifically-literate populace (Coll & Taylor, 2004; Laugksch, 2000). As a result, all Botswana education policy documents continue to emphasize this.

As indicated earlier, the significance of science is expressed in several educational policy documents and this comes out quite broadly in the upper primary school science syllabus that stipulates that:

> Science is a vehicle on which the world is dependent for its survival. Furthermore, the technology, which is the application of science concept and scientific discoveries to solve problems, uplifts the living standards of humanity. (Ministry of Education [MoE], 2005, p. 106)

CONTEXT ISSUES MILITATING AGAINST THE TEACHING OF SCIENCE

Though it is acknowledged that education in Botswana has come a long way and that generally significant growth has been achieved since independence, there are, however, context-specific issues which militate against the successful implementation of science curriculum in the country. In fact, the teaching of science runs parallel with that of mathematics such that the two experience the same implementation problems. The Trends in International Mathematics and Science Study (TIMSS) for 2003 undertaken in Botswana schools clearly captures problems that characterize the teaching of science. The report indicated that there is a strong relationship between the inadequacy of resources, and student performance. A shortage of calculators for science instruction, shortages of audiovisual resources for science and mathematics instruction, the absence of science laboratory equipment together with a shortage of computers for science instruction, all are noted in the report. Other impediments include a shortage of special equipment for disabled students. The problems highlighted here are a result of the economic context existing in the country, which consequently make the teaching of science extraordinarily difficult.

BASIC FACTS ABOUT BOTSWANA

A description of the physical, demographic, and economic features of Botswana will help in understanding the country.

Physical Characteristics

The Botswana National Commission for UNESCO Information Magazine of 2005 indicated that Botswana is a land-locked country dominated in geographical terms by the Kalahari Desert - a sand-filled basin, averaging 1,100 meters above sea level (Republic of Botswana [RoB], 2005). The country is bordered by Zambia and Zimbabwe to the northeast, Namibia to the north and west, and South Africa to the south and southeast. Botswana has consistently dry conditions and rainfall is very unreliable. Such a situation is not good for a country that over the years has been predominantly an agrarian society. Consequently, some parents would not have enough agricultural produce, which they could sell to pay for their children's school fees, and other basic educational needs. Further, failure to have adequate farm produce would not allow them to buy their children supplementary reading materials, particularly for challenging subjects like science.

Political and Demographic Characteristics

Though there may be a few grey areas, Botswana has been hailed as a leader in African democracy. Since independence in 1966, the country operates on the governing principle of democracy (RoB, 2005). All changes in power have been peaceful and democratic and followed the guiding principles of the Botswana constitution. Further, the country's democratic process is enriched by the plurality of political parties such that though the political landscape has for a long time been dominated by the Botswana Democratic Party (BDP), smooth working relationships with opposition parties exist. It is a given that such a calm political climate would allow for proper management of resources

The Botswana population census of 2001 indicated that the country has a population of 17 million people, thereby making it a thinly populated country by African standards (RoB, 2001). The population has a reasonably fast growth of 3.4% per annum, and statistics for the same year indicated that 48% of the population is below 15 years in age. That is, in common with other developing countries, the young predominate (RoB, 2001), indicating that the majority of people are of school age. Such a situation suggests that the education system faces a challenge to provide adequate educational resources including those for science teaching.

Economic Characteristics

Botswana's economy is one of the strongest and best managed in the developing world. At the time of independence in 1966, Botswana was one of the poorest countries in the world (RoB, 2005). The economy was predominantly agrarian, and 40% of GDP came from agriculture. However, by 1985/86, the structure of the economy had changed profoundly. Mining, which was virtually non-existent in 1966, had become the dominant economic activity, accounting for 50.7% of Botswana's GDP. Botswana is now classified as an upper middle-income country with US$9,200 in purchasing power parity. It is interesting to note that, though the country has such a sound economic background several challenges continue to constrain the teaching of science. This will be described below when those factors that shape the curriculum are discussed in more depth.

FACTORS SHAPING THE CURRICULUM: THEIR ORIGINS

Since independence, Botswana, like other African countries, went through several curriculum innovations, all of which were intended to improve the quality of education as well as develop education relevant to the country's needs and aspirations. The improvements targeted a variety of educational stakeholders, the learners, teachers and school administrators. In some cases, the innovation(s) in question meant a modification of curriculum content in one way or the other including for science. Though this chapter aims to discuss contextual factors that have over the years affected the teaching of science, first, an attempt will be made to present a general picture of the factors that often shape the curriculum in many settings. First, the history of the factors and where they come from in Africa will be discussed.

Literature that focuses on the interactions between curriculum development and implementation point to similar factors that are viewed as capable of shaping and or modeling the curriculum in a particular fashion. The factors apply to all subjects, including science. Several research reports point to a myriad of these factors including cultural influences, economic influences political factors (Bayona, 1995), major developments in the society, as well as the teacher competency in implementing the syllabus. Other factors may include official legislation in the form of governmental directives, white papers and standing orders, all of which explain how the education system is intended to operate. However, here for the teaching of science is undertaken in Botswana, the five most important issues will be dealt with here. Specifically, the political context, HIV/AIDS and wellness issues, economic contexts, gender equity, and equality issues.

Historical Trends in the Political Context of Africa

It is essential to note that political context, a factor that may shape the teaching-learning of any subject, has a long history in several African countries. Bayona (1995) noted that Africa's independence decade (1960-70) brought to many African nations' education systems changes at all levels. In Tanzania and Botswana for instance, this is reflected in policies of Education for Self-Reliance and Education for Kagisano (Social Harmony). Specific activities undertaken to achieve this include the introduction of national languages in schools, the introduction of vocational subjects, centralization of education and curricula development, and centralized control of national examinations.

In what may be viewed as a continuation of the spirit expressed above, in Botswana, there have been since about 1996, several such innovations across all levels of education (primary, junior and senior secondary). The primary reason for bringing the innovations was that several African governments felt that the education offered did not address the needs of their countries. They believed that the education did not generate relevant human resource and economic benefits needed by their countries. Strategies designed to address this resulted in the inclusion of traditional subjects such as social studies, science, English language, Setswana language. The innovations were introduced with a view to offering education considered relevant to prepare school leavers for the world of work as well as for smooth transmission into secondary and post-secondary education.

Additionally, in regard to politics and their relationship with education, as noted above, Botswana has had one party in power since independence, that is, well over four decades. Perfection is not an easy thing to achieve; however, having one government with reasonably sound leadership and a calm political climate over the years provided an environment that allowed reasonable development of education in Botswana.

Economic Influences

Economic factors exert considerable influence on the curriculum, not only in Africa but also in all developing countries. In many countries, for instance, high rates of school dropout at the end of each level have mainly been reported because nations cannot afford to provide all learners with education at successively more advanced levels. Further, schools in such countries have a shortage or have used poor quality materials. However, in Botswana there has been a different situation. Between 1986 and 1987, the Government abolished school fees in all public primary and secondary schools. Since 2004, due to other pressing needs, the Government felt that it could no longer fund secondary education wholly. School fees were introduced in all public secondary schools as a cost-sharing measure. Naturally, some individuals and organizations did not take this kindly. Such concerned groups believed that some children whose parents could not afford fees would drop out of school because they did not have funds to pay for their children. However, the Government set up some 'safety nets' showing clearly that while all parents are expected to make efforts to pay for their children, no child would fail to go to school because his or her parents could not afford to pay school fees.

HIV/AIDS and Wellness Issues

One major challenge the country faces is the devastating impact of HIV and AIDS, such that the pandemic has become a serious development and education issue. The first case of HIV-related illness in Botswana was diagnosed in 1985. Since then the virus has spread sharply from urban to rural areas leaving no district spared. According to the UNAIDS/World Health Organization Report on Global AIDS Epidemic for 2006, Botswana is one of the countries hardest hit by the HIV/AIDS pandemic in the sub-Saharan Africa, with an estimated 24% of its residents living with the virus. The report estimated the adult (15 to 49 years) prevalence rate for 2005 of 24.1%. The National AIDS Coordinating Agency reported that their figure for the same age group was 25.3% in 2005. According to the Principal Public Relations Officer for NACA, 17.1% of the population is HIV positive.

Between 1993 and 2002, HIV spread rapidly among all age groups and the prevalence rate more than doubled, reaching 35.4% at the end of the period. In the entire context within which science subject is taught, HIV/AIDS has become devastating. The high prevalence rate of HIV/AIDS continues to reduce the capacity of some of the most productive members of society, thus affecting both employment and economic growth. In this instance, the productive workforce referred to includes science teachers. This has started to reverse some of the hard-earned economic and social gains of the past decades, as reflected in the lowering of life expectancy rates. The Government has been compelled to divert the meager resources available into the provision of medication and other health needs of the sick. Consequently, other economic sectors tend to suffer because they are somewhat neglected.

Inadequacy of Resources and Technology

It is also worth noting that the teaching of science in Botswana also takes place in settings where certain resources that may be considered essential for its teaching are inadequate. The shortage encompasses a wide range of such resources and this includes: shortage of infrastructural amenities, computers and other instructional materials.

Shortage of infrastructural amenities. Due to ever-increasing school enrolments in all levels of education, infrastructural amenities are seldom adequate, with primary schools particularly being the hardest hit by a shortage of classrooms, teacher accommodation, storerooms and other buildings. In regard to such amenities, secondary schools are considerably better off than the primary schools. When it comes to science laboratories, the situation is similar. For most schools, even where a recommendation has been made in regard to the number of laboratories per school based on the school's enrolment, science laboratories are never adequate. For example, for most senior secondary schools whose enrolments remain around 1,500 or so students, the Ministry of Education recommends a minimum number of laboratories to service such a school, commonly 10 to 12 laboratories. But one often finds that a school functions with significantly less than 10, thereby making the teaching of science much more difficult. This naturally impacts on the quality of science teaching.

Shortage of computers and other instructional materials. The Trends in International Mathematics and Science Study (TIMSS) report on Botswana for 2003 indicated that there is a strong relationship between the inadequacy of resources and technology and student performance, for example, the shortage of calculators needed for both science and mathematics instruction. For science in particular, there are shortages of audio-visual resources, laboratory equipment, computers with appropriate software; relevant library materials, instructional materials such as textbooks, and qualified teachers. All such situations compromise quality in the teaching of science.

Gender Equity and Equality

While Botswana has taken significant steps in recognizing and addressing gender equity and equality in education, there are indicators that the country has not done enough. Matsoga, Lekoko, Tsheko and Garegare, (2006) note that internationalization and globalization of education have come to inform the world that regardless of sex, each individual needs to be educated. This reminds one of conventions such as the 1990 World Conference on Education in Jomtien, Thailand; the 1995 World Conference on Women held in Beijing, China; the 1997 Humburg International Conference on Adult Education; and the 2000 World Education Forum held in Dakar, Senegal. In the latter, 2015 was targeted as the year by which gender disparities should be eliminated in primary and secondary education. Also, 2015 was targeted as a year in which gender equality should be achieved with a focus on ensuring girls' full and equal access to and achievement in quality basic education (World Education Forum, 2000). Because Botswana is a signatory to the World Education assemblies, she is responsible to such declarations to provide free basic education and re-admit those who left school because of pregnancy. Matsoga et al. (2006) further explain that traditional Botswana society driven by its patriarchal attitudes, has seen it fit to educate boys and leave girls out, hoping that they would get husbands to fend for them. As these traditional beliefs are changing and new developments are felt, more people are realizing that education is equally important for boys and girls. The Government too recognizes this and consequently, the significance of science and the need to popularize it have been reflected in several policy documents. As a result of this in regard to science and technology education the Revised National Policy on Education (RNPE) of 1994 emphasized the need to:

> Intensify measures to popularize science amongst students and to develop an interest and positive attitudes towards science and technology and that this should be developed through science and mathematics fairs and other competitive activities, special awards, establishment of a science and technology park and so forth. (RNPE, 1994, p. 25)

As for the girl child and her position in as far as the teaching and learning of science is concerned, the Policy states that "special measures should be developed to increase the participation and performance of girls in science, mathematics and technology" (RNPE, 1994, p. 25). Against this understanding, one essential question to pose is; since there are so many policy documents that strongly advocate for the welfare of the girl child, what are the teaching and learning realities of science in Botswana in relation to gender equity and equality?

DISCUSSION

Several positive points exist in relation to gender equity and equality issues in the education system in Botswana; for instance, learning materials such as textbooks are prescribed on the basis of their being gender sensitive. Further, the Government has set up a policy that allows for re-admission of children who drop out of school (both primary & secondary) for various reasons. However, the policy has not yet been taken advantage of by parents or students, since not many of them know about its existence and its details. As a result the policy is underutilized.

In summary, it is interesting to note that the Government strongly advocates for the promotion of science among students and that "special measures should be developed to increase the participation and performance of girls in science, mathematics and technology" (RNPE, 1994, p. 25) Further, the senior secondary education program stated that "the school curriculum will be designed to be gender sensitive and make efforts to positively address existing biases in order to help all learners fully develop their potential" (RNPE, 1994, p. 3).

Contrary to all this, realities in schools indicate a different scenario altogether. Teacher education programs have very little or no traces of content on gender education and, as a consequence, it is the teacher who to some extent contributes to gender stereotypes seen in school. Some teachers still hold to a view that science and mathematics are for males. Matsoga et al. (2006) note that "in secondary schools, high performance in traditional male dominated subjects (like pure science, mathematics, design and technology) is skewed in favor of boys" (p. 48).

It has been noted earlier that the country has two very significant areas that allow for appropriate management of the available resources: a reasonably stable political climate, coupled with sound leadership. Following this, reasonable efforts have been made in the channeling of resources into education, though this is not enough; much more needs to be done. Exacerbating the situation is the prevalence of HIV/AIDS and other wellness issues. The HIV/and AIDS pandemic and its other forms of related illnesses render Government endeavors ineffective. Many resources that could otherwise be used to improve education are channeled into medication and other forms of caring for the sick. The system keeps losing a substantial number of the teaching workforce through deaths, absenteeism and among these science teachers are included. The diversion of big chunks of money into the cure and control of HIV/AIDS leads to other challenges such as shortages of computers, and other ICT tools that could lead to improved teaching of science in schools.

CONCLUSIONS

It may be concluded that despite its macroeconomic success, Botswana has serious problems of poverty and high unemployment compared to other countries at a similar level of economic development. Many have argued that the poverty situation in Botswana appears to be more of a structural problem, attributable to a rather narrow economic base largely mining and agriculture, that tends to limit opportunities for employment. This is exacerbated by a small and sparsely distributed population, and a very small and fragmented domestic market. Issues of gender inequities and how they impact on the promotion and teaching of science in schools have been discussed. It is natural a few grey areas still exist in the system; however, significant potential in regard to education improvement is still there and concerted effort is needed from all stakeholders to improve on the quality of education offered. Botswana has been hailed as a leader of African democracy. There is need for all to take advantage of the country's governing principle of democracy which has been in existence since independence in 1966. Such a context would be conducive for the teaching and learning of science in all schools, primary and secondary.

REFERENCES AND BIBLIOGRAPHY

Bayona, E.L. (1995) *Curriculum design and development*. Gaborone: Print Consult.

Botswana AIDS Impact Survey. (2004). Gaborone, Botswana: Government Printer.

Coll, R.K., & Taylor, N. (2004). Probing scientists' beliefs: How open-minded are modern scientists? *International Journal of Science Education*, *26*(6), 757-778.

Curriculum Blueprint (Revised 2002). *Botswana Senior Secondary Education Program.* Gaborone, Botswana: Government Printer.

Curriculum Blueprint. (Revised 2002). *Basic Ten-Year Education Program.* Gaborone: Government Printer.

Education for Kagisano (Social Harmony). (1977). *Report of the National Commission on Education.* Gaborone, Botswana: Government Printer.

Faculty of Education. (2005/2006). *Department of Educational Foundations: Handbook* Gaborone, Botswana: Government Printer.

Laugksch, R.C. (2000). Scientific literacy: A conceptual overview. *Science Education*, *84*, 71-94.

Matsoga, J.T., Lekoko, R.N., Tsheko, G.N., & Garegare, K. (2006). *A review of gender equity and equality in primary and secondary education: The case study of Botswana.* Gaborone, Botswana: Botswana Educational Research Association and UNICEF.

Ministry of Education. (2003, April). *National plan of action for education for all.* Paper presented at a national stakeholder forum. Gaborone, Botswana.

Ministry of Education. (n.d.). *Curriculum development and evaluation: Core text evaluation score sheet.* Gaborone, Botswana: MoE.

Ministry of Education. (n.d.). *Checklist for selecting books and other support materials: Guidance and counseling.* Gaborone, Botswana: MoE.

Ministry of Education. (2005). *Upper Primary school syllabus: Standard five to seven.* Gaborone, Botswana: Curriculum Development and Evaluation Department, Ministry of Education, Government Printers

National AIDS Coordinating Agency Newsletter. (2007). Gaborone, Botswana: Government Printer.

National Commission on Education. (1993). *Report of the National Commission on Education (1993).* Gaborone, Botswana: Government Printers.

Presidential Task Force. (1997). *Vision 2016: Towards prosperity for all.* Gaborone: Presidential Task Group.

Republic of Botswana, Botswana National Commission for UNESCO. (2005). *Information Magazine No. 7*. Gaborone, Botswana: Government Printers.
Republic of Botswana, Presidential Task Force. (1997). *Long term vision for Botswana, Vision 2016.* Gaborone, Botswana: Government Printers.
Republic of Botswana. (2001). *2001 Population and housing census: Preliminary result brief.* Census Office: Gaborone, Botswana.
Republic of Botswana. (2003). *National development plan 9*. Gaborone: Government Printers.
Republic of Botswana. (1994). *Revised national policy on education*. Gaborone: Government Printers.
Republic of Botswana (2004). *Botswana millennium development status report 2004: Achievement, future challenges and choices*. Gaborone, Botswana: Tiger Design and Graphics.
Trends in International Mathematics and Science Study. (2003). Gaborone, Botswana: Government Printer.
UNAIS/World Health Organization. *Report on Global AIDS Epidemic for 2006*. Retrived 10 September 2007, from www.kit.nl/exchange
University of Botswana. (2007). *Horizon,* Vol. 1.

AFFILIATION

Joseph Thoko Matsoga
Faculty of Education
University of Botswana
Botswana

KOOSIMILE AND PROPHET

SCIENCE TEACHING IN CONTEXT IN BOTSWANA

INTRODUCTION

The level of investment in education and science education in Botswana would be the envy of many comparable nations worldwide. However, developing a sound education system involves even more than substantial financial investment. The model of curriculum development, and mode of implementation also are determining factors. It seems that educational organizational structures, and that the centre-periphery model used for curriculum reform and implementation in Botswana have limited the quality, if not the quantity, of science teaching, and resulted in science teaching that is strongly teacher-dominated and fails to address individual learners' needs and capabilities. In this chapter the authors argue that a new organization structure and culture associated with curriculum reform is needed if we are to improve the quality of science education in Botswana.

Botswana, is classified as a middle-income country, and is a former British Protectorate that has made significant progress in education since her independence in 1966. After a series of structural changes during the 1980s and 1990s, Botswana currently runs a tripartite education system comprising seven years of primary school education followed by three and two years respectively of junior and senior secondary school education. Schools are co-educational with 6-13 year olds in primary schools, 14-16 year-olds in junior secondary schools and 17-19 year olds in senior secondary schools. The primary and junior secondary schools in Botswana make up the 10-year Basic Education cycle that embraces an egalitarian perspective. Running parallel with the government school system is a private, English Medium School system that is the choice for the majority of middle income parents. With smaller class sizes and better resources, both human and physical, these schools guarantee better examination results for students thus ensuring progress to higher levels of education.

Within government primary schools, a generalist teacher, who handles all subjects making up the curriculum, teaches science. The primary school curriculum, which has undergone a number of significant changes since independence, is currently split into two phases. From Standards 1 to 4, known as Lower Primary, students follow an integrated curriculum in which science forms part of the subject known as Environmental Science. At Upper Primary (Standards 5 to 7), the focus is on the development of pre-requisite skills for junior secondary curriculum and science is taught as a separate subject. Emphasis is placed on the basic principles, skills and processes of science and students are expected to

Richard K. Coll, Neil Taylor (eds.), Science Education in Context: An International Examination of the Influence of Context on Science Curricula Development and Implementation, 187–198.

recognize the nature and limitations of science. Teachers are encouraged to be ingenuous and improvise with basic everyday materials in their science classes. However, for the most part, students copy copious notes from the board into notebooks and few practical activities take place. Little or no attention was paid to language issues or to the level of students' cognitive development as the primary school syllabus was being developed and thus the rote regurgitation of science concepts continues to be emphasized.

The Primary School Leaving Examination (PSLE) had served to select the best students for junior secondary schooling but in 1985, government accepted the principle of universal access to 10 years of basic education. At this time, only about 40% of Standard 7 leavers entered Form 1 but this had increased to 95% by 1993. The move was not welcomed by many junior secondary teachers who were required to change their teaching approach to cater for the whole ability range of students leaving the primary schools

The whole ethos of teaching and learning science changes at the junior secondary school level. There is now a specialist teacher, textbooks, and a moderately equipped science laboratory. The current science syllabus has a strong life skills and vocational orientation giving pupils some foundational knowledge that prepares the majority of them for life as adolescent school drop-outs at the end of their ten-year Basic Education.

Performance in terminal examinations at the end of the basic education cycle, known as the Junior Certificate Examination (JC), continues to control enrolment into senior secondary education. The 27 government senior secondary schools take relatively few students from the 206 junior secondary schools resulting in a high attrition rate of students. As of 2007, about 60% of junior secondary students move into senior secondary schooling although the Government aims to guarantee 100% access to senior secondary schools by the year 2016 when the country will celebrate 50 years of independence. The current progression rate, while still far from satisfactory, is a recent innovation. Admission to senior secondary schooling was very selective and even in the early 1990s the progression rate was only around 25%. So in just over a decade, the admission rate to senior secondary education has more than doubled. This has had a massive impact on the ability range of students moving forward to Form 5 examinations, a trend not always favorably received by teachers, and certainly not one for which they received any further training.

The science syllabuses offered since 1996 in the senior secondary schools are adapted versions of the University of Cambridge's International General Certificate in Secondary Education (IGCSE). The syllabuses and examinations, known as the Botswana General Certificate in Secondary Education (BGCSE), and are offered in collaboration with, and under accreditation from, IGCSE. The curriculum package now claims to be relevant to the local socioeconomic contexts and experiences of the youth in Botswana, despite the design and content, which mimics that of the IGCSE. The curriculum rhetoric for the science syllabuses advocates a learner-centered approach with the local environment used to provide context. There is an over-arching emphasis on science process skills, problem-solving skills and the acquisition of hands-on experience, together with diverse assessment strategies that include formal examinations and continuous assessment, including assessment of practical work to ensure that learners attain the set aims.

STATUS OF TEACHING SCIENCE IN BOTSWANA

The literature suggests that most curriculum changes in developing countries, generally driven and promoted by trends from metropolitan nations, result in some form of first world hegemony which constrains and limits the autonomy of critical persons (Gray, 1999). These curriculum changes have been shown to fail in developing nations as their aims, philosophy and instructional methods are not questioned deeply (Lewin, 1990), and they remain as what Ingle and Turner (1981) call 'cultural misfits'. Despite significant advances in science education as a whole since 1990, the status of science teaching has not undergone a similar improvement across the world (Hodson, 2003; van den Akker, 1998). In Botswana, even with a strong political will and the diamond-powered economy that sustains the commitment of the Government to education, there also has been little positive development in science teaching. Consequently, the peculiarity of Botswana as a developing country is that it has in place every important component in curriculum development, implementation, and institutionalization.

The general state of teaching science in Botswana is not encouraging. Science teaching remains teacher-centered, lecture-based with passive participation of students, textbook-driven and examination-driven. These trademarks have become prima facie evidence for decrying the quality of science teaching not only in Botswana, but in all developing countries in general. The prevailing assumption by the Government of Botswana is that resource availability should readily translate into effective implementation. Consequently, other key contextual factors that make up science teaching have been, and continue to be, neglected. Research on science education by university academics has so far failed to have any positive impact on the situation. The effect of increased access to secondary schooling on the process of science teaching has been ignored. Little attention has been paid to the quality of teaching and to the learning experiences provided to students. Furthermore, relatively little has appeared in the literature on how teachers themselves define, translate and implement the curriculum. Thus we would contend that the study of science education in Botswana lacks a cutting edge, has had a low impact factor, and has failed to bring about change at school level.

The foregoing scenario, however, does not imply a lack of capacity of schools and teachers to implement various science curricula. Rather we would wish to suggest some deep-seated problems in the structural, organizational, corporate and societal culture that mediate science curriculum matters. Hence the term 'science teaching in context' is used to refer to the wider structural and organizational framework that mediates and gives meaning to activities in curriculum implementation and the quality of teaching in science. Thus, the chapter does not focus on operational issues at school and classroom level, such as classroom interactions, but on the wider structures and policies that give meaning to what happens in the science curriculum. Consequently, the issues in science teaching that are considered in this chapter are in two interrelated areas: curriculum control within a centre-periphery model of educational management and the science teacher and curriculum innovations. The discussion considers a number of possibilities which have the potential to improve the quality of science teaching in Botswana.

CENTRE-PERIPHERY MODEL

The Government of Botswana runs a centralized education system whereby all critical decisions regarding the curriculum, curriculum support materials, physical facilities, human resource management (recruitment of teachers), and examinations are handled at the Headquarters of the Ministry of Education in the capital city, Gaborone. The information cascades from the centre to the periphery of schools and teachers. The relationship between the centre and periphery is timid, to say the least, and is punctuated by frequent directives issued from the centre on curriculum matters. The system is rigid, not allowing for a diversity of possible developments in curriculum development and there is no alternative platform for exploring and maximizing opportunities arising in the context of curriculum implementation. Ironically, the centre-periphery model symbolizes the prevailing dissonance in curriculum implementation, with each unit (Department of Curriculum Development & Evaluation, Division of Examination Research & Testing, Department of Teacher Training & Development, schools and teachers) bent on independently fulfilling its own mandate in its own way.

One fundamental issue that emerges in the context of Botswana is the failure of the revised science curricula to embed in the education system in general, and in school contexts, in particular. This contradicts a strongly held belief in the fallacy that the uniform design of schools and teaching facilities promotes a uniform compliance with government's intention to reform the science curriculum at school level. For example, all laboratories built in the junior secondary schools during an era of massive expansion during the late 1980s and 1990s, conformed to one design with uniform provision of equipment and furniture. This physical structure was intended to give meaning to and provide the context in which the teaching of science would take place. The Government believed that now all schools would implement, without deviation, the revised (1996) junior certificate science curriculum with its emphasis on process skills and problem-solving skills. However, there is no evidence that any change took place at school level and the laboratory simply became the venue for the same teacher-centered handing down of scientific facts.

The development of the localized Botswana General Certificate of secondary Education (BGCSE) syllabuses in the mid-1990s provides another example. The decision to break away from the reliance on the University of Cambridge Overseas School Certificate (COSC) was made in 1994 as a result of a recommendation of the 1993 National Commission on Education. The new examination would be based on the International General Certificate of Education (IGCSE) model and the process would be overseen by the Cambridge Examinations Syndicate. A curriculum blueprint for the senior curriculum was quickly developed and the decision was made that the syllabuses for the three science subjects of biology, chemistry and physics would take priority.

For each subject, a task force was set up to 'develop' the syllabus but it was far from clear what their role would be. The Curriculum Blueprint set a very tight framework for the development, the IGCSE syllabuses in the sciences provided sample content and University of Cambridge advisors were present to 'vet' proposed changes. Given the very short time frame available to submit the final syllabus, any form of critical debate and analysis was very limited. It should be of no surprise that the final syllabus for each subject bore a remarkable resemblance to the original IGCSE syllabus, and that teachers in the field felt that there had been no change.

The pure sciences were intended for the most able students in science who were likely to progress to science-based tertiary education programs. However, the expansion of the senior secondary school system meant a much wider ability range of students, the majority of whom were not destined for science-based careers or simply did not have the ability to cope with the academic sciences. Since some form of science was a compulsory part of the curriculum for all, the Task Forces were now charged with developing science curricula for the majority of students. There appears to have been little debate on the issue of what sort of science would be most suitable for these students and a Double Science Syllabus (comprising of a selection of two thirds of each pure science syllabus) and a Single Science Syllabus (comprising a selection of one third of each pure science syllabus) were quickly developed. Again, that teachers were given little or no guidance on implementation and the inadequacy of, for example, the Single Science Syllabus is shown by the pass rate of about 11% which has remained consistent over the past four years.

More recently, the decision to implement the continuous assessment of practical work has come into force. From January 2007, science projects and practical investigations resting on the 'processes of science' became part of the assessment process in all government senior secondary schools. In spite of inconclusive trialing of this mode of continuous assessment in a small number of schools, it is currently being implemented, and schools do not have the opportunity to make a case for their readiness and willingness to participate or otherwise. By contrast, the IGCSE allows continuous assessment only in schools that meet its stringent requirements for accreditation in running the program. In this way, schools have the privilege to engage or withdraw on professional grounds, as a quality assurance measure, from adopting certain aspects of the curriculum.

Clearly, the Government has often proved woefully inadequate in its ability to respond to concerns on science curriculum matters to a point that problem areas become entrenched in the system, frustrating the development of science education in general. This leads to questions about the philosophy and vision of the Government of Botswana for science education and what it hopes to achieve through the teaching of science in schools. We appear to have a curriculum driven by the centre in a poorly conceptualized framework that is not meant to marginalize teachers and schools per se, but which results in an incomplete feedback loop and weak articulation between the various organizational units in curriculum implementation. Under these circumstances, curriculum implementation and its meaning, rather than being an outcome of critical thought and action by the practitioners themselves, become incidental to the process of teaching. Thus in terms of goals, objectives, effect and outcomes, the quality of science education and the process of science teaching in classrooms has become somewhat doubtful. Important questions on how best to balance the control of curriculum and how to institutionalize and embed the curriculum in the education system have been overlooked or ignored.

SCIENCE TEACHERS AND CURRICULUM INNOVATIONS

In bringing about any change in the science curriculum, the teachers and the curriculum are inseparably linked. Botswana manages her curriculum change using a centre-periphery model in which the teachers are part of the Government's 'resources' (Davies, 1988). However, instead of serving as partners in curriculum change, the teachers are largely powerless functionaries in curriculum matters, learning of curriculum innovations through workshops, documents and directives. The teachers are the major instruments for curriculum implementation, although they often lack the criticality, drive and enthusiasm that would enable them to ask meaningful questions and proactively drive curriculum innovations. It appears that teachers are viewed as technicians who can unravel the complexities of any curriculum with relative ease, except in areas where the Government is of the view that they might be deficient or incompetent. This view easily passes as a 'needs analysis' for in-service teacher training and development. While the learner is at the focus of the latest curriculum innovations which promulgate the doctrine of learner-centered education, the teachers are generally neglected. Their professional development in terms of the development of critical thinking skills and conceptual tools for interpreting the curriculum vis-à-vis teaching is paid little or no attention.

Nevertheless, the Government appears to take the whole issue of teacher development very seriously as indicated by the fact that a whole department of the Ministry of Education is dedicated to these activities. The main vehicle for addressing the developmental concerns for teachers is the workshop which may be held at local, regional or national level. However, the Government keeps a tight control on these in-service activities for teachers, commissioning only workshops which address some aspect of curriculum they consider important at that time or in which the teachers are considered 'deficient'. Workshops on the various curriculum innovations are simply in pursuit of the fulfillment of policy initiatives and curriculum rhetoric. The workshops fail to help teachers attain the breadth of knowledge, the creative skill and the critical awareness needed to enhance the implementation of the science syllabuses and the impact of these workshops on teachers' lives and work in the science classroom is relatively unknown.

Examples of such workshops held in recent years include, among others, mixed-ability teaching, the processes of science, learner-centered teaching, remedial teaching, problem-solving strategies, and constructivism in science teaching. Workshops focus on each of these discrete components of the science syllabuses at the expense of a holistic and broad-based consideration of curriculum and teaching. Teaching strategies to be employed by the teachers are perceived and presented as some kind of a recipe which is to be followed in clearly defined stages thus standardizing the whole process and making it, as far as possible, teacher proof. This contrasts sharply with the recent emergence of innovative thinking about teaching in science, presenting it as an eclectic educational movement drawing from philosophical, psychological, cognitive and instructional theories in order to build a *vision* that usually includes beliefs, values and ideology about education (Lazarowitz & Hertz-Lazarowitz, 1998).

This poorly conceptualized vision of in-service development is underpinned by an even more problematic mode of dissemination. This rests on a cascade model of dissemination in which workshop participants are expected to sponsor and drive the innovation amongst colleagues at school level. This is not a priority for science teachers in their day-to-day working lives in the classroom. The cascade model is based on certain assumptions about the professional work of the teacher which are generally not shared by teachers. In this sense, the cascade model challenges the science teachers' perceptions of their positions, roles and identities in a centre-periphery system and this alien perception is readily ignored.

The failure of the cascade approach in science teacher development underscores the truism that change does not happen spontaneously, but must be firmly grounded in the context of teachers' lives and identities. There is no reasonable evidence of success that the contemporary ideals and practices embedded in the various syllabuses have been adopted in the classroom. Rather, the schools and teachers futilely grapple with the Euro-centric conception of teaching and curricula change that is expected, with little or no professional assistance. Whole-class teaching, rote learning and regurgitation of science facts and concepts with little or no understanding are the norm, and remain strongly entrenched in the education system. Issues such as culture, language, large class enrolments, examinations with local content, and how best to incorporate the experiences of learners in teaching, are largely ignored. This remains a great paradox as it would be reasonably expected that these should be priority issues in the implementation of a localized curriculum that seeks to relate education to the social, political, cultural and economic circumstances of the nation (National Commission on Education, 1977).

These issues and challenges in science education remain stubborn, resistant to change, and Botswana continues to fail to come up with her own definition of a relevant science curriculum and how best to teach it through which the process of science teaching could be anchored in schools and the wider society. In a brave, but seemingly futile, attempt to transform the teaching process, the Government embarked on a radical overhaul of assessment strategies throughout the education system from primary to senior secondary levels. In 1994, assessment policies were put in place to promote the change from norm referenced to criterion referenced testing and examinations (Republic of Botswana, 1994). The primary schools were the initial focus of attention and after receipt of a handbook to explain what was expected and one or two short workshops, teachers were expected to develop student profile records, and criterion-referenced testing was to be fully utilized to assess the students for diagnosis, remediation and selection. The concept was extended into the secondary school sector with the revised syllabuses in the late 1990s.

However, in spite of the changed assessment policy, comprehension-based tests and examinations in which rote learning is rewarded remain the norm in schools. The marking of examinations continues to be a secretive activity and marking keys are confidential and classified documents that are never made public. Workshops hosted for those teachers selected for the live marking of examinations are intended to train them on using the marking keys, but not on the principles of developing assessment instruments. The expectations and comments of examiners should be a valuable resource for teachers which could lead to an improvement in the quality of teaching and assessment at school level. However, the hidden nature of the assessment process negates any possible positive effect. The end result of this is that the curriculum does not realize its optimum potential to force a change in examination questions and assessment strategies, this in turn feeding back to change and develop the curriculum itself.

In spite of the kinds of problems outlined above, changes in the curriculum continue to be passed down unabated from the centre to the periphery. The frequency with which these ideas are launched gives cause for concern, given their gestation periods and the blind faith and idealism that serve as a platform for selling innovative ideas to science teachers. This is an indication of the mechanistic, objectivist, and positivist view of curriculum implementation held at the centre which overlooks teaching as a social construct and the teacher as the meaning-maker in defining and implementing the curriculum. It is suggested that it is this divide in the whole conception of curriculum between centre and periphery that is failing science teaching in Botswana.

DISCUSSION

The major thesis of this chapter is that Botswana has a relatively advanced education system, characterized by significant advances in both human and physical resource provision for science education, yet operating well below expectation and its optimum level. In spite of the large investment in science education, well-meaning and innovative ideas are failing to embed in the system. This chapter explores a number of possible reasons, the main one being that it is the Government's stranglehold on the science curriculum that is restricting the development of values such as problem-solving, critical thinking and teamwork skills in the teachers and schools. We contend that the bureaucratization and centralization of curriculum and its support structures excludes teachers from active participation in the reinterpretation of their roles and positions in the teaching profession and in curriculum development. In-service training and development seems to be premised on a deficiency model rather than seeking to engage the science teachers in a creative and critical approach to curriculum development and implementation. Questions about the relationship between a teacher's own values and those that they are expected to promote in teaching/implementation of the curriculum are unavoidable. We contend that the interplay between the Government's perceived areas of need and teachers' values, and perceptions of curriculum implementation is more problematic. It is almost impossible to detach curriculum decisions from teachers' attitudes to authority and the legitimacy of the pressures it brings within a centre-periphery regime of curriculum management.

The centre-periphery model of curriculum management has led to the development of a powerful bureaucratic structure with a clear divide between managers and teachers. In their professional work, teachers have been acculturated into looking to the centre for answers to their problems with curriculum implementation rather than taking personal responsibility at school and classroom level. It is not that teachers feel constrained to discuss and innovate and develop in their teaching, although it may be that they do not perceive curriculum development and implementation as part of their role in being a science teacher, rather there is some evidence that they creatively and strategically use resources such as in-service workshops to resist pressure to adopt innovative ideas. In this way, change that is perceived as being in the interest of the Ministry of Education rather than in their interest as science teachers is resisted and stifled. This inevitably undermines the quality of science teaching in schools in Botswana. As noted by Reynolds and Skilbeck (1976), "so long as curriculum analysis is a dominantly armchair activity, it is possible to filter out and ignore inconvenient facts (about organizational complexities, and teachers' and pupils' interests, for instance) by interposing a preconceived framework of ideas or theory" (p. 28).

This prompts a question of the sort "how do we *reshape governance and administration* of the education system in order better to achieve the genuinely educational goals" (Husen, 1980, p. 180), and enhance the quality of curriculum implementation, and perhaps, science instruction? How then can an organizational structure and culture be developed in Botswana that will uplift the teaching of science, encourage and motivate pedagogical innovation, and bring into the public domain a critical debate on science education and teaching? Three interrelated strands underpin the response to these questions. One strand is that the huge investment in education thus far by the Government is a significant leverage point in as far as it enables consideration of *what is missing* for an enhanced delivery quality of education by the existing structures and teachers. The second strand is that the bureaucratized and centralized control of curriculum has been at the expense of *collegiality* - defined as some form of collaborative partnership (see Fullan, 1995; Hoyle, 1975) - amongst stakeholders in science education. The third strand is that an increased awareness of the issue of quality in science education is possible and can be facilitated if science education is accepted as a value-laden political activity open to public scrutiny through a well-informed debate involving all stakeholders.

If the mandate of science education is to be fully realized, how do we assert the supremacy of collective responsibility and professional development of teachers and other stakeholders over the regulated control of science education and associated resources by the Government? Two areas emerging from our debates in the chapter are worth considering in some more detail: in-service education provision, and the development of a professional science education association.

The in-service education of science teachers should be playing a crucial role in their professional development, helping them to critically analyze curriculum and their own teaching practice and values in order to better understand and assimilate good practice into the culture of their workplace. This possibly requires an interpretive view of curriculum, as opposed to a technicist one, where dialogue and debates on aspects of curriculum can be articulated and appreciated within the existing wider organizational structures for science education. The Department of Mathematics and Science Education (DMSE) of the University of Botswana has an in-service wing for science and mathematics teachers, and DMSE-INSET had the initial vision to play this very central role in the professional development of teachers. However, over the 15 years of its existence, it has failed to have much positive impact in science laboratories and classrooms. A major part of the problem has been the institutionalization of DMSE-INSET into the bureaucratic structures of the Ministry of Education in-service activities. Because DMSE-INSET needed the approval of the Ministry in order for teachers to be released for workshops, it became assimilated into accepting the deficiency model of in-service and simply became a provider of resources to help overcome these perceived deficiencies. Thus, DMSE-INSET lost her independence and became seen by the teachers as an agent of the bureaucratic centre which should be resisted.

However, the continuing presence of DMSE-INSET provides an important site for interventions in changing the consciousness of science teachers, vis-à-vis curriculum development and implementation. It has to be realized, though, that this change will not be brought about without struggle as an in-service provider working outside the immediate control of the Ministry's Department of Teacher Training and Development would pose a powerful threat to the authority of the centre. What is needed is the development of a vision for in-service education that places teacher knowledge, awareness and analytical skills about science curricula at centre stage. Only in this way will teachers be professionally developed in a way that encourages them to assume the role of partners in curriculum change and development and to assume responsibility for embedding the aims and objectives of the curriculum in the context of their teaching. There is also an urgent need to disengage DMSE-INSET from the auspices of the Ministry of Education. Unless this can be achieved, DMSE-INSET will not be perceived as independent and working in the best interests of the teachers themselves.

The second area concerns the development of a vision of science education as a profession with the ability to coordinate and control all the activities of the members of the profession. In the United Kingdom, for example, a powerful body known as the Association for Science Education (ASE) plays a key role in all matters pertaining to the teaching of science, especially in curriculum development. This professional association membership is drawn from primary and secondary teachers, technicians, science educators involved in teacher education and students. The major mandate of ASE is to provide an authoritative medium through which the opinions of teachers of science can be expressed at all levels, including policy statements submitted to Parliamentary Select Committees on Science and Technology. In this way, science teachers feel they have a voice in curriculum matters and experience some sense of ownership of their work.

This sense of ownership and voice is currently absent in Botswana. There is an organization known as the Botswana Science Teachers Association (BOTSA) which is nominally in existence but this association is essentially moribund and plays little or no part in representing the voice of teachers in matters concerning their professional work. This has to change if science teachers are to become partners with the Ministry of Education in the development of science curricula and science teaching in Botswana. However, while this professionalization of science teaching in Botswana is an imperative, it is not clear at present where the drive to encourage and implement such an association will come from.

CONCULSIONS AND IMPLICATIONS

This chapter has focused on organizational structures and culture as the key underpinning factors in the context of science teaching in Botswana. Within this framework, several aspects of science education in Botswana have emerged including, among others, the bureaucratic and centralized governance of curriculum and all activities that support it, with teachers serving as powerless functionaries. This has had a negative impact on the quality of science education in Botswana. We contend that the system of governance in Botswana has blinded the Government and educators to the possibilities and opportunities for improvement in the quality of science education that should have been inherent in the system, given the massive investment in human and physical resources and the advances made thus far in curriculum.

It is clear that in the context of Botswana, a new organizational structure and culture is required if the whole enterprise of science education is to achieve its full potential. The culture has to balance the interests of all stakeholders involved in science education: any one partner should not dominate. The culture has to inspire and sustain the new priorities identified by the stakeholders and in the process give a sense of ownership and professional accountability to the science teachers themselves who have the responsibility of implementing changes in the classroom. In fact, the process of enhanced teacher participation in science education matters takes on even greater significance when considering the usually vacuous phrase 'improving the quality of science teaching in schools'. Teacher self-actualization and professionalization become more of a possibility when the freedom to engage in shaping and interpreting curricula increases. However, it has to be said that one might have to contend with the passivity and inertia of the teachers and that of various institutions in the system when they are challenged to take substantial responsibility for curriculum decisions.

REFERENCES AND BIBLIOGRAPHY

Davies, L. (1988). Contradictions of control: Lessons from exploring teachers' work in Botswana. *International Journal of Educational Development*, *8* (4), 293-303.

Fullan, M. (1995). *Change forces*: *Probing the depths of educational reform*. London: Falmer.

Gray, B.V. (1999). Science education in the developing world: Issues and considerations. *Journal of Research in Science Teaching*, *36*(3), 261-268.

Hodson, D. (2003). Time for action: Science education for an alternative future. *International Journal of Science Education*, *25*(6), 645-670.

Hoyle, E. (1975). Planned organisational change in education. In A. Harris, M. Lawn & W. Prescott (Eds.), *Curriculum innovation* (pp. 291-311). London: Croom Helm with the Open University Press. (Reprinted from *Research in Education*, 1970, *3*, 1-22).

Husen, T. (1980). *The school in question*: *A comparative study of the school and its future in Western societies*. Oxford, UK: Oxford University Press.

Ingle, R.B., & Turner, A.D. (1981) Science curricula as cultural misfits. *European Journal of Science Education*, *3*(3), 357-371

Lazarowitz, R., & Hertz-Lazarowitz, R. (1998). Cooperative learning in the science curriculum. In B.J. Fraser & K.G. Tobin (Eds.), *International handbook of science education* (Part 1, pp. 449-470). London: Kluwer.

Lewin, K. (1990). International perspectives on development of science education: Food for thought. *Studies in Science Education*, *18*, 1-23.

National Commission on Education (1977). *Education for Kagisano*. Gaborone: National Commission on Education

Republic of Botswana. (1994) *Revised national policy on education. White paper No. 2 of 1994*. Gaborone, Botswana: Government Printer.

Reynolds, J., & Skilbeck, M. (1976). *Culture and the classroom*. London: Open Books.

van den Akker, J. (1998). The science curriculum: Between ideals and outcomes. In B.J. Fraser & K.G. Tobin (Eds.), *International handbook of science education* (Part 1, pp. 421-447). London: Kluwer.

AFFILIATIONS

Anthony T. Koosimile
Department of Mathematics & Science Education
University of Botswana
Botswana

Robert B. Prophet
Department of Mathematics & Science Education
University of Botswana
Botswana

KASANDA

IMPROVING SCIENCE AND MATHEMATICS TEACHERS' SUBJECT KNOWLEDGE IN NAMIBIA

BACKGROUND

Namibia is one of the youngest democracies in Africa, having gained independence from apartheid South Africa in 1990. During the colonial era, the education system was segregated, fragmented and based upon race and ethnicity. Each one of the 11 Tribal Administrations ran its own education system, resulting in 11 different education systems in the country.

The education system for the white community was much superior to that of the non-whites. The school structures were of superior quality; the teachers were highly qualified and had more teaching and learning materials than was often needed, often for a very small number of learners. The contrary was true for the tribal administrations, even though they catered for a larger segment of the population. The colonial education system placed less emphasis upon mathematics and science education for non-whites. Ramananandan (1995, p. 139) quotes Prime Minister Verwoerd's speech of 1945 thus: "What is the use of teaching a Bantu child mathematics (and science) when it cannot use it in practice?...Education must train and teach people in accordance with their opportunities in life". This official view kept the blacks in an inferior position and as such provided cheap and unskilled labor for the colonial power (World University Services [WUS], n.d.).

Separate development entailed peoples of different races living separately from each other and having their own facilities. Accordingly, the segregated and fragmented education system could not be expected to cater for the educational needs of the majority of the Namibian people nor for an independent Namibia (Angula, 1993). It was also wasteful in terms of human and material resources. It failed in educating the majority of black Namibians to such an extent that 50% to 65% of the black Namibian adults were illiterate (Mshigeni, Duncan, Dzinotyiweyi, Martin & Zikonda, 1993; WUS, n.d.).

Hoey (1989), Turner (1990, 1999), Cunnington (1989) and Clegg (1989) have documented the state of mathematics and science education in Namibia before independence. They describe the teaching and learning of science and mathematics subjects as boring and uninspiring. Rote learning was the norm rather than the exception in schools. Cunnington observed that the training of teachers in Namibia during the colonial era did not help in fostering creativity in student teachers, because it reinforced chalk-and-talk teaching strategies. Consequently, Namibia in 1990 had a large number of both unqualified and under-qualified teachers in the system (UNESCO/UNDP, 1990).

Richard K. Coll, Neil Taylor (eds.), Science Education in Context: An International Examination of the Influence of Context on Science Curricula Development and Implementation, 199–209.

Many of the schools in northern Namibia (comprising more than half the population) did not offer all science or mathematics subjects to students. Clegg (1989) notes that the lack of emphasis on science and mathematics subjects during this period, was due to the implicit assumption by the colonial authorities that these subjects were inherently difficult for blacks. The high failure rates among the few blacks taking these subjects did not help matters at all. Furthermore, a lack of adequately prepared indigenous science and mathematics teachers, and poorly equipped laboratories made it even more difficult for black Namibians to study mathematics and science to any higher levels. Kasanda (1995) notes that this mixture of passive and active discouragement of blacks to study mathematics and science subjects up to matriculation level contributed to the unpopularity of science and mathematics education in the country.

EDUCATION REFORM IN NAMIBIA

Major education reform in Namibia took place immediately after independence. The 11 education authorities were merged into a single unified education ministry catering for all Namibians. The education system policy emphasized equity, quality, equality and democracy, and included life-long learning (Ministry of Education & Culture [MEC], 1993a, 1993b; National Institute for Educational Development [NIED], 2003). This reform also prescribed and endorsed the use of learner-centered methods of teaching instead of the teacher-centered methods that had been in use during the colonial era (MEC, 1993a, 1993b).

Curriculum Changes

Immediately after independence, a vigorous curriculum reform in education took place country wide. The Cape Education Syllabus for Senior Secondary Education was replaced by the Higher/International General Certificate of Senior Secondary Education (H/IGCSE) which was administered by the Cambridge International Examinations Syndicate in collaboration with the Namibian Directorate of National Examinations and Assessment (DNEA). According to Geingob (1993) this was necessary to avoid perpetuating the elitist undertones of the Cape Education system, and hence address the needs of the newly independent country and its people (Angula, 1993). The H/IGCSE was replaced in January 2006 with the Localized Namibia Senior Secondary Certificate Syllabus Ordinary and Higher Levels (NSSCO/H) (Ministry of Education [MoE], 2005), a version of the former H/IGCSE.

Until 2006, curricula were subject based, and not based on the core competencies to be mastered by the learners. NIED, in collaboration with interested stakeholders, is involved in developing the learners' core science and mathematical knowledge, skills and essential competencies to be acquired for the primary, junior secondary and senior secondary phases (i.e., Grades 1 through 12).

The identified competencies and skills are what are expected to be displayed by students upon completion of Grade 12 science and mathematics in Namibia.

Science and mathematics education play an important role in the social and economic development of any given country (Angula, 1989). Mshigeni et al. (1993) observed that developed countries spend more money on science and its related subjects, including mathematics and technology, than developing countries. Without such infusion of funds, little could have been achieved in terms of real development. The scarcity of science and mathematics teachers the world over has been well documented (Turner, 1990; Mshigeni et al., 1993; Omari, 1991). Omari (1991) cites Canada, USA, the UK, and Sweden as some highly developed countries, which lack teachers in the sciences and mathematics. Even countries with strong science and mathematics systems complain about the dwindling numbers of mathematics and science teachers in their classrooms. Consequently, many countries are being forced to put in place strategies to keep teachers in the classroom. Science and mathematics teachers are often in high demand in industry and since schools often struggle to compete in terms of remuneration and conditions of service, teachers are leaving the classroom in large numbers for the private sector (Mshigeni et al., 1993; UNESCO, 2004).

Lack of laboratory equipment, chemicals and textbooks are other causes of ineffective and theoretical treatment of science and mathematics subjects in developing countries, including Namibia. This contributes to teacher frustration, leading to their mass exodus from the teaching profession (Kasanda & Nkosi, 1994). Mshigeni et al. (1993) describe the lack of equipment, facilities, textbooks and so on, as a major hindrance affecting the teaching of these subjects in the Southern African Development Community (SADC) countries. Kitchen (1995) identifies almost the same factors as affecting the delivery of effective instruction in a Guatemalan school.

The quest for effective teaching of science worldwide is probably better illustrated by the Maryland State Department of Education (1991). They initiated the Building Effective Teaching Through Educational Research (BETTER) project in science. They have summarized their research findings in non-technical, and non-threatening language, thereby enabling classroom practitioners to make use of it. This shows that better instruction in science and science-related subjects is not the concern of developing countries only, but of all countries expecting to achieve economic and technological advancement.

POST INDEPENDENCE EDUCATION IN NAMIBIA

At present Namibia follows a 7-3-2 education model. That is seven years of primary education (Grades 1-7), followed by a three year junior secondary education (Grades 8-10) and a two year senior secondary education (Grades 11-12). After Grade 12, learners may secure admission to tertiary institutions in the country and elsewhere. The Government has the responsibility of providing free basic education (Grades 1 to 10) to all its citizens (MEC, 1993a, 1993b).

There is no leaving examination at the end of Grade 7 for admission to Grade 8. A junior school-leaving examination, though, is administered at the end of Grade 10, with successful candidates continuing their studies through Grades 11 and 12. At the beginning of senior secondary school, candidates select and follow subjects leading to the Namibia Senior Secondary School Certificate (NSSC) which is a localized version of the former (Higher) International General Certificate of secondary Education [(H)IGCSE] curriculum.

The learning of science and mathematics subjects in Namibia is compulsory up to Grade 10. After Grade 10 students are allowed to make selections of fields of study, which may not necessarily include a science or a mathematics subject. It should be pointed out that learners who opt not to study science and mathematics subjects at the senior secondary level, do select themselves out of science-related career fields. This is due to the fact that passes in the senior secondary school in science and mathematics subjects are prerequisites to the further study of science and mathematics subjects at the university level. Without passes in these subjects, a student will often not be admitted to science faculties even though he/she obtained higher grades in other subjects in their Grade 12 examinations.

SCIENCE AND MATHEMATICS EDUCATION

Science and mathematics education enjoys a high priority in Namibia. Several attempts to popularize the study of these subjects with the majority of Namibians who were denied the chance to study these subjects have been put in place. The establishment of the President's Endowment Fund Bursary Scheme for the study of science and technology at the tertiary level for deserving learners is a further testimony to the importance the country attaches to the study of science and mathematics subjects.

As a consequence of this deliberate policy of encouraging the study of science and mathematics subjects in secondary schools, the number of candidates studying science subjects has increased. For example, while only 1,188 and 726 candidates studied mathematics and physical science at Grade 10 in Namibia in 1988, these figures increased to 23,983 in 1993 for both subjects. These increases are a reflection of Namibia's commitment to the study of science and mathematics.

But, the amount of money and effort being spent on the study of science and mathematics education in the country notwithstanding, the effectiveness of science and mathematics education continues to be adversely affected by a number of factors; these are discussed in more detail below. Nonetheless, suffice it to say that unprecedented numbers of students taking science and mathematics subjects at secondary school level since independence has also contributed to the unsatisfactory teaching of these subjects in some schools, and has resulted in derailing the Government's efforts to make Namibia a knowledge-based society.

TRAINING OF SCIENCE AND MATHEMATICS TEACHERS

In Namibia, teacher training for basic education takes place at four colleges of education (three in the densely populated North East, North and the Caprivi regions, and the fourth in Windhoek), while that for senior secondary school teachers takes place at the University of Namibia.

The candidates at the four teachers' colleges follow a three year Basic Education Teacher Diploma (BETD), which was started in 1993. During their three years of study, candidates are generally required to follow common foundation studies in the first year with mathematics and integrated natural science each accounting for 12.5% of the learning time. In the third term of the second year, the students are expected to select a subject area of specialization related to the chosen phase of Basic Education (Grades 1-7 or 5-10). During this period mathematics/science takes up 45% of the study time. Unfortunately, the mathematics and science specialization is optional, and a number of candidates opt not to specialize in it because of their low grades in these subjects in the H/IGSE examinations and the fear of failing. In the third year, candidates continue with the specialization started in the second year. In the third year, the specialization subject now accounts for 50% of the time. This is to enable the teachers to know and understand their teaching subjects thoroughly (MEC, 1994a). Teachers who graduate with the BETD often teach either Grades 1-7 or Grades 5-10.

Since these graduates are expected to be the backbone of popularizing science and mathematics education at the primary and junior secondary levels, their 'in-depth knowledge' should be equal to or above that they are required to teach. Until recently, the emphasis in the BETD was on 'how' to teach and not on 'what' to teach, making it difficult for those graduating with a junior secondary specialization to teach mathematics and science subjects at the senior secondary school in times of teacher shortage. Erwee and Kasanda (2006) observed that the subject content knowledge of the final year BETD mathematics trainee teachers was very low. Indeed, many failed to perform at a level commensurate with their specialization. Further, Grade 6 mathematics learners and teachers' low subject knowledge was attested to by the study of the primary school teachers and learners in 14 countries by the Southern Africa Consortium Monitoring Education Quality (SACMEQ) that placed Namibian learners last and teachers second last (MBESC, 2004). To be effective and confident mathematics and science teachers, more subject content understanding is necessary.

FACTORS HINDERING THE EFFECTIVE TEACHING OF MATHEMATICS AND SCIENCE SUBJECTS IN NAMIBIA

The importance of science and mathematics education for the social and economic development of a given society can not be over-emphasized. Nonetheless, for Namibia, a number of problems appear to be hindering the effective teaching and learning of science and mathematics subjects including the use of learner-centered teaching. Some of these are addressed here.

Inadequate Numbers of Qualified Science and Mathematics Teachers

One of the major problems experienced in offering effective instruction to students in science and mathematics education is the lack of adequately qualified and professionally prepared teachers at secondary level. The World University Services (WUS, n.d.) and UNESCO/UNDP (1990) observed that 80% of the more than 12,000 schoolteachers in Namibia were untrained. Given such a large number of unqualified or ill-equipped teachers in schools with the majority of the school population, one fails to see how effective instruction could be carried out (Clegg, 1989; Cunnington, 1989). The situation as portrayed in 1990 has to a large extent improved with the introduction of the BETD and BEd programs for the preparation of teachers.

Even though a significant number of the newly trained teachers are still in schools in Namibia, a number of the experienced teachers have been leaving the profession for 'greener pastures' in the private sector, or have been promoted to non-teaching administrative positions in the Ministry of Education. Provision of a higher salary and related fringe benefits may forestall the exodus from the teaching profession before it becomes endemic and renders insignificant the gains achieved so far in providing qualified science and mathematics teachers for the schools.

Lack of Laboratory Space and Laboratory Equipment

In most schools, especially in the rural areas, laboratories lack even the most basic equipment for carrying out simple experiments (Mshigeni et al., 1993; UNESCO, 2004; West, 1995). Even though practical/experimental work in science does not necessarily ensure understanding or meaningful learning, lack of such equipment reduces the teaching and learning of science to an abstraction. The theoretical teaching of science and mathematics subjects pushes them beyond the students' realm of experience and understanding. They invariably perceive the subjects as being useful for passing the examination only rather than as part and parcel of their daily lives. Such a view of science and mathematics will likely only fuel the negative attitudes students hold towards the study of these subjects in our schools.

Lack of Textbooks, Other Teaching and Learning Materials, and Teaching Methods

Namibia has endorsed a learner-centered approach, which places the child as the center of all learning while the teacher plays the role of facilitator, learner and advisor (Branch, 1995; MEC, 1993a, 1993b). Swarts (1995) notes that a number of teachers appear to be familiar with a learner-centered approach, some of whom had earlier either passively or actively opposed its implementation. NIED (2003) notes that teachers are familiar with a learner-centered method but do not implement it as envisaged, while van Graan (1998) notes that the implementation of learner-centered teaching was slow and that for most teachers learner-centered teaching simply implied group work. It should be pointed out that learner-centered teaching is material and equipment intensive, and requires confident and content versed teachers for successful implementation in the classroom. Therefore, the lack of these essential materials and equipment may reduce learners' active participation in the learning of science and mathematics and result in rote learning. Further, a learner-centered approach to teaching may not be embraced by an overworked teacher (Kasanda, 1981; 1993).

Inadequate Funding for the Purchase of Educational Requisites

Financing education is another factor that appears to affect the provision of better instruction in science and mathematics subjects. This is mainly due to the fact that science subjects often require equipment and materials to enable hands-on experience. Namibia spends approximately 20% of the national budget on education (Mshigeni et al., 1993). But, West (1998) noted that most of this goes to pay salaries and not on infrastructure or consumables. The limited budget for the purchase of consumables and provision of infrastructure for the teaching and learning of science and mathematics in schools reflects the existence of competing societal demands and in the face of these demands education is most likely to come out second best, since its 'products' are not observable in the short term (Omari, 1991). The effects of such cuts are not immediately obvious in the products of the education system. Nonetheless, cuts in education expenditure invariably result in impoverished teaching, especially for the sciences; because chemicals, equipment and other needed resources will not be available in adequate quantities in the schools to cater for practical work which is necessary for enhancing learners' understanding of the subject content. It should be noted, though, that practical work in science subjects in many Namibian schools does not take place due to lack of equipment, large class size, examination pressure, lack of suitably qualified teachers, lack of self confidence by the science teachers, especially older teachers who trained during the colonial era (UNESCO, 2004; van Graan, 1998).

IMPROVING SCIENCE AND MATHEMATICS TEACHERS' CONTENT KNOWLEDGE

The problems encountered in the teaching and learning of science and mathematics subjects has led to a variety of attempts to address them. In Namibia, the introduction of the Mathematics and Science Teachers' Extension Programme (MASTEP) in 2003 has been one of the latest efforts in attempts to improve the subject content knowledge of the mathematics and science BETD trained teachers, a recognition of the fact that teachers make a difference in the learning process (NIED, 2003).

As noted above, the paucity of well-trained teachers is more acutely felt in the areas of mathematics and science education, especially at the higher secondary school level in almost all countries in Africa and the outside world (Haambokoma, 2003; Pabale & Dekkers, 2003). Benson (1994) notes that many science and mathematics teachers in South Africa are under-qualified, and are themselves the product of a ‘disadvantaged education system’. In Namibia, the lack of adequately trained mathematics and science teachers can also be attributed to the colonial legacy of not encouraging the study of mathematics and science subjects among the black majority.

As indicated earlier, the Mathematics and Science Teachers’ Extension Programme (MASTEP) was created to improve the quality of science and mathematics education at secondary level in the country through improving the content understanding and teaching methods of the teachers in service. The program is an example of a practitioner-led development, defining, stimulating and implementing an innovative practice in the preparation of senior secondary school mathematics and science teachers in Namibia. The content and activities of the MASTEP initiative have been tailored to produce an effective mathematics and/or science teacher for Namibian schools. A variety of delivery methods are used in this program and include distance learning, face-to-face tuition and the Professional Development Placement. In addition, the MASTEP has also successfully introduced the participants to the use of the learner-centered approaches favored by the Ministry (MEC, 1993a, 1993b). Nonetheless, the structures in the Namibian education system may be discouraging many newly trained mathematics and science teachers from applying learner-centered methods (van Graan, 1998).

At the inception of the MASTEP program, the aims and objectives to be achieved were clearly stated. The program’s curriculum comprises devoting 66% of the time to content study; 22% for professional study with emphasis on learner-centered methods and 12% on the Communication and Study Skills content. The emphasis on content is a deliberate move to give the BETD graduates enough subject content, and hence self confidence, to teach mathematics and science subjects at the senior secondary school level.

MASTEP uses successful practicing teachers in secondary schools and the education lecturers at the University of Namibia to teach both the content and the methods. Thus, effective ways of how to deliver content are shown to the participating teachers and they are expected to put these skills and knowledge into practice during the Professional Development Placement period.

The MASTEP teachers tend to use more everyday context in their teaching of mathematics and science because of the materials (modules) used during training and given to each in-service student. These materials were prepared by College of Education staff, University of Namibia lecturers and University of York staff in such a way as to drive teachers to use their local contexts to teach the subject content. Further, the use of experienced host teachers provides further help in terms of appropriateness of content to be taught and how to effectively teach it back in their schools. As Benson (1994) notes, lack of professional support for teachers may produce a sense of isolation and a sense of powerlessness in teachers to solve some problems they face at school on their own. Accordingly, the support provided by the host teachers in the MASTEP program gives these teachers the assurance that they are not alone and that help is always available to them.

PROFESSIONAL DEVELOPMENT PLACEMENT

The Professional Development Placement (PDP) is four weeks (two separate fortnights) long, during which period the MASTEP teacher observes and teaches in a senior secondary school, often away from his/her actual school, which might be an upper primary school or combined school. The PDP is different from the usual school teaching practice often required of pre-service teachers. Under the watchful eye of the host teachers, the MASTEP teacher takes over and teaches an IGCSE or senior secondary mathematics or science class. An evaluation of the PDP experience (van der Zwan, 2003) found that the PDP makes a positive difference to the experiences of the MASTEP teachers.

USE OF THE IMMEDIATE ENVIRONMENT

The use of the immediate environment in the teaching and learning of mathematics and science in Namibian schools is currently under-utilized by many Namibian teachers. Even though during their training the student teachers are introduced to improvisation of teaching and learning media, most tend not to use these upon graduation. For many mathematics and science teachers, the use of commercially produced teaching and learning materials is the norm. Absence of these in the schools results in the overdependence on textbooks and, tends to result in teacher-centered teaching. The pressure of the examinations and the need to prepare learners for these examinations appears to further reduce the use of examples from the immediate environment to enhance learners' understanding of the subject matter.

CONCLUSIONS

In conclusion, it should be emphasized that education in general, and science and mathematics education in particular, play an important role in raising the general living standards of the people. It is probably this realization which has compelled nations the world over to spend more money in improving the quality of the teaching and learning of mathematics and science education. In Namibia, mathematics and science teachers are being provided with the subject content that should give them the self confidence that they need to teach this content effectively to their learners. The use of learner-centered methods based on the constructivist theory (NIED, 2003), the use of local contexts as a vehicle for the learning of science and mathematics at both the primary and secondary levels, although not fully embraced by all teachers, will in the long run popularize the learning of these subjects in Namibia. Indeed, the teacher's knowledge and ability to explain the content clearly to his/her learners are key to learners' desire to learn mathematics and science subjects beyond the compulsory 16 years of basic schooling.

REFERENCES AND BIBLIOGRAPHY

Angula, N. (1989, September). *Teacher education and national building; Namibia in transition.* Keynote address, International Conference on Teacher Education for Namibia. Lusaka, Zambia.

Angula, N. (1993). Senior secondary education in independent Namibia - Policy issues. In *Proceedings of the National Conference on IGCSE and HIGCSE. Swakopmund.* Windhoek, Namibia: Ministry of Education and Culture.

Benson, H. (1994). *A history of CASME.* 1985-1994. Retrieved 20 June 2004, from http://www.hula.org.za/article_detail.asp?Article_ID=159.

Branch, R.C.M. (1995, August). *A paradigm for learner-centered instruction.* Paper presented at the sixth BOLESWA Symposium, Gaborone, Botswana.

Clegg, A. (1989). *Science and mathematics education in Namibia. Report of a Specialist Tour to Namibia from July 26 to August 13th.* Paper presented at the International Conference on Teacher Education for Namibia. Lusaka, Zambia.

Cunnington, J.L. (1989). *Science in Namibia: With particular emphasis upon the preparation of teachers for the primary age phase.* Paper presented at the International Conference on Teacher Education for Namibia. Lusaka, Zambia.

Erwee, S.P., & C.D. Kasanda (2006, May). *Assessment scores in mathematics JSC examination in Namibian schools: Do they tell the whole story?* Paper presented at the Mathematics Congress. Swakopmund, Namibia.

Geingob, H. (1993). Opening address. *Proceedings of the National Conference on IGCSE and HIGCSE.* Swakopmund, Namibia.

Haambokoma, C. (2003). Feasibility of providing training in teaching through distance mode to untrained graduate teachers of mathematics and science. In B. Putsoa, M. Dlamini, B. Dlamini & V. Kelly (Eds.), *Proceedings of the 11th annual SAARMSTE Conference.* Waterford, Kamhlaba VWC.

Hoey, S. (1989). *Mathematics education in schools and colleges in Namibia.* Paper presented at the International Conference on Teacher Education for Namibia. Lusaka, Zambia.

Kasanda, C.D. (1981). The plight of mathematics teachers in Zambia. *ZAME Bulletin, 13*(2), 2-10.

Kasanda, C.D. (1993, July). *School organization and problems encountered in teaching mathematics in Zambian secondary schools.* Paper presented at the Lesotho Educational Research Association Conference. Maseru.

Kasanda, C.D. (1995, August-September). *The science and mathematics imperative in the new curriculum in Namibia: A change of focus.* Paper presented at the SASE Conference. Windhoek College of Education, Windhoek, Namibia.

Kasanda, C.D., & Nkosi, M. N. (1994). Factors contributing to burnout of mathematics teachers in Zambian secondary schools. *Southern Africa Journal of Mathematics and Science Education, 1*(2), 25-36.

Kitchen, R. (1995). Mathematics pedagogy in the 3rd world: The case of a Guatemalan teacher. *International Study Group on Ethnomathematics Newsletter, 10*(2), 1-4.

Maryland State Department of Education. (1991). *BETTER Science.* Baltimore, MD.

Ministry of Basic Education, Sport and Culture. (2004). *The SACMEQ II report: A study of the conditions of schooling and the quality of primary education in Namibia.* Windhoek, Namibia: MBESC.

Ministry of Education. (2005). *Namibia senior secondary certificate* (NSSC). Okahandja, Namibia: NIED.

Ministry of Education and Culture. (1993a). *Toward education for all. A development brief for education, culture and training.* Windhoek, Namibia: Gamsberg Macmillan.

Ministry of Education and Culture. (1993b). *The pilot curriculum for formal secondary education.* Windhoek, Namibia: MEC.

Ministry of Education and Culture. (1994). *The broad curriculum for the Basic Education Teacher Diploma.* Windhoek, Namibia: MEC.

Mshigeni, K.E., Duncan, T., Dzinotyiweyi, H.A.M., Martin, C.A., & Zikonda, A.R. (1993). *Unpublished report of the study of science and technology of the Southern African Development Coordination Conference.*

National Institute for Educational Development. (2003). *Learner-centered education in the Namibian context.* Okahandja, Namibia: NIED.

Omari, I.M. (1991). *Higher education at crossroads in Africa*. Nairobi, Kenya.
Pabale, M.F., & Dekkers, P. (2003). Science teachers' views on outcome-based education: Partial evaluation of UNIN's Advanced Certificate in Education Programme. In B. Putsoa, M. Dlamini, B. Dlamini & V. Kelly (Eds.), *Proceedings of the 11th annual SAARMSTE Conference*. Waterford. Kamhlaba VWC.
Ramananandan, K. (1995). My IGCSE experience as a teacher of science. In C.D. Kasanda & F.A. Phiri (Eds.), *Proceedings of the (H)IGCSE colloquium on teacher education* (pp. 139-145), Windhoek, Namibia: University of Namibia in collaboration with the University of Cambridge.
Swarts, P. (1995). The new senior secondary system in Namibia: Achievements and challenges. In C.D. Kasanda & F.A. Phiri (Eds.), *Proceedings of the (H)IGCSE colloquium on teacher education* (pp. 6-20). Windhoek, Namibia: University of Namibia in collaboration with the University of Cambridge.
Turner, J. (1990). *Education in Namibia. Report of a consultancy*. Windhoek, Namibia: Ministry of Education and Culture.
Turner, J.D. (1999). *Higher education in Namibia. Report of a Presidential Commission*. London: Commonwealth Secretariat.
UNESCO (2004). *Partnership for relevant science and technology*. Paris: UNESCO.
UNESCO/UNDP (1990). *Namibia in-service teacher training project. Project formulation framework*. Windhoek, Namibia: UNESCO.
Van Graan, M. (1998). Learner-centered education: Equal to group work? Findings from Namibian classrooms. In T. Squazzin & M. van Graan (Eds.), *Education reform and innovation in Namibia: How best can changes in classroom practice be implemented and supported?* (pp. 52–65). Okahandja: NIED.
Van der Zwan, F. (2003). *Professional Development Placement: Does it make a difference?* Unpublished Master of Education thesis, Vrije Universieit, Amsterdam.
West, R.C. (1995). Teacher preparation, in-service training and resourcing. In C.D. Kasanda & F.A. Phiri (Eds.), *Proceedings of the (H)IGCSE colloquium on teacher education* (pp. 21-32). Windhoek. University of Namibia in collaboration with the University of Cambridge
West, R.C. (1998). Teacher preparation, in-service training and resourcing. In K. Legesse & B. Otaala (Eds.), *Performance at (H)IGCSE in Namibia, 1995-1997: Implications for teaching and learning*, (pp. 27–35), Windhoek, Namibia: UNAM.
World University Services. (n.d.). *Namibia: Education in conflict*. Nottingham, UK: Russell.

AFFILIATION

Choshi D. Kasanda
Department of Mathematics, Science and Sport Education
University of Namibia
Namibia

SCIENCE AND MATHEMATICS EDUCATION IN ETHIOPIA: POLICY, CURRICULUM AND IMPLEMENTATION

INTRODUCTION

Ethiopia has an area of 1.1 million square kilometers, with an estimated total population of 70 million consisting of diverse languages, culture and topography. The vast majority (85%) of the population of Ethiopia is living in rural areas and earns their living from rain-fed agriculture which constitutes 52% of GDP. The country is one of the poorest in the world, with per capita annual income less than 100US$, and with some 44 % of the population living below the poverty line. The male and female proportions of the population are about the same, with 50.1% of the population male and 49.9% female (Ministry of Education [MoE], 2005).

HISTORY AND CONTEXT OF EDUCATION IN ETHIOPIA

Ethiopia had its own indigenous education and curriculum prior to the introduction or emergence of modern education and curricula. Though concepts of the terms such as 'curriculum' and 'policy' were not well known or widely used at government level, there were social institutions dedicated to the development of these ideas in traditional Ethiopia (Areaya, 2007). The first attempt to unify curricular offerings at primary level of Ethiopia was made in 1947, and the Elementary School Curriculum: Grade I-VI developed by a committee consisting largely of foreign staff of varying nationalities (Areaya, 2007) was published in that year. This first formal curriculum had a 6-6 structure with six years of elementary school followed by six years of secondary school. Among other subjects, the curriculum included science and arithmetic.

Since the establishment in 1908 of the first modern school in Ethiopia, three attempts at educational reform have been made by the three most recent governments. Early in the 1950s, the Government sensed general dissatisfaction with the education system of the imperial regime and initiated a study called the Education Sector Review (ESR) in 1971 (Engida, 2002). It was suggested at this time that the rural population be made the target of educational policy. However, this reform process was abandoned in September 1974 when The Dergue (a military group led by Mengistu Haile Mariam) began to rule the country. The Dergue argued in 1982 that "the education policy of the Imperial regime was elitist (favoring some regions and urban areas) and that the curriculum did not take into account the concrete conditions in the country" (Negash, 1996, p. 107).

Richard K. Coll, Neil Taylor (eds.), Science Education in Context: An International Examination of the Influence of Context on Science Curricula Development and Implementation, 211–222.

Accordingly, this Government then launched a reform study called Evaluative Research of the General Education System in Ethiopia (ERGESE) in 1983. Its influence, however, was masked by the Ten Year National Perspective Plan (1984-1994) that was launched by the Government in 1984, and that set policy statements for education (Tefera, 1996).

The current Government of Ethiopia also noted the inadequacy of the education system to prepare learners for useful participation in the community (The Transitional Government of Ethiopia [TGE], 1994a). The Government subsequently developed a new Education and Training Policy that states, among other things, that the education system "is entangled with complex problems of relevance, quality, accessibility and equity" (TGE, 1994b, p. 2). Within the framework of this policy, Ethiopia launched the first Five year Education Sector Development Program (ESDP-I) in 1997. According to the Ministry of Education:

> The main thrust of ESDP was to improve educational quality, relevance, efficiency, equity and expand access to education with special emphasis on primary education in rural and underserved areas, as well as the promotion of education for girls as a first step to achieve universal primary education by 2015. (MoE, 2005, p. 4)

Ethiopia has now launched the Education Sector Development Program III, which will span five years (2005/06 to 2009/10). The ESDP-III (MoE, 2005) states that the education sector's vision is to see that all school age children gain access to quality primary education by the year 2015, and the creation of trained and skilled human power at all levels who will be driving forces in the promotion of democracy and development in the country. The mission statements derived from the above stated vision are to:

- Extend quality and relevant primary education to all school age children and expand and standardized education and trainings at all levels to bring about rapid and sustainable development with increased involvement of different stakeholders (community, investors, NGOs, etc.);
- Ensure that educational establishments are production centers of all rounded, competent, disciplined and educated human power at all levels through the inclusion of civic and ethical education with trained, competent and committed teachers;
- Take affirmative action to insure equity of female participation, pastoral and agro-pastoral and those with special needs in all education and training programs and increase their role and participation in development; and
- Expand non-formal education training programs to out of school age children and adults to enable them actively participate in poverty reduction.

HARMONIZING POLICY, CURRICULUM AND IMPLEMENTATION IN SCIENCE AND MATHEMATICS

There is considerable consensus among policy-makers, educationists, and curriculum experts in Ethiopia that there is a natural and dynamic relationship within issues and practices inherent in policy, curriculum, and implementation process (Areaya, 2007). In other words, the implementation of a policy or curriculum is always influenced by the nature of that policy or curriculum.

There is also a worldwide view that improving science education should be regarded as a priority for developing countries in order to promote long-term economic development. To this effect, there are a number of initiatives aimed at improving science education in developing countries. However, all too often, the focus of such initiatives is limited to the *development* of science curricula, while the details of how the curricula will be *implemented* at school level are often neglected (Rogan & Grayson, 2003). It also is argued that although there are many well-intentioned and well-designed curriculum reform programs in the developing world, many fail to take root. One of the most important reasons for this failure seems to be a lack of clearly thought out implementation strategies that take into account the local context, including diversity that may exist within that context and psychological factors that influence learning and change.

At present, Ethiopia does not have a separate science and mathematics education policy. It has, however, a general education and training policy that is expected to guide the country's overall education and training practices. This policy was developed by the Transitional Government of Ethiopia (TGE) in 1994. The Education and Training Policy (ETP) argues strongly that the objectives and relevance of education in Ethiopia in the last 30 years have become questionable. The policy further points out that: "The science and cultural components are weak and inadequate to prepare the learner for useful participation in the community" (TGE, 1994b, p. 1). The training of science teachers was also reported to be inadequate in terms of equipping student teachers with the kind of realities they face in Ethiopian schools (Engida, 2000).

Currently, science and mathematics are offered throughout the Ethiopian educational system. For instance, the general objective of the Ethiopian primary school science is to make the curriculum relevant to the everyday life of the learners. Local curriculum experts have attempted to implement this general objective by introducing separate science (biology, chemistry and physics) courses starting from Grade 7 (about age 13 years). Mathematics is offered to all students starting from Grade 1 (about age 7 years). The emphasis of the objectives of the primary school science curriculum on relevance maybe considered as a response to the longstanding criticisms of African science and mathematics education programs. However, the question is: *To what extent is this aim translated into appropriate contents and instructional strategies?* Educational research in Ethiopia suggests that little has been achieved in this regard. Context-specific issues seem under-represented in the curriculum - to the advantage of hard science (Engida,

2002). Ignoring the context-specific issues may, in turn, hinder the pupils' attempt to link school science with everyday experiences.

Areaya (2007) argues that, as a result of the current Education and Training Policy of Ethiopia, the quality of secondary education in general, and science and mathematics education in particular, has declined to the extent where the majority of secondary school graduates are not fit for university level education. One reason for this may be that the Policy is too ambitious. For example, it proposes that the four year secondary education is completed within two years, and the four year university program in three years. Although policy statements advocate the development of problem-solving skills, there is little evidence this is achieved in the implementation processes. According to Engida (2002), for instance, the chemistry curriculum is still organized under the major theoretical concepts. The implicit objective here, as opposed to the stated curricular/educational objective, is thus to teach chemistry to beginners as a 'true' subject of facts, and not as a means of properly understanding their immediate environment. The content is thus a better fit to the objective of the old high school chemistry curriculum - providing a solid foundation of chemistry to students. The aim of a culturally-relevant curriculum/policy should not be to impart basic factual science to children, both in its statements of objectives and contents and approaches (Engida, 2002).

Areaya (2007) also argues that it is not difficult to see resemblance to traditional educational practices among what are supposedly modern and contemporary educational practices in Ethiopia, particularly with regard to curriculum organization and methods of teaching. Areaya's (2007) in-depth study of two regional secondary schools (Debub & Oromia) identified a number of gaps, which seem to have existed since the inception of the ETP. There is a gap, for instance, between what the policy/curriculum aspires to, and what actually happens at regional, district, and school levels. There is also a gap between central and regional governments' perceptions of their role and role relationships, particularly with regard to secondary education.

It seems, then, that the formulation of an educational policy needs to anticipate and work on strengths, weaknesses, opportunities, and threats so that the educational system harmonizes the resulting curriculum development and implementation processes.

REVISITING PLASMA TELEVISION SATELLITE INSTRUCTION IN ETHIOPIAN SECONDARY SCHOOLS

After eight years of implementation of the new secondary school curriculum, the Government of Ethiopia decided in 2004 to use information and communication technologies (ICTs) as an alternative means for implementing the curriculum. The intention was to resolve quality and equity issues that the education system had faced for some time (Areaya, 2007). Furthermore, its intention was to give all secondary schools in the nation opportunities for uniform and standardized instruction consistent with the curriculum. Subsequently, the MoE designed six secondary school subjects; English, mathematics, physics, chemistry, biology, and civic education, all of which were to be taught via satellite and appear on plasma televisions. The reason for giving priority to mathematics and science subjects was that mathematics, for example, because of its theoretical nature, could be presented using computer animations - especially for geometry - therefore was well suited to such instructional technologies; something most mathematics teachers fail to do in class. Likewise, science instruction in most Ethiopian secondary schools, for a variety of reasons, seldom involves students doing experiments. But satellite technology could allow teachers to demonstrate quality, interesting, experiments (EMA, 2003, cited in Areaya, 2007).

Educational reforms worldwide often face resistance, particularly at the early stage of inception. Sometimes resistance is based on valid theoretical and practical concerns; other times, not. In the same way, there has been resistance to the use of the instructional television system in Ethiopia all around the country. This led the Ethiopian Government to launch an internal evaluation of the instructional system by the Ministry of Education. In addition, the Ministry approved an external evaluation of instructional television system. Although both the internal and external evaluations of the television instructional system are expected to come up with official and feasible recommendations about the fate of the system in the near future, it is also useful to consider the findings of other local independent research. As might be expected, recent research identified pros and cons of instructional television.

Research into teaching biology in a secondary school of eastern Ethiopia suggests that lessons were "completely teacher-centered or more specifically television, lecture-centered and completely unidirectional" (Lemma, 2006, p. 80). The major differences between the instructional television-based lessons and the regular teaching are that in television-based lessons:

- The lectures were prepared in a foreign land;
- The instructor on the TV was a white lady;
- The lessons were done on the TV and broadcast via a satellite system with state-of-the-art technology;
- Lessons were communicated through very good, but alien, South African English;
- The students were completely passive throughout the program;
- The teacher at one school observed in eastern Ethiopia was passive, except for instances at the beginning and end of TV lessons, i.e., only when it was possible to be active; and
- Most students have started to become punctual because of the TV lessons. It looks like they are aware of the risk that once they miss the TV lessons they miss them forever. (Lemma, 2006, pp. 80-81)

Overall, it seems that biology instruction using television "seriously scales down the involvement of teachers in curriculum development and implementation, their place in the community and the ethical values they should represent and guide in their communities" (Lemma, 2006, p. 86).

Similarly, recent research into the instructional television practice in science and mathematics suggests that the use of television-based instruction to *supplement* normal teaching activities, rather than using it as the sole teaching tool, is the more acceptable to secondary school communities (Areaya, 2007). It also seems there is often inconsistency between the policy implication for expected instructional processes and the actual practices in secondary school lessons. For example, the policy asserts that the essence of learning is the *engagement* of students in meaningful experiences, and that the instructional process in classrooms should shift from passive learning, to active problem-solving. However, neither instructional television classroom practice nor conventional teaching seems to be consistent with this policy in general and the curriculum in particular.

Research into teachers' and students' views of the benefits and difficulties of teaching via instructional television reveals mixed views about such instruction. On a positive note, instructional television allows teachers to demonstrate experiments when they lack facilities (Areaya, 2007). That most science laboratories in Ethiopian secondary schools are short of qualified and committed science teachers, justifies the use of instructional television as one medium of instruction. As noted above, it affords the use of computer animations for demonstration of some experiments making them accessible to all schools regardless of their geographic location.

On the other hand, science and mathematics teachers point to a number of problems in using instructional television for lessons. The main issues raised are that instructional television can make teachers 'idle'; just watching instructional television-based lessons, students find the mode of delivery too fast, and it employs language beyond their capability (Areaya, 2007).

Overall it seems that the use of information and communication technologies in the form of instructional television can support meaningful learning when used appropriately. One enabling factor would be to give the local teacher control of the 'what', 'when' and 'how' of instructional television lessons. If teachers could, for example, have some control of the program, be provided with transcripts so they can help the students before lessons start, or have video or DVD recordings so that teachers can stop, start, and interact with students, this would help with a) language difficulties, and b) may serve to revitalize teachers - so that they do not see themselves as being confined to people who turn the television on and off! A particularly important issue is that of the language of instruction; partly due to students' weak proficiency in English. It seems the use of English language as a medium of instruction in the instructional television-based lessons has not helped the majority of students to learn meaningfully and benefit from the program, but rather it has served as a distraction. This suggests it is appropriate to reconsider the use of English as a medium of instruction in general, in television instruction in particular.

CONTEXTUALIZING CURRICULA THROUGH A SCIENCE-TECHNOLOGY-SOCIETY (STS) APPROACH

The last decades represented turbulent times in the history of science education (Ajeyalemi, 1990). Societal, scientific and technological changes have forced a rethinking of what is 'basic' in science education. While earlier science programs have been criticized for their emphasis on the structure of the discipline, more and more aims of science education are formulated to include messages that indicate that students of science should be aware of the practical applications of the science content as well as the resulting social implications.

Examination of the content of old and new in primary and secondary Ethiopian school chemistry textbooks for inclusion of STS themes shows some interesting trends (Engida, 2005). Based on Piel's classification of STS themes, (Chiang-Soong & Yager, 1993) it seems STS appears in all such textbooks (see Table 1, Figures 1 & 2).

Table 1. Analysis of Ethiopian Chemistry Textbooks for STS Themes

Grade	*Year of Publication*	*Type*	*Author/s*	*Publisher*
7	1999	New	R14EB[1]	Mega PE[2]
8	1999	New	R14EB	Mega PE
9	1999	New	R14EB	Mega PE
9′	1989	Old	ICDR[3]	EMPDA[4]
10	1989	Old	ICDR	EMPDA
11	1993	Old	ICDR	EMPDA
12	1983	Old	ICDR	EMPDA

[1]Region 14 Education Bureau, [2]Mega Publishing Enterprise, [3]Institute of Curriculum Development and Research, [4]Educational Materials Publication and Distribution Agency.

Although there are certain STS issues in all Ethiopian textbooks, most of the issues are limited to describing the uses of materials (e.g., chemical substances such as metals, nonmetals & oxides) in industry and, in some cases, in everyday life. This is particularly true in the case of the new Grade 8 chemistry textbook, which contains the highest proportion of STS issues.

For developing countries like Ethiopia, due consideration might be given to food production, family planning, energy sources and uses, utilization of natural resources, and so on. Unfortunately, however, there is little about such locally-relevant topics in Ethiopian chemistry textbooks. In other words, the space devoted

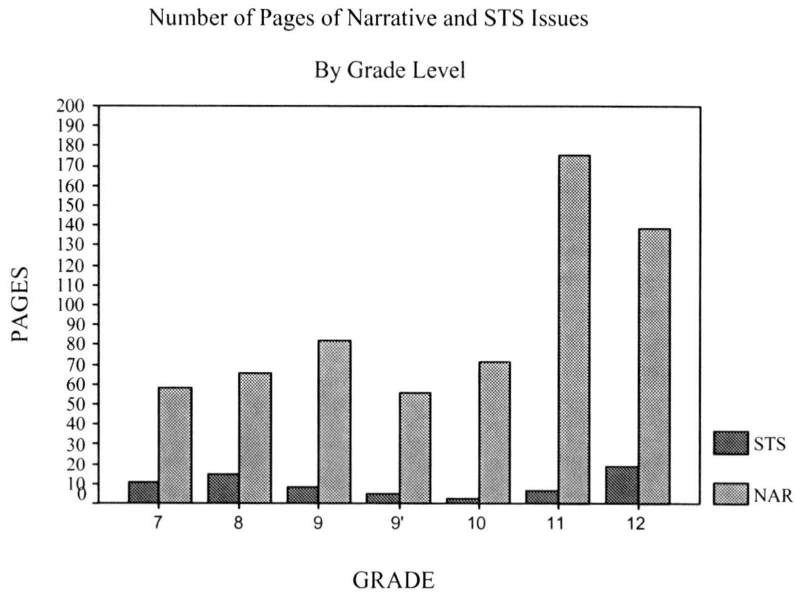

Figure 1. Pages of narrative (NAR) and science-technology-society (STS) issues and topics according to Piel's classification

For developing countries like Ethiopia, due consideration might be given to food production, family planning, energy sources and uses, utilization of natural resources, and so on. Unfortunately, however, there is little about such locally-relevant topics in Ethiopian chemistry textbooks. In other words, the space devoted to STS topics, though large overall, is very small for issues that might encourage students to interact with their local environment and local technology (Engida, 2005). It should be noted, however, that the old Grade 12 chemistry textbook has a reasonable distribution of STS topics.

With regard to the STS themes across grade levels, it seems that there is no clear trend as one goes from Grade 7 through 12 (Figure 2). What is seen is that the primary school chemistry textbooks contain more STS themes than the secondary school texts, irrespective of the type of specific STS themes. On the other hand, although there is a gap of about 10 years between the old and the new Grade 9 chemistry textbooks, there are few differences in the inclusion of STS issues. Considering the attention given to the STS movement in science education in the last two decades internationally, this is surprising. Using Piel's STS categories, irrespective of proportion, sociology of science is the only topic covered by all Ethiopian textbooks at all grade levels, with the new Grade 9 textbook having the maximum representation.

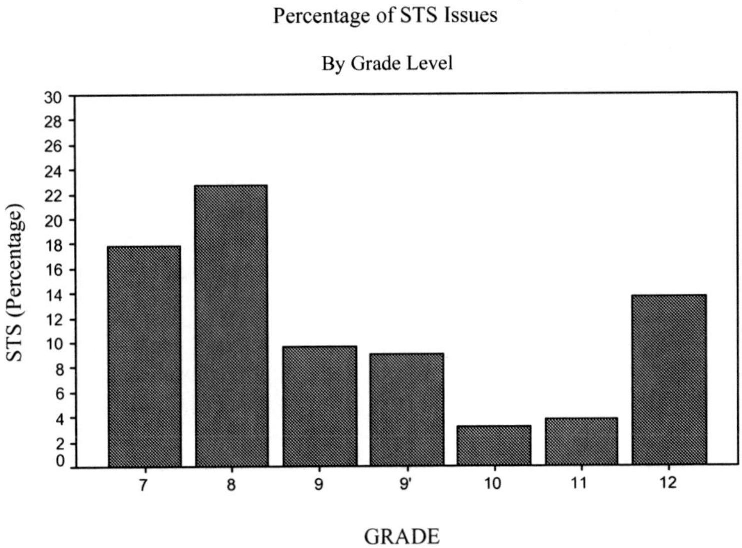

Figure 2. Proportion of STS topics (%) by grade level

If we are to realize the notion of STS, then it needs to be localized to the educational context in which the learner confronts scientific concepts and local technology. How this might be done in the case of Ethiopian school chemistry is presented below. Inclusion of STS issues in school chemistry could thus involve three major components.

Identifying and organizing society-related issues and indigenous knowledge for local school science. Such issues help science curricula/textbooks to begin from where the learner is. The foundation and passage to the learning of school science should be the local knowledge or natural phenomena. The preparation of local beer (*tella*), the manufacture of charcoal, and so on, are examples of issues that have local relevance. Focus should be given to Piel's categories such as Energy, Population, and Environmental Quality and Utilization of Natural Resources. All concerned personnel, such as policy makers, science educators, science curriculum experts, and science teachers should identify sources of STS issues. International journals and magazines that publish articles about African science education are primary sources. Local journals on the environment and its protection also are potential sources. For instance, AKIRMA published an article entitled, *Mirt* for the Household! (*Mirt* is a fuel-efficient biomass Injera stove, see Koenmund, 2001), pointing out that 50% of the total primary energy consumption is utilized in baking *Injera* (local bread). The article shows that among other advantages the *Mirt* uses half the amount of fuel compared with an open fire (18-23% efficiency as compared to 8-10%). Including such issues of local relevance is likely to help students see science learning as meaningful and authentic. Such issues would also help to make concrete otherwise abstract concepts like energy.

Revising, or at least enriching, science textbooks in line with STS themes. It is common in Ethiopia for teachers to rely heavily on textbooks. Textbooks even dictate teachers' teaching strategies, and incorporating the STS themes simply as body of facts will little generally not help students' understanding of their environment. Rather, the textbooks need to encourage students to construct meaning by themselves. In this regard, the STS issues might usefully draw upon a constructivist view of learning.

Increasing the amount of STS materials in the science textbooks progressively across grade levels. As students grow and are exposed to more science concepts and skills, they will be better able to identify and deal with the interaction of science concepts and skills with society and available technology: "This assumes that content coverage prepares students to be good decision makers regarding these issues" (Chiang-Soong & Yager, 1993, p. 348).

MINIMIZING THE GAP BETWEEN MATHEMATICS CURRICULUM AND ITS IMPLEMENTATION

As noted above, the implementation of any curriculum depends on many factors; people, the program, and the organization. However, the implementation of any curriculum predominantly depends upon the daily activities of the organizational members such as teachers, principals, and administrators who are responsible for its implementation. If these members respond to influences other than the intentions of the developers, then even the most carefully planned and supported curriculum is unlikely to be implemented as intended (Kolko, 1965, cited in Fullan & Pomfret, 1977). Evaluation of the implementation of the Ethiopian mathematics syllabus using Stake's (1967) Congruence-Contingency Evaluation Model suggests that the mathematics syllabus has not been implemented as intended (Areaya, 2000). Mathematics teachers in most Ethiopian classrooms tend to teach the facts of mathematics, rather than teaching the skills required to do mathematics. As a result, most classroom time is spent writing and copying lessons from the blackboard, and not doing mathematics. Although the mathematics curriculum requires the instructional process to be as student-centered as possible, the reality in the classroom is that teaching is, by-and-large, teacher-centered.

It also seems many Ethiopian mathematics teachers are unhappy with the process of mathematics curriculum development, and most teachers believe that little consultation occurred. This perception of lack participation in the mathematics curriculum development process and textbook preparation is likely to have contributed to teachers' lack of a sense of ownership of the reforms. Examination of mathematics achievement tests suggests that the majority of the students failed to obtain the expected outcomes for the subject. For instance, of the seven curricular objectives intended to be achieved in one semester (16 weeks of lessons), for only two did the majority of the students acquire basic mathematics knowledge and skills (Areaya, 2000).

It, therefore, seems that there is a need to revisit the process of mathematics curriculum development and textbook preparation with particular regard to teacher participation and professional development. This might help mathematics teachers not only to have sense of ownership for curricular reform, but also to contextualize and properly implement the mathematics curriculum in their classrooms.

CONCLUSIONS AND IMPLICATIONS

Ethiopia has been an independent country for many centuries. It has a population of over 70 million people, 85% of whom are living within a traditional agricultural economy. The country consists of many ethnic/language groups and the Government is structured as a federal state, composed of ethnic-based regional states. The Ethiopian educational system has changed in the last four decades, following the political ideology of the three most recent governments.

The current Government of Ethiopian has been developing and implementing educational policy, curricula and textbooks for science, mathematics and other subjects of various levels, including teacher education programs. Such policy demands that the educational system should produce graduates who possess strong problem-solving capacity. It is, however, clear that Ethiopian schools are poorly equipped for science and mathematics education. The science and mathematics curricula and textbooks are not designed in a way that encourages context-based or student-centered learning. The recently introduced television instruction for secondary schools has in some ways aggravated these problems in the sense that student-centered learning detailed in education policy documents has been inhibited, and instead, instruction by this mode of delivery consists largely of teacher-centered, direct instruction received passively via television. There is current concern in the country that the science and mathematics education has thus been reduced to eyes-on and ears-on learning at the expense of hands-on learning. Science and mathematics teacher training also is a challenge, since the television instruction was developed in a foreign country and took little consideration of local context, although it does allow for the demonstration of some science experiments.

Many stakeholders in the educational system, including the Government, now agree that there are some weaknesses in the current curriculum. Attempts at reform are therefore necessary, and these need to strive to achieve a match between policy, curriculum, and implementation of science and mathematics curricula.

REFERENCES AND BIBLIOGRAPHY

Ajeyalemi, D. (Ed.). (1990). *Science and technology education in Africa: Focus on seven sub-Saharan countries*. Lagos, Nigeria: University of Lagos Press.

Areaya, S. (2007). *The process of curriculum development and implementation in Ethiopia*. Unpublished PhD thesis, University of East Anglia, Norwich.

Areaya, S. (2000). *The evaluation of the implementation of mathematics syllabus*. Unpublished MA thesis, Addis Ababa University, Addis Ababa.

Chiang-Soong, B., & Yager, R.E. (1993). The inclusion of STS material in the most frequently used secondary science textbooks in the US. *Journal of Research in Science Teaching*, *30*(4), 339-349.
Engida, T. (2005, August). *Experiences in science and technology education.* Paper presented at Science and Technology Education Policy in Ethiopia Workshop. Nazareth, Ethiopia.
Engida, T. (2002). Reflections on African science education for the new millennium: The case of the Ethiopian chemistry curriculum for beginners. *International Journal of Science Education*, *24*(9), 941-953.
Engida, T. (2000). Student teaching in non-western science classrooms: Analysis of views from potential participants in the program. *Journal of Science Education*, *1*(2), 86-93.
Fullan, M., & Pomfret, A. (1977). Research on curriculum and instruction implementation. *Review of Educational Research*, *47*(2), 335-397.
Koenemund, K. (2001).'*Mirt*' for the Household. AKIRMA, March (5 & 6), 18-20.
Lemma, B. (2006). Plasma television teachers: When a different reality takes over African education. In L. Dahlstrom & J. Mannberg (Eds.*), Critical educational visions and practices in neo-liberal times* (pp. 71-90). Umea: Global South Network Publishers.
Ministry of Education. (2005). *Education sector development program III (ESDP-III)*: *2005/06-2010/2011 (Program Action Plan).* Addis Ababa, Ethiopia: Ministry of Education.
Negash, T. (1996). *Rethinking education in Ethiopia*. Uppsala, Nordiska: Afrikainstitutet.
Rogan, J.M., & Grayson, D.J. (2003). Towards a theory of curriculum implementation with particular reference to science education in developing countries. *International Journal of Science Education*, *25*(10), 1171-1204.
Tefera, S. (1996). Attempts at educational reform: A top-down or a bottom-up reform. *The Ethiopian Journal of Education*, *16*(1), 1-37.
Transitional Government of Ethiopia. (1994a). *Education Sector Strategy*. Addis Ababa, Ethiopia: EMPDA.
Transitional Government of Ethiopia. (1994b). *Education and Training Policy*. Addis Ababa, Ethiopia: EMPDA.

AFFILIATIONS

Temechegn Engida
International Institute for Capacity Building in Africa
UNESCO
Addis Ababa
Ethiopia

Solomon Areaya
College of Education
Addis Ababa University
Ethiopia

CHIKASANDA AND MBENDERA

TECHNICAL EDUCATION REFORMS IN MALAWI

INTRODUCTION

Malawi is one of the world's poorest countries, with low per capita income, an economy based on subsistence agriculture, and heavily impacted upon by globalization. There is an urgent need to up-skill Malawi citizens if the nation is to reduce its poverty, and increase its standard of living. Technical education thus assumes major importance in the Malawian educational system. This chapter describes reforms in technical, vocational and further education in Malawi and the impact these reforms have exerted on Malawi and its citizens. It seems the reforms have largely failed because they neglected to take into account important contextual factors such as the size of the industrial labor market. There also is a need to link technical and vocational education to the general school curriculum to enhance the status of technical education in Malawi.

Malawi's per capita income is estimated at US$170 per capita and 65% of the population living below the poverty line (World Bank, 2004). A number of international economic development bodies such as the World Bank, UNESCO, and UNDP (2006) have consistently asserted that education is critical for human development in countries such as Malawi because of the links with health, equity, and empowerment. The Policy and Investment Framework of the Government of Malawi also observes that "increased investment in education can yield broad economic and social benefits. Evidence shows that education is associated with increased agricultural productivity, higher incomes, and lower fertility rates and with improved nutrition and health" (Government of Malawi [GoM], 2000, Section 1.2.1). A healthy and educated population is necessary to achieve sustainable economic growth (GoM, 2006). Therefore, Malawi needs to invest in education in order to enhance literacy, numeracy, and employability of students when they leave the school system.

The current technical education curriculum in Malawi has proved to be economically unsustainable due to a requirement for heavy machinery, and associated maintenance - which few schools can cope with. As a result, technical education is offered in fewer than 13 secondary schools, and many primary schools opted out of technical education because of a lack of teaching and learning resources. The school technical educational curriculum places emphasis on examinations and because of the comprehensive nature of the syllabus, low enrolments, and disparities in participation based on gender, few students progress into further and higher education in this area.

Richard K. Coll, Neil Taylor (eds.), Science Education in Context: An International Examination of the Influence of Context on Science Curricula Development and Implementation, 223–234.

MALAWI PROFILE AND EDUCATIONAL AIMS

Malawi is a sub-Saharan African country often called 'the warm heart of Africa' because of its friendly people. It is bordered by Tanzania to the northeast, Zambia to the west and Mozambique to the south. It has a total land area of 118,484 square kilometers, of which 20 percent is water. In 2004, the population of Malawi was 12.6 million and it is expected to grow to about 16 million by 2015 (UNDP, 2006). About 51% of the population are female (GoM, 2002) and the economy is predominantly based on agriculture with tobacco, tea and sugar accounting for about 60% of the country's exports. There have been attempts to diversify the economy in order to transform the country from being a predominantly importing and consuming economy, to a predominantly manufacturing and exporting economy (GoM, 2006).

Malawi was a British colony up until 1964 when it attained political independence, and as a consequence of its colonial history, the education system is modeled on the UK system. However, despite significant education reforms in the UK, Malawi has retained the traditional UK system with few subject changes. Craft- and skill-based technical subjects in the general education curriculum have remained much the same. In order to attain the goals of the Science and Technology Policy for Malawi, Malawi Poverty Reduction Strategy Paper (2006/11), Malawi Growth and Development Strategy (MGDS), Vision 2020 and the Millennium Development Goals, technical education reforms are imperative for technological literacy. This chapter discusses the framework for embarking on curriculum reforms for the education system to embrace technical education for sustainable economic and social development.

White (2006) asks "Why else are we bringing up and teaching children, unless it is to equip them to lead a flourishing life themselves and help others?" (p. 30). In the context of Malawi, we might well note that if just one child from a poverty-stricken family flourishes, with Malawi's extended family structure, then others such as nephews or cousins also may escape poverty. Poverty is 'contagious' and a curriculum that empowers the youth to be able to 'fish rather than waiting to be fed', may go some way towards alleviating poverty in Malawi. Here education is seen as crucial. However, the curriculum needs to reflect the context and, for example, a curriculum that integrates personal relevance, intellectual processes and social reconstruction (Johnson, 1992; Petrina, 1992; Zuga, 1992) is necessary for the personal growth of students, and to help them become responsive to society values and beliefs.

EDUCATION SYSTEM IN MALAWI

Malawi has an 8-4-4 education system comprising eight years of primary, four years of secondary and four years of university education. Within this framework there are limited opportunities for pre-school and post-secondary education (GoM, 2000). At the primary level children start school at the age of six, completing Standards 1 to 8 at the end of which pupils sit the Primary School Leaving

Certificate examination (PSLCE). The PSLC Examinations are administered entirely by the Ministry of Education and the Malawi National Examinations Board (MANEB). These examinations form the entry point into the secondary schools in Malawi. Prior to 1994 the only alternative for the bulk of students who failed to gain entry into secondary school was to engage in secondary education by means of distance education, or to attend private schools. Theoretically, since 1994, primary school education has been compulsory. However, in reality Malawi has neither the resources nor the capability to enforce or support such a compulsory system.

Secondary school education in Malawi covers four years, being Forms 1 to 4. There are over 900 public secondary schools in Malawi and these are open to all students who pass the PSLCE examinations. There also are a large number of private secondary schools run by the private sector. The quality of secondary education varies widely for both private and public schools. The secondary school system has two examinations, the Junior Certificate Examination (JCE) at Form 2, and the Malawi School Certificate Examination (MSCE) at Form 4. Both of these examinations are set by MANEB but they are administered jointly by MANEB and the Ministry of Education. Pass rates at MSCE have been falling sharply in the last few years reaching a low of around 16% in 1999 - prompting a Presidential Commission of Enquiry into this rather dismal performance. The MSCE is the prerequisite for selection to various post-secondary school institutions, including the two public universities, the University of Malawi and Mzuzu University, and various teacher training colleges and vocational and technical colleges. Some other technical training colleges offer programs in such fields as forestry, marine science, social welfare and hotel management, as well as in various trades. Such programs lead to certificates after studies lasting between six months and four years. There are a few private secondary schools - mostly in Kasungu, Blantyre and Lilongwe - that run a UK style A-Level programs and examinations. However, most of these demand fees well beyond the reach of most Malawians.

The University of Malawi opened in 1965, and is a typical post-colonial institution. The initial aim of the University was to train Malawians to take over the administrative duties of the outgoing colonial administrators. Hence, in its early days it offered a variety of bachelor level degrees and a selection of technician type programs. The University has five colleges each offering some specializations in the liberal arts and natural sciences, science and technology, agriculture, medicine or nursing. The recently opened Mzuzu University situated in the northern region, offers education degree programs, and plans to offer training in other areas in the near future.

THE HISTORY OF TECHNICAL EDUCATION IN MALAWI

Technical education in Malawi was initiated by missionaries and this was followed by the introduction of craft type subjects at several government and missionary institutions. A more comprehensive national plan put into effect in the post-independence period resulted in the introduction of a secondary school technical curriculum, and this was followed by the expansion and upgrading of government technical schools and colleges. This stage was significant because it resulted in the introduction of sub-degree and degree level programs, and training of technical teachers at the Malawi Polytechnic.

According to Aipira, (1988) Scottish missionaries in Malawi believed that education was a sober, practical, business and this view captures the origins of the development of technical education in Malawi as it was first introduced by Scottish Presbyterian missionaries. With such UK origins, technical education consisted of discrete academic subjects and the aims of technical education at this early point was to train local artisans to help erect mission buildings and maintain machinery that the missionaries brought to Malawi. The main areas of training were thus things such as carpentry, joinery and brickwork. Other missionaries such as Roman Catholics and Adventists emulated the Presbyterian missionaries at Khondowe, and introduced technical education in other parts of the country. The colonial government did not join in the provision of technical education until the 1950s when they accelerated activities in technical education with schools opening at Dedza Secondary school, Mpemba, and later on Kanjedza in Blantyre. This flurry of activity continued in the period just before independence with Soche Technical and Lilongwe Technical schools opening in the early 1960s.

The lack of interest by the Government in any other type of education outside general education at primary level was attributed to the fact that the colonial government needed clerks and typists to join in the administration of the Government. As a result, it never felt compelled to become involved in technical education for a long period. At the time of independence, therefore, the country had two Government and three grant-aided mission technical colleges. Two Government technical colleges were established and other mission technical schools were subsequently established. Currently Malawi has 13 public and private technical colleges.

THE MALAWI POLYTECHNIC

Technical education at tertiary level became possible in Malawi with the opening of the University of Malawi in 1965. The opening of the Malawi Polytechnic as a constituent college of the University of Malawi was unique because it was mandated to offer both technician level and tertiary level programs. According to Cartwright (1990), the Polytechnic was able to offer tertiary technical courses at diploma and degree levels in engineering, construction, business studies, commerce, laboratory technology, public health inspection and secondary school technical teaching. Technician apprenticeship courses were offered in motor vehicle mechanics, general fitting, construction, and electrical and electronics courses, while craft level courses were offered at the technical schools or colleges. The Polytechnic in particular was unique in the sense that the founders made a deliberate decision to include vocational and technical education within the university system.

An important feature of the Polytechnic was its mandate from 1969 to produce technical teachers who would teach in secondary schools in Malawi. With the support from International Development Agency's (IDA) first and second projects, the Government of Malawi was able to introduce technical subjects to 13 secondary schools in Malawi (Cartwright, 1990). The Polytechnic again was expected to produce diploma teachers who would teach woodwork, technical drawing and metalwork in the newly created technical departments in 13 secondary schools in the country. The first students entered the Polytechnic in 1969 and from these beginnings the Polytechnic developed and upgraded its diploma to a degree which was launched in 1982. In 1988, the Polytechnic hosted students from Lesotho, Swaziland, Namibia, Zimbabwe, Zambia and Malawi under the Southern African Development Community (SADC) capacity-building project. Currently, the department offers three degrees: a Bachelor of Science in Technical Education, a Bachelor of Education (Technical), and a Bachelor of Education (Business Studies).

However, over the past 10 years, teacher training at the Polytechnic has remained stagnant to such an extent that the institution has missed opportunities for diversification and expansion into equally important technical and technology subjects. For instance, secondary schools in Malawi have introduced computer studies but the Polytechnic has failed to introduce aspects of computer education in its curriculum. Another weakness of the current teacher training program is over-emphasis on secondary school technical subjects of woodwork, metalwork and technical drawing, at the expense of equally important stakeholders such as the technical colleges who expect the Polytechnic to prepare instructors for the varying trades. The technical college curriculum is more diverse in the sense that it includes areas such as construction, motor vehicle mechanics, and electrical courses among others.

TECHNICAL AND VOCATIONAL EDUCATION AND TRAINING IN MALAWI

The labor market in Malawi is very small due to lack of a vibrant industrial and manufacturing base. Manufacturing accounts for 22% of GDP while the agriculture sector accounts for 33% of Malawi's GDP (GoM, 2004). The agriculture sector employs about 80% of the workforce, and contributes over 80% of foreign exchange earnings but the sector has, over the years, dwindled due to natural calamities such as drought, little irrigation development, and low uptake of improved farm inputs (GoM, 2006). The MGDS sees manufacturing as a key area for economic growth and considers this should occur by improving the quality of products and productivity of both labor and capital, enhancing skills through better integration of science and technology into vocational training, and improving standard certification capacity. The implementation of formal technical and vocational education and training (TVET) was shared by the Ministry of Education and Ministry of Labour up until 1998 when the Government developed a new TVET policy. The policy focuses on strategies outlined in Vision 2020 which centers on promoting sustainable development and poverty reduction including diversifying the economy through industrialization. Greater emphasis was to be placed on appropriate technical/vocational and entrepreneurship education and training. Technical education reforms were implemented using financial and technical support provided by the German Development Assistance Agency (GTZ) and Danish International Development Agency (DANIDA) through the Canadian College Partnership Program (CCPP), the Saskatchewan Institute of Applied Science and Technology (SIAST), and Cambrian College of Applied Arts and Technology with local partners; the Ministry of Education/Ministry of Labour and the Malawi Polytechnic.

The new system was entrusted to an authority called the Technical, Entrepreneurial and Vocational Education and Training Authority (TEVETA) and this authority was mandated to promote an integrated demand-driven competency-based modular education system. Through its Competency Based Education (CBE) programs TEVETA was expected, as detailed in the MPRSP (GoM, 2001), to become the engine of economic development and poverty reduction through the provision of effective formal and informal skills training for job and self-employment. The TEVETA is therefore charged with implementing formal training programs (Chafa, 2005a): apprenticeship schemes, private sector training programs; informal training programs: skills development; small enterprise development, and on-the-job training. The TEVETA's main focus is on curriculum development and assessment and it has developed the CBE and the National Qualifications Framework (NQF), career guidance and other outreach programs. The TEVETA has since developed CBE modules for most of the programs offered in technical colleges (except secretarial studies), launched a qualification framework and manages curriculum delivery and implementation, and student recruitment in all public colleges. The NQF was launched, and now provides national standards for learners, trainers and employers within the TEVET system.

The first project, the Technical Education Reform Project (1998-2003), implemented by SIAST (as the lead partner) and Cambrian College and funded by CIDA through CCPP, focused on curriculum development and new education concepts like CBE, DACUM (for Develop A Curriculum), Standards, and Certification. These issues were brought into the technical and vocational education arena, and are issues which required enhancement. Hence, the Enhancing TEVET Outcomes (ETO) project was developed to deal with these issues. This project was launched in August 2003. The objective of the project was to address the needs of the TEVET systems such as policy and management development, teacher training, entrepreneurship and development of model technical colleges - one rural and one urban. The project also addresses common issues such disparity in participation based on gender, environmental education and HIV/AIDS education. The project was managed by Canadian partners: SIAST, Red River College (RRC) and Bow Valley College - with support from the Association of Canadian Community Colleges (ACCC). Local partners were; the Ministry of Labour, the Ministry of Education and Vocational Training, TEVETA and the Malawi Polytechnic. Under this project about 100 instructors from public and private technical colleges were trained in pedagogic principles and practice, using a modular-based education program from RRC.

Despite such progress in TEVET activities, there has not been any notable change in economic growth that can be attributed to TEVET's role in the sector. Poverty has continued to increase with minimal microeconomic growth, and rising unemployment the norm in both rural and urban areas. The TEVET system, therefore, needs further reforms in order to make a meaningful contribution towards rapid and sustainable poverty alleviation for the large number of Malawians living below the poverty line. While the formal training programs are probably worthwhile for the industrial and manufacturing sectors of the economy, the informal sector needs urgent attention since that is where there are a large number of unemployed youths who might benefit.

The TEVET system has also not done much to raise the status of technical and vocational education, or show how formal college training can be linked to the rest of the education system in Malawi. Technical and vocational training has always been associated with students of low academic stature and is not seen as an alternative route for tertiary training. In Malawi, such students are classified as 'unintelligent' compared with their university counterparts. Often they are not considered for white collar jobs, and changes to the status of such students has far-reaching consequences, and is often resisted by those who see such moves as disadvantageous (Paechter, 1993). According to Paechter, TEVET teachers are seen as "... thick jocks with brawn but little in the way of brains ... or as people who are more comfortable working with their hands and not with heads" (p. 352).

TEVET also is seen as a male domain. According to Layton (1994), "Women have never lived without technology ... have barely a toehold in the discourse and direction of it ... have been robbed of the history of female technical initiative, imagination and invention" (p. 15). Hence, any TEVET reforms need to be done in a manner that develops a coherent and unified framework that addresses the needs of all social groups while, at the same time, providing flexible, sustainable, and gender-neutral pathways.

TECHNICAL EDUCATION CURRICULUM

Secondary School

In the midst of international debate on the benefits of vocationalizing the school curriculum (see, e.g., Foster, 1977; Goodwin, 1990; Sanderson, 1993; Watson 1994), Malawi via funding from the IDA, introduced technical subjects in 13 secondary schools soon after gaining political independence from the UK in 1964. The aim at the time was to attract able students into engineering-related study and to provide them with skills for jobs and self-employment. These assumptions were largely based on the premise that a comprehensive curriculum would help in adaptation and increase social and occupational mobility (Urevbu, 1988). Opponents of a vocational-based curriculum argue that such a curriculum inhibits further study, and thus reduces future socioeconomic attainment. Proponents, in contrast, argue that it helps students avoid unemployment and increases their chances of becoming skilled workers (Arum & Shavit, 1995). Goodwin (1990) comments that many countries in Africa also hoped that diversification of the curriculum would ease unemployment, and enhance growth and productivity. The impact of vocationalizing the curriculum in Malawi has not been measured, but, suffice to say, that poverty and youth unemployment has continued to escalate - despite the promises of such a curriculum.

As in most African countries that were former British colonies, the integration of vocational and liberal education was largely based on the system prevalent in the UK at the time, and the influence of donor organizations (Sanderson, 1993; Urevbu, 1995; Watson, 1994). As a result, Malawi's indigenous technologies, culture, values and beliefs are not reflected in the curriculum. In contrast, in some European countries, local cultures and industries form part of the curriculum. For example, in Germany, schools draw upon a range of local handcrafts - clocks, toys, woodcarving, and wrought iron work, while students in French schools develop their design skills by engaging with distinctive French luxury craft trades such as glass, porcelain, fashion couture and furniture (Sanderson, 1993). In Malawi, pupils may not easily transfer more generic skills learnt in an industrial-based classroom into their local context - since there are no industrial-related activities in rural villages where most pupils come from. Despite professional development initiatives undertaken in the early 1970s in collaboration with Bolton College and Huddersfield University of UK, the Malawian curriculum remained 'alien' and implementation of technical curriculum was confined to the 13 pilot schools that were provided with heavy and sophisticated machinery. As a consequence, enrolments in technical subjects continue to dwindle, and some schools closed their technical departments. Although policy and political will may well overturn events, the curriculum still needs to be reviewed so that the curriculum considers current demographic indicators and emerging issues. Malawi needs a curriculum that promotes social, economic and environmental awareness and development - and at the same time enhances the beliefs and values of society.

Primary School

As a signatory to the Education for All (EFA) initiative at the Jomtien Conference in Thailand in March 1990, the Government of Malawi committed to EFA goals. One result was the introduction of free (compulsory) primary education in 1994 (GoM, 2005). With this, enrolments in basic education increased from 1.9 million in 1994 to 3.2 million in 2000, thereby creating enormous pressure on school infrastructure, teachers, and teaching and learning materials (GoM, 2001).

The primary school curriculum has undergone a variety of reforms in terms of subject content, pedagogy and assessment. The reforms were necessitated by emerging issues such as gender issues, the HIV/AIDS pandemic, rapid population growth and environmental problems (Nyirenda, 2005). The most recent reforms in 2001, called the Primary Curriculum and Assessment Reforms, were designed to address major shortfalls in educational attainment by primary school children - while at the same time addressing current national policy initiatives such as the Malawi Poverty Reduction Strategy Paper (MPRSP), Vision 2020 and the 2001 Policy and Investment Framework (Kambankadzanja, 2005). Besides improving internal efficiency, the reforms were also implemented in order to reflect curriculum developments taking place within the Southern African Development Community region and internationally where, for example, in Canada, Australia, Mozambique and South Africa, the education systems had embarked on outcomes-based education (Nyirenda, 2005).

Rather depressingly, the quality of primary school education in Malawi has gone down despite huge investment by the Government and its multilateral partners such as the World Bank, African Development Bank, UNESCO and various NGO's (GoM, 2004; Nsapato, 2005; World Bank, 2004). Low public expenditure per pupil, under qualified teachers, and poor teaching and learning conditions, are but some of the major causes of poor returns in primary education leading to low literacy and numeracy rates estimated at less than 0.6% by Grade 6 (Kambankadzanja, 2005; World Bank, 2004). The World Bank (2004) also reports that between 1992 and 2000, primary school completion rates, a major factor in achieving the goals of education for all of 100% primary completion rates by 2015, are below 40% - the lowest in the region. Out of every 10 pupils, 6 drop out before completing primary school and join the ranks of the unemployed engaged in manual labor.

The primary school curriculum has undergone a number of reviews to address emerging issues. However the plight of dropouts, many orphaned due to HIV/AIDS, has not been adequately considered. The generality of the curriculum is an essential feature for pupils to be able to transfer classroom activities to real life work (Cognition & Technology Group at Vanderbilt, 1990; Hennessy, McCormick & Murphy, 1993; Jones, 1996, 1997; Perkins & Salomon, 1989). For example, most curricula promote academic reading where transferability of knowledge to social and practical situations is hardly possible at the lower levels of schooling. With poverty, an unprecedented dropout rate, unemployment and limited opportunities for secondary education, the curriculum needs to empower students to engage in gainful activities after schooling. The review of the primary curriculum that took place in the 1990s included the introduction of creative arts as a subject, but no appropriate professional development program was put in place (Chafa, 2005b). Creative arts here included some technical subjects like carpentry, building, pottery, tin-smithing, and welding as learning units, and teachers were required to have all these skills or utilize local artisans. The curriculum was heavy and required relevant human and material resources development and investment. The problem was also compounded by the fact that the subject was not examinable, and meaning most teachers preferred teaching subjects which involved their pupils sitting national examinations.

CONCLUSIONS AND IMPLICATIONS

Malawi still offers traditional technical subjects as part of its education system while the UK has changed in a number of ways, reflecting changes in the local context, in terms of societal trends and conditions. For example, Design and Technology were made compulsory in the UK National Curriculum in the 1990s (Layton, 1993; Lewis, 2000; McCormick, 1993). Lewis also notes that the USA likewise published a framework - *Technology for All Americans* - a curricular manifesto for the teaching of technology in American schools. New Zealand (Jones, 2003), Australia (Treagust & Rennie, 1993), Finland (Rensburg, Ankiewicz, & Myburgh, 1999), The Netherlands (Rasinen, 2003), and South Africa all have reformed their technical curricula, shifting from their craft and construction origins towards technology, something seen as more holistic in nature. Although Malawi is still a developing country with few technologies commonplace (Lewis, 2000), it needs to take note of world trends to reform technical subjects more towards technology. Technology transfer, diffusion and creation should be enhanced by the curriculum so that children as future leaders are offered these opportunities. Hence the urgent need for reform so the nation can make that much needed quantum leap into technological modernity.

Context related technical education reforms in Malawi are necessary in order to enhance technological literacy, reduce overdependence on imported materials and reduce Malawi's vulnerability as a dumping ground for substandard and counterfeit products. There is also a need to link general education with technical and vocational education and training, in order to enhance the status of technical education in Malawi and provide alternative opportunities for the large number of students that drop out of school.

REFERENCES AND BIBLIOGRAPHY

Aipira, O.S.L. (1988). *Curriculum design for the technical teachers training and education programme at the Malawi Polytechnic*. Unpublished MEd Thesis, Manchester University.

Arum, R., & Shavit, Y. (1995). Secondary vocational education and the transition from school to work. *Sociology of Education, 68*(3), 187-204.

Burns, J. (1992). Students' perceptions of technology and implications for an empowering curriculum, *Research in Science Education, 22*, 72-80.

Cartwright, R. (1990). *Secondary technical teacher education in Malawi: A test project country study 1985-1988*. Bolton, UK: Bolton Institute of Higher Education.

Chafa, J.W. (2005a, March-April). *The role of technical, entrepreneurial vocational, education and training in social economic development in Malawi*. Paper presented at the National Education Conference. Retrieved 27 December 2006, from http://www.sdnp.org.mw/~hosea/workshops/march2005-conference/papers.htm

Chafa, J.W. (2005b, March-April). *Curriculum for entrepreneurship skills and practical subjects*. Paper presented at the National Education Conference, March 29-April 1, 2005. Retrieved 17 December 2006, from http://www.sdnp.org.mw/~hosea/workshops/march2005-conference/papers.htm

Cognition and Technology Group at Vanderbilt. (1990). Anchored instruction and its relationship to situated cognition. *Educational Researcher, 19*(6), 2-10.

Foster, P.J. (1977). Vocational school fallacy in development planning. In J. Karabel & A.H. Halsey (Eds.), *Power and ideology in education* (pp. 356-365). New York: Oxford University Press.

Government of Malawi. (1998). *Vision 2020. National Economic Council*. Retrieved 4 January 2007, from http://www.sdnp.org.mw/~esaias/ettah/vision-2020/

Government of Malawi. (2000). *Policy and investment framework. Ministry of Education: Lilongwe*. Retrieved January 2, 2007, from http://www.sdnp.org.mw/~phindu/min-education/whole.htm#background.

Government of Malawi. (2002). *1998 Malawi population and housing census: Analytical report National Statistical Office of Malawi, Zomba*. Retrieved 31 December 2006, from http://www.nso.malawi.net/

Government of Malawi. (2003). *Malawi DHS EdData Survey 2002*. Calverton, UK: The National Statistical Office ORC Macro. Retrieved 30 December 2006, from http://www.nso.malawi.net/

Government of Malawi. (2004). *Malawi integrated framework: Diagnostic trade integration study*. Retrieved 4 January 2007, from http://www.integratedframework.org/ files/ Malawi_finaldtis_20Feb04.pdf

Government of Malawi. (2005). *National education sector plan*. Lilongwe, Malawi: Ministry of Education.

Government of Malawi. (2006). *Malawi growth and development strategy 2006 – 2011*. Retrieved 31 December 2006, from http://www.malawi.gov.mw/News3/MGDS%20November%202006%20-%20MEPD.pdf

Goodwin, D. (1990). Technical education in sub-Saharan African countries: A review of development and strategies. *The Vocational Aspect of Education*, *17*(113), 91-99.

Hennessy, S., McCormick, R., & Murphy, P. (1993). The myth of general problem solving capability: Design and technology as an example. *The Curriculum Journal*, *4*(1), 73-89.

Johnson, S.D., (1992). A framework for technology education curricula which emphasises intellectual processes. *Journal of Technology Education*, *3*(2), 26-36

Jones, A. (2003). The development of a national curriculum in technology for New Zealand *International Journal of Technology and Design Education*, *13*, 83-99.

Jones, A. (1996, January). *Recent research in learning technological concepts and processes*. Paper presented at the second Jerusalem International Science and Technology conference. Jerusalem.

Jones, A. (1997). Recent research in learning technological concepts and processes. *International Journal of Design and technology Education*, *7*(1), 83-96.

Jones, A., & Carr, M. (1990). Teachers' perceptions of technology education. In *Towards technology education: Working papers from the first phase of the Learning in Technology Education Project* (vol. 1, pp. 49-83). Hamilton, New Zealand: Centre for Science & Technology Education Research, University of Waikato.

Kambankadzanja, (2005, March-April). *The Malawi primary curriculum & assessment reform: The process, the challenges and the way forward*. Paper presented at the National Education Conference. Retrieved 27 December 2006, from http://www.sdnp.org.mw/ ~hosea/workshops/march2005-conference/papers.htm

Layton, D. (1993). *Technologies challenge to science education*. Buckingham, UK: Open University Press.

Lewis T. (2000). Technology education and developing countries. *International Journal of Technology and Design Education*, *10*(3), 163-179.

Malawi DHS EdData Survey. (2002). *Educational data for decision making*. Zomba, Malawi: National Statistical Office.

McCormick, R. (1993). Technology education in the United Kingdom. In R. McCormick, P. Murphy & M. Harrison (Eds.), *Teaching and learning technology* (Part 1, pp. 15-27). Wokingham, UK: Addison-Wesley.

Nsapato, L. (2005, March-April). *Factors affecting quality of basic education in Malawi*. Paper presented at the National Education Conference. Retrieved 27 December 2006, from http://www.sdnp.org.mw/~hosea/workshops/march2005-conference/papers.htm

Nyirenda, D.C. (2005, March-April). *Malawi secondary school curriculum reform: Issues and challenges*. Paper presented at the National Education Conference. Retrieved 27 December 2006, from http://www.sdnp.org.mw/~hosea/workshops/march2005-conference/papers.htm

Paechter, C. (1993). What happens when a school subject undergoes a sudden change of status? *Curriculum Studies*, *1*(3), 349-363.

Perkins, D.N., & Salomon, G. (1989). Are cognitive skills context bound? *Educational Researcher, 18*(1), 16-25.
Petrina, S. (2000). The politics of technological literacy. *International Journal of Technology and Design Education, 10*(3), 181-206.
Petrina, S. (1992). Curriculum change in technology education: A theoretical perspective on personal relevance curriculum designs. *Journal of Technology Education, 3*(2), 37-47.
Rasinen, A. (2003). An analysis of the technology education curriculum of six countries. *Journal of Technology Education, 15*(1), 31-47.
Rensburg, S.V., Ankiewicz, P., & Myburgh, C. (1999). Assessing South African learners' attitudes towards technology by using the PATT (Pupils' Attitudes Towards Technology) questionnaire. *International Journal of Technology and Design Education, 9*(2), 137-151.
Sanderson, M. (1993). Vocational and liberal education: A historian's view. *European Journal of Education, 28*(2), 189-196.
Treagust, D.F., & Rennie, L.J. (1993). Implementing technology in the school curriculum: A case study involving six schools. *Journal of Technology Education, 5*(1), 38-51.
UNDP. (2006) Human development report. Retrieved 31 December 2006, from http://hdr.undp.org/hdr2006/pdfs/summaries/HDR2006_English_Summary.pdf
Urevbu, A.O. (1988). Vocationallizing the secondary school curriculum: The African perspective. *International Review of Education, 34*(2), 258-270.
Watson, K. (1994). Technical and vocational education in developing countries: Western paradigms and comparative methodology. *Comparative Education, 30*(2), 85-97.
White, J. (2006). Education for well-being. *Education Review, 19*(1). 28-33.
World Bank. (2004). Cost of financing and school effectiveness of education in Malawi. *Africa region human development working paper series*. Retrieved 27 December 2006, from http://siteresources.worldbank.org/AFRICAEXT/Resources/No_78.pdf
Zuga, K.F. (1992). Social reconstruction curriculum and technology education. *Journal of Technology Education, 3*(2), 48-58.

AFFILIATIONS

Vanwyk Khobidi Mbubzi Chikasanda
Center for Science and Technology Education Research
University of Waikato
New Zealand

Ida K. Mbendera
Department of Technical Education
University of Malawi-The Polytechnic
Malawi

O-SAKI

SCIENCE EDUCATION IN TANZANIA: THE CHANGING CONTENT AND FORM OF SCIENCE TEACHING AND TEACHER EDUCATION

INTRODUCTION

The United Republic of Tanzania is located in East Africa and has an area of 945,087 km^2, the 12th largest of the 53 countries of Africa. The population in 2006 was estimated to be 38.5 million[i]. It is surrounded by Kenya in the north, Uganda, Rwanda and Burundi in the northwest, the Democratic Republic of the Congo (DRC) in the west, and in the southwest by Zambia and Malawi. In the south there is Mozambique, and in the east it has a coastline bordering the Indian Ocean. It has a heterogeneous population made up of over 95% indigenous Africans of about 120 ethnic groups; the rest is a mixture of Asians, Arabs, European and other nationalities. Humans are thought to have originated in the Olduvai Gorge in Tanzania, where human fossils discovered by the Leakeys in 1959/79 were found to be 1.75 million years old. Stoneage tools and pre-historic rock paintings are tourist attractions found in Tanzania. Early humans are thought to have evolved in Tanzania, dispersed at some time and returned later as travelers, farmers, warriors, settlers, invaders, explorers and so on. Over 40% of the land is conserved as game reserves and national parks where large herds of mammals such as wildebeest, zebra, giraffe, antelope, lions, leopards, hyena, wild dogs and many birds, reptiles, and invertebrates roam. Famous tourist attractions include Mount Kilimanjaro (5,895 m), the Ngorongoro Crater, the Serengeti, Selous, Lake Manyara and Gombe Stream National Parks, and Zanzibar.

The GNP in 2005 was estimated be US$12.1 billion and expected to grow by 6.8% in comparison to the previous year. The per capita income in 2005 was estimated to be around US$340 (see Table 1). Tanzania has been governed by a multi-party democratic government since 1995. In terms of the environment, Tanzania is rich in natural resources including natural gas, gemstones, gold, coal, and phosphates, though the full potential is far from having been exploited.

EDUCATION SYSTEM

Tanzania follows a 7-4-2-3 education structure. Children attend seven years of compulsory primary education, and upon completion up to 35% are able to join secondary school. They attend four years of junior secondary after which they sit

Richard K. Coll, Neil Taylor (eds.), Science Education in Context: An International Examination of the Influence of Context on Science Curricula Development and Implementation, 235–250.

Table 1: Basic facts about Tanzania: 2006 (Ministry of Planning Statistics, 2006)

Area: 945,087 km2 [including Zanzibar] (59,050 water) -12th largest in Africa.
Population (2006 estimate): 38,500,000.
Population density: 42 per sq km.(64% rural).
Annual population growth: 1.83% p.a (2006 est.).
Highest point: Kilimanjaro Uhuru peak: 5,895 m.
Lowest Point: Indian Ocean.
Economy:
GNP growth (2005): 12.1 billion US $ (up 6.8% from 2004
GNP per capita (2005): US $340
Agriculture 50%
Trade: 16%
Business Services: 10%
Mining: 2% (growth 15% in 2005 compared to 2004)
Tourism: growth 8% in 2005 compared to 2004.
Main exports: Coffee, tea, cotton, tobacco, cashew nuts, gemstones, gold.

the Certificate of Secondary Education Examination (CSEE). Those who succeed proceed to two years of senior secondary which end with the Advanced Certificate of Secondary Education Examination (ACSEE). After this level, students can choose to join higher education institutions to study for degrees, or tertiary education colleges to undertake professional diplomas in business education, technical and vocational education, teacher education, health education and other fields including agriculture, fishery, social studies, and so on. In 2007 the transition rate from primary to secondary school was around 35% (up from 7% in 2004).

Tanzania inherited a rather elitist system of education from its colonial ruler, Britain, in 1961 with primary, as the main structure but with very small secondary and tertiary education sectors. The number of children attending school was very low and much lower for those going further than primary school level. The curriculum was also very academic and Western-oriented especially from secondary school level and beyond. Primary education was semi-vocationalized following the 1930s 'Education for Adaptation' Policy of the British colonial government (Cameron & Dodd, 1966). By 1961, secondary education was academic and designed for the minority who would fill the positions then existing in government. The transition rate from primary to secondary level was below 15% up to 2003 when steps were taken to increase enrolment at secondary school level. The current transition rate ranges between 30-40%, with universal primary education (UPE) almost fully achieved.

EDUCATION FOR SELF RELIANCE

In 1967, the Government introduced the *Education for Self Reliance* (ESR) Policy paper which re-defined the purpose of education in the country. The main goal of education was defined as being to develop the following four qualities in learners: an enquiring mind; the ability to learn from others and make a relevant judgment on what to adopt or adapt; the ability to think critically; and the development of confidence and mental liberation.

The philosophy of ESR evolved from the larger philosophy of Socialism and Self Reliance enshrined in the Arusha Declaration. ESR was designed to review the previous curriculum and the experience of schooling in order to produce children who were more confident, curious, inquisitive, and with critical, liberated minds. The text of this philosophy was not directly translated into operationalized school practices and hence was left open to various forms of interpretation by curriculum developers, head teachers, district education managers and classroom teachers (see Komba, 2006). At the time, curriculum planners based at the Institute of Education adopted science and mathematics curricula that were imported from abroad and attempted to use these to promote an enquiring mind and critical thinking skills. Three curriculum packages were selected in science and mathematics.

The first was the African Primary Science Project materials for primary schools which focused on promoting thinking, imported from the USA. This was adapted into a form called: ‘Thinking Science: Ask the Ant Lion” because it was based on an investigative study of the behavior of a common bug called the ant lion. It was translated into the local Kiswahili language: *Elimu ya Kufikiri: Mwulize Fukufuku.* Primary school teachers continued to be recruited from among school leavers until after 1980 when it was made mandatory for primary school teacher trainees to be recruited only from among secondary school graduates.

The School Science Project (SSP) and School Mathematics Project (SMP), which were adapted from Nuffield Science materials, were introduced in around 1968. Selected school teachers were trained to rewrite and adapt the Nuffield and SMP into a suitable form called the School Science Project (SSP) which covered Physics, Chemistry and Biology and School Mathematics Project (SMP) of East Africa. The thrust of these materials was mainly experimental, following an experimental approach that also touched on the historical development of the ideas in each topic and a great deal of laboratory and outdoor activity. Biology learning, for example, involved bird watching, ecological sampling, collecting and identification of specimens and experimental write-ups, while chemistry had practical work of analyzing substances, doing titrations, visiting local industry, mines and quarries.

Many schools also introduced an extra dimension into the curriculum which was often described as Education for Self Reliance (ESR) activities or *Elimu ya Kujitegemea* (EK). These included an opportunity for each child to engage in a productive activity that involved manual work, designed to encourage them to respect work and to relate classroom learning to economic activities practiced in their communities. At the peak of these activities, this author was manager of such a program in one school. The school divided the pupils into several groups and they took part in the following activities on selected days of the week: farming, animal husbandry, beekeeping, shop-keeping, maintaining the school laboratories, and caring for themselves.

Farming. The students grew maize, millet and grapevines, which were the common crops of the area.

Animal husbandry. There was a project of around 300-400 chickens, fully taken care of by a selected group of students, a shed with eight dairy cattle run by another group, and a hut with no less than 20 pigs managed by another group. The job of the students included ensuring that the shed was clean, animals were fed, and eggs collected, or cows milked , and so on.

Bee keeping. A bee hive project was introduced in 1979 with about 10 experimental hives and in one month all had bees in them. The students had to monitor the bees, harvest the honey, and market it.

Other students ran the school shop/mini supermarket and, for others, their work included maintaining the school laboratories. Students also cleaned their dormitories themselves, and washed and ironed their own clothes. Each student was expected to be attached to at least one project, and each project had a team of teachers to supervise it in addition to the normal academic program (O-saki, 1996, 2000a, 2000b).

CHALLENGES OF ESR

Though exciting for learners, these ESR programs faced three main challenges as schools expanded.

First, except for ESR activities which were somehow productive, the other goals were based on expensive laboratory equipment targeting only a small number of schools. The equipment, bought through the UNESCO funded project in 1968-70, could not meet the needs of expanding enrolments, nor be replaced regularly, due to a shrinking budget and rising costs. Second, many secondary school science teachers were leaving the schools; in case of expatriates either to find jobs elsewhere (as contracts expired and were not renewed) or, in case of local graduates, they were appointed into government and parastatal managerial positions which had better pay and social status than teaching. Third, around 1972, when Cambridge overseas syndicate were examining, there was lack of communication between the local school managers and the Cambridge Overseas Examination Board on the curriculum then on trial in some schools. A traditional examination was set for students who had followed the inquiry curriculum for four years, leading to failure because it demanded knowledge of formulae and facts rather than processes. The subsequent poor performance of the students doing the experimental SSP and SMP curricula in comparison to the control group was interpreted as a weakness on the part of the inquiry curricula; although in fact they had been given the 'wrong' examination. Because of this, teachers, parents and other conservatives were angry and thought it was the new syllabus that was weak, causing students to fail examinations. This pressure group, aided by lack of effective teachers, forced the closure of the inquiry science and mathematics curriculum trial in the mid-1970s (see O-saki, 1991). The Institute of Education, therefore, abandoned the emphasis on inquiry science, and continued to write textbooks that focused more on remembering facts and formulae than inquiring skills and experimental work. A survey found that science and mathematics teaching had seriously deteriorated and been reduced to copying and memorization of facts for examinations (Chonjo, O-saki, Possi & Mrutu, 1995). Practical work had virtually been stopped, and an 'alternative to practical examination' introduced since the mid-1980s had been interpreted by teachers to mean that there was no need for practical science (Kibga, 2004).

Pre-Service Teacher Education

In terms of teacher training, the introduction of the SSP and SMP was accompanied by the intensive training introduced at the University of Dar es Salaam from 1968, funded by UNESCO. This was a three-year undergraduate training program which led to the award of the BSc with education degree (and a BA with education) and thereafter the graduates were directly posted into schools. These graduates studied two academic subjects (e.g., physics & mathematics, physics & chemistry, chemistry & biology, botany & zoology), and were prepared to teach in secondary schools). However, there was a parallel program called Bachelor of Education (BEd) which produced a few graduates who went to TTCs to train teachers or work in adult education institutions. These did one academic subject (mathematics, chemistry, physics, geography, a language, etc.) and studied more courses in educational psychology, educational administration, curriculum or adult education to increase their knowledge of professional education issues. In 1990, it was decided that the three-year programs were too short to produce an effective teacher and they were revised and elongated to four years. The BEd students did more optional subjects in the fourth year, while the BA(Ed) and BSc(Ed) students did extra courses in their academic subjects to prepare for better teaching of that subject at advanced levels.

TEACHER EDUCATION ASSISTANCE IN MATHEMATICS AND SCIENCE (TEAMS) PROJECT

In 1995, the University of Dar es Salaam, cooperating with three Dutch universities, established the Teacher Education Assistance in Mathematics and Science (TEAMS) project, funded jointly by the governments of Tanzania and the Netherlands, through a bilateral program known as the MHO (Cooperation with Dutch Universities). The project had three goals: i) to revise and improve pre-service and in-service teacher education, ii) to develop capacity through training of human resources, and iii) promote research capability.

Revision of Pre-Service Undergraduate Training

The project reviewed the quality of undergraduate programs in science education, namely BSc(Ed) and BEd at the University of Dar es Salaam, and found that despite the good work going on, there was a great weakness in the quality of teacher preparation and this was contributed to by an over-emphasis on content development at the expense of pedagogical training. The science education staff restructured the undergraduate program and introduced one new course in each discipline and one compulsory pedagogy course to try and match university level science and school curriculum. The five new courses introduced then were called: Physics for Education Students; Chemistry for Education Students; Mathematics for Education Students; and Biology for Education Students, all of which addressed pedagogical issues specific to those subjects teaching and syllabus requirements. Science and Mathematics Education was an advanced pedagogy course designed to reflect on the history and philosophy of science, students' conceptions, practical work, selection and use of textbooks.

These courses ran for five years and produced graduates who are now working in the schools. Unfortunately, some courses were shelved in 2005 following a restructuring of the four-year program to three years to cut costs. The issue of 'how long a professional program should be' has been a hot issue for discussion in Tanzania. Some have claimed that once a person knows the subject matter, teaching skills will come automatically through experience, and hence funds should not be invested in lengthy pedagogical training. This is, of course, a fallacy held by many, and still a subject of research in all professions. Others believe that a lengthy pedagogy program will help raise the teacher's ability to deliver content effectively, hence it is most important. We now know that both components are important, and a lifelong professional development in content development and pedagogy expertise is the ideal model.

Staff Development and Capacity Building in Science Education

The project also runs parallel training programs for selected academic staff, whereby four staff completed PhDs in pedagogy in sandwich programs run jointly by Dutch Universities and the UDSM, and one at the University of Pretoria South Africa. A Master of Education (MEd) program in science education was also designed and established from 2001. By 2007 it had produced over 30 graduates, while in 2007, 18 candidates were registered. The program consists of two semesters of course work introducing students to research methods, basic issues in curriculum design, ICT, professional development, and science and society, followed by design of a research project and writing of a dissertation. Optional courses focus on areas of interest such as Assessment, ICT, Science and Society, Environmental Education. It is now becoming popular, with more students applying each year. In 2006-7, over 60 students applied but the faculty was able to select only 18 based on capacity. The goal is to strengthen this program so that it can produce more qualified master teachers who can work as mentors, curriculum developers, teacher educators and professional development staff in the education sector. There are, of course, other education masters programs running in the universities, including the general MA in Education, Masters in Applied Social Psychology, MA in Educational Administration at UDSM and others in the new emerging universities in Tanzania such as the Open University, St. Augustine University and Tumaini University, though many of these are at the initial stage. Doctoral research produced work of various natures ranging from ICR use in education (McKenney, 2001; Tilya, 2003), mathematics pedagogy (Kitta, 2004), and microlab chemistry (Mafumiko, 2006).

In-service Education

TEAMS also ran an in-service education prototype development program which developed lesson materials that were used in professional development, and later improved and exported to other projects, including Education II Project, Zanzibar Trainers' Project, and Induction Program for licensed teachers. The main thrust of these lessons was to encourage interactive and transformative learning strategies in schools and promote the use of outdoor activities, and practical laboratory work. As Ottevanger, Feiter, O-saki and van de Akker (2005) comment, the project might have been more sustainable if following project conclusion in 2004, the proposal to create a university-wide centre for science and mathematics education was taken up. However, internal struggles led to shelving of the proposal, and like other similar projects on the continent, equipment and materials acquired during the project were left unattended. Staff trained during the project have been assigned other duties and their team broken up. This provides a good lesson for development partners to focus more on mainstream development than on the project approach in development assistance.

SCIENCE EDUCATION IN SECONDARY SCHOOL PROJECT (SESS)

SESS was a project designed to improve classroom performance in secondary science and mathematics started by the Ministry of Education and Culture in 1996. The project identified 15 pilot schools country-wide and improved the condition of textbooks (at least ensuring that three children shared one science or mathematics textbook) and laboratory facilities by purchasing a minimum laboratory working apparatus kit for each school. The project also launched a Training of Trainers Program which was based in zones and was supported by a national resource person training team. An internal review in 2000 showed that while textbooks and laboratory facilities had been improved in project schools, classroom practice changed very little (O-saki, 2000a, 2000b). Following recommendations of the review, project management was improved and by end of the project in 2003, the project produced high caliber of trained resource persons, who are working in the secondary school sector, and some of whom are exemplary school inspectors. The project is now mainstreamed into the secondary education department of the

Ministry of Education and Vocational Training, and its resource persons continue training whenever they called upon to do so.

THE SCIENCE TEACHER IMPROVEMENT PROJECT (STIP)

A third project, the Science Teacher Improvement Project (STIP), supported by the German government organization GTZ, operated in schools run by the Christian Social Services (CSSE) organization had main goals which were similar to those of SESS, but had a specific focus on a starter experiment before any new science or mathematics topics were taught. This project ran several trial workshops in church schools and ended in 2003, but its ideas were mainstreamed into SESS and are now part of the Ministry of Education INSET training focus.

EDUCATION II PROJECT

In 2002 the Ministry of Education received financial assistance from the African Development Bank to improve the science and mathematics situation in schools in three ways: by the purchase of relevant textbooks, construction of science laboratories, and design of science INSET courses and materials. The Faculty of Education at the University of Dar es Salaam was charged with the task of designing INSET courses and materials and train resource persons, which it completed it in 2006. The main products of the project included modular training materials for specific groups in the secondary sector: science teachers (physics, chemistry, biology, mathematics); science/mathematics teacher educators; school inspectors; school heads on the management of science teaching and learning resources.

These modules, which focus on interactive teaching, experiments and field work, were completed, tried out and revised, and have been used for running workshops in newly established schools during 2005 throughout the country. No evaluation has been done so far to determine their impact.

In language education, the British Government through the British Council ran an English language support program from 1990-2000 to improve secondary science teachers' language skills and also to produce teaching materials titled 'Baseline English'. These materials have been used in schools since then, but over all, the quality of English teaching is still viewed as unsatisfactory (Bevan, 1992). Those who say English standards have fallen may be comparing recent school leavers with those of the 1970s and 1980s, when there were only a few students in schools, and very good teachers, both the best local graduates who then joined teaching and the expatriate teachers who came from various parts of the world (UK, Canada, Australia, India etc.). This era is gone, unfortunately. We shall return later to the challenges created by the exodus of the good teachers and the recruitment of weaker graduates into the teaching profession. For some, the issue should now be to return to the national language as a medium of instruction (Brock-Utne, 2002; Qorro, 2004).

SECONDARY EDUCATION DEVELOPMENT PROGRAM 2005-2009.

When the TEAMS project closed in 2004, the Ministry of Education launched the Secondary Education Development Program, which implies a large expansion of the secondary sector to take in more students. As Ottevanger et al. (2005) and the Education Minister's Annual Report (Tanzania Government, 2006) note, from January 2005 to 2007 the secondary school intake more than doubled, and hence the programs designed by TEAMS are of crucial importance in producing teachers and advisers for the system. The demand for secondary school teachers has also more than doubled and the Government has taken several measures to ensure that the system functions effectively.

The Government has increased the capacity of universities to admit more undergraduates into the teaching degree programs. The University of Dar es Salaam main campus, for example, has doubled its intake of BSc(Ed), BA(Ed) students since 2006 and maintained the BEd student intake, while also encouraging the Postgraduate Diploma in Education program to increase enrollment.

Two new constituent colleges were formed in 2005, with a total annual intake of 2000. Other public universities are also admitting education students and private universities and colleges have been opened up in Moshi, Zanzibar, Mbeya, Dodoma, and Mwanza.

To increase the speed of producing teachers, the four year university degree program was reduced to three years under instruction from the Ministry of Education. Thus, from the 2005-6 academic year the new admission cohort will graduate in three years. There may be an impact on quality of teachers, but the Government is less concerned with this at the moment. Thus the undergraduate program revised by TEAMS in 2001 has been affected by the restructuring, and some of the new pedagogy courses have been dropped by the faculties of science and education due to reduction of allocated time to three years.

There is also a teaching diploma program offered in 35 teachers' colleges run by the Ministry of Education and Vocational Training which admit lesser qualified advanced secondary students, and offers a two-year diploma program, then sends them to teach in lower secondary school classes. This program is also being revised and for the first time, the University of Dar es Salaam and its colleges are also offering a Diploma in Education. These students take two teaching subjects (e.g. physics & mathematics, chemistry & physics, chemistry & biology, etc.) and education courses and graduate after two years. The present group will be allowed to proceed and complete the undergraduate program in two years (instead of three) if they pass the diploma with distinction.

Finally, the Ministry of Education also launched in 2005, a 'crash training program' for advanced level finalists who prefer to join teaching immediately after completing senior secondary school. They undergo a one-month intensive training program and are given a license to teach thereafter, but have an opportunity to undergo on-the-job professional development, either by joining the diploma or degree program part time or full time later on. This group worries some because if they do not upgrade, they could be a very weak teaching force in schools. There are various forms of curriculum review taking place to meet the diversity of student population enrolled in schools. All the school syllabuses have been revised as well as certificate and diploma teacher training programs.

REFLECTIONS

There are a few lessons we have learned in this process.

There is no one right formula on how long a teacher education program should run, though there is a need for balanced training in subject matter, professional studies (psychology of learning, pedagogy, etc.), field experience with professional feedback and on-the-job coaching. Tanzania has come a long way, restructuring and running pre-service teacher training programs. But these have not often been formally evaluated and revised, and this is unlikely until foreign assistance is available. The three-year program designed in 1968 went to four years duration in 1990, and was returned to one of three years duration in 2005 without any evaluation. There is now the need to evaluate mainstream programs before experiments become too long and costly.

The small secondary sector until 2004, led to the production of very few highly qualified graduate teachers in the education system. The expanding sector needs experienced professionals. The challenge is to produce more of them and learn from others education systems.

Since the expansion in 2005, the demand for science and mathematics teachers has grown fast and it will peak in 2009, when over 50% of primary school leavers are expected to join secondary school at Form 1. Currently about 35% are being promoted to secondary. With an estimated 500,000 children joining first year secondary education in 2009 (currently it is 243,359), and over 50% of these doing science, the demand for well trained teachers is likely to be very high. However, signs of diminishing interest in science careers are threatening the profession, and perhaps the only way out, is to hire good expatriate teachers albeit for three to four years to stimulate interest and provide support.

There is a very big challenge to improve the quality of teacher training in universities and the Ministry of Education teachers' colleges, because school class sizes are now very large, and expansion is at a critical level. Class sizes of up to 70-80 in secondary schools are now common. Much of this may require expansion of in-service education and training.

The lessons learned in implementation of TEAMS, SESS, STIP and Education II projects are a great support for the expanding secondary sector. These include the following:

- That increased laboratory supplies do not necessarily make teachers more inquiry oriented (Chonjo et al., 1995, O-saki, 2000a, 2000b);
- That the supply of textbooks must be accompanied with addressing *local relevance* and teacher training on how to encourage textbook use in classrooms and for homework;

- That the activity approach, laboratory experiments and other innovations taught in workshops may not be practiced in school if there is no follow up, including peer coaching, mentoring and effective school management (Knamiller, O-saki & Kuonga, 1995); and
- That as enrollment expands, student diversity increases, and teachers need to be prepared for a more constructivist approach to teaching and handling of individual student problems.

Recent research underscores the need to promote reflective practice and lifelong professional development, rather than 'one shot' pre-service training (Schön, 1986). The view that professional skills develop from knowledge accumulated from progressive practical experience through such stages as novice, competent, and expert (Dyerfus, 1986) are gaining momentum. Alba, Dell and Sandberg (2006) propose a combination of vertical development (of skills that accompany experience) and horizontal development (understanding of practice within the practice in question); our challenge is to research and develop a model that balances these two aspects of professional training.

The postgraduate education programs need to be expanded in order to produce more leaders and researchers to serve the education system. The University of Dar es Salaam had proposed to expand enrollment and also create a larger Graduate School of Education which would run postgraduate programs in science and mathematics education, language education, counseling, and management, and so on. This proposal was shelved in favor of a single graduate school for the whole university.

Finally, research into teacher education, classroom management and pedagogical practice are areas people are working on in collaboration with others in the Institute of Education of the University. They have, so far, also published materials that reflect the research collaboration with colleagues within southern Africa (see, O-saki, Hosea & Ottevanger, 2004; O-saki, van den Akker, Uiso & Ottevanger, 2002), and hope to continue through the graduate education team. These need the support of government and international organizations.

EMERGING ISSUES IN SCIENCE TEACHING IN TANZANIA

In the process of implementing the changes that have taken place since achieving its independence, Tanzania has made four bold moves uncharacteristic of countries in similar situations. It continued with the colonial curriculum; then, reviewed it, focusing on self reliance; adapted the inquiry curriculum from Nuffield, which was subverted; and then reviewed this and reverted to the original Nuffield system.

Like other countries, Tanzania first continued with the old colonial curriculum with its differentiated learning opportunities for the rural African, the urban commercial Asian and the European Western strata, for a short period of six years. These curricula used materials adopted from the West for Africans contained materials originally designed for less able children in Western countries, or specially written for the colonies.

Then, it reviewed the old system and decided to focus on a reform into the education for self-reliance (ESR) focus, where the main goal in all subjects (including science) was to promote an inquiring mind, encourage learning from other societies, and develop critical thinking and respect for work, including manual work (which was discouraged in elite schools). In implementing the ESR goals, Tanzanian politicians pushed educators to lean more towards encouraging activities such as rearing livestock, farming, commerce and the like.

Third, science and mathematics curriculum designers adapted inquiry-based curricula from Nuffield sciences and elsewhere to introduce inquiry-like activity-based science and mathematics learning. But, due to a lack of trained teachers and innovators, the materials failed to make any impact on pedagogy and by 1990 the schools had reverted to a predominantly transmission pedagogy.

A review by Chonjo, O-saki, Possi and Mrutu (1995) saw the danger to science education nationwide, and proposed a return to activity-based approach which was implemented through TEAMS, SESS, SIP and Education II projects.

NATURE OF SCIENTIFIC INQUIRY

Despite these reform efforts in Tanzania and elsewhere in post-colonial societies, three issues plagued the curriculum development process during the period. In most of sub-Saharan Africa, there is a tradition of indigenous science activities in the countryside that have served communities for centuries, but whose mode of inquiry is slightly different from Western science. There are indigenous technologies of beer-brewing, food preservation and storage, medicine extraction from various plant materials, work with soil and chemicals, animal husbandry, and veterinary services, indigenous surgery, hematology, astronomy, and so on (O-saki, 1991, 2002). Children growing up in rural (and even urban) communities in Africa experience these technologies at home, but are rarely given opportunity to learn them at school. At home few get the opportunity to learn informally, and no reinforcement is given in school (O-saki, 2002; O-saki et al., 2004). Three examples are given here.

First, take fermentation of alcohol. In rural areas of Tanzania there are various brewing technologies depending on the type of raw materials locally available. Where fruit products are abundant, fruit-based local brew is made. From bananas is made *rubisi* (West of Lake Victoria), and *mbege* (mixed with finger millet) is made in the Kilimanjaro area. Different varieties of banana are involved in each case. These two drinks are different in taste. From cashew nut fruits they make a brew called *uraka*, from pineapple, *tembo* pawpaw, and other fruits, they make other local brews. From plant juices they make *tembo* (palm wine) from coconut palms and from bamboo juice *ulaasi*. Likewise, from grains such as maize, millet, sorghum, and seeds such as 'beans', come brews such as *komoni, kangara, mtama, mapuya* and so on. Children know the process used in their local community.

Since the colonial curriculum was introduced in schools, the only method of fermentation taught in the topics of *anaerobic respiration* and *fermentation* is that of adding brewers yeast to a solution of sugar and fermenting for a day or two. The village processes used for decades or centuries have never used externally added yeast. Science books have, until now, only encouraged teaching the Western idea of sugar/yeast fermentation, and hence few of those who have attended school can make the indigenous beer, or improve the processes. Yet majority of ordinary people drink their local brew, it is cheap and locally available. The science curriculum here thus *alienates* indigenous practices from modern science, and indigenous lifestyle from what school teaches. This suggests science is irrelevant, and is just a subject done to pass examinations.

A second example relates to theories of evolution. There are many indigenous explanations about how the modern species may have come about. There is, for example, an interesting theory about the origin of animals emerging from one species and diversifying later, taught in village tales, but ignored by all schools. Textbooks say nothing about these, nor attempt to link them to Darwinism, Lamarckism, and so on.

A similar situation occurs with regards to classification of living organisms. The local criteria of separating species, families in some communities, is left out and children are compelled to memorize the Linnaean system of Binomial nomenclature, cladistics, or dichotomous keys without relating to anything they have learned from, or is happening, at home.

The second issue relates to the way indigenous scientists generate and test hypotheses (Odhiambo, 1981). If you have a headache and you go to an indigenous healer he/she will not take a blood sample to test for malaria parasites. If he/she suspects malaria he will refer you to a Western hospital or give you a herb. However, if he suspects hysteria, schizophrenia or any aspect of mental illness, asthma or diabetes the approach may be totally different from that of a Western-trained doctor. What this approach is, few people know, but indigenous healers are much more successful in dealing with mental illness in Tanzania. And yet their approach or any other indigenous processes are not addressed in school. In view of this, there is a very big gap between the dual world of science in African communities and the unitary approach taken in science textbooks and classrooms (O-saki, 1991, 2002). This issue has rarely been seriously addressed.

The third issue has to do with language and medium of instruction (Brock-Utne, 2005). This may be related to the issue of the processes of science. The activities done in the local environment, and discussed in the vernacular, build on scientific concept development in children. When children are compelled to use a foreign language to understand and explain phenomena, there arise serious problems of translation and re-interpretation of past experience. Whether Chagga children are taught in Kiswahili, Arabic, English, French, they remain Chagga in their novel experiences, of stargazing, banana growing and sampling, beer-brewing, or rock sampling (O-saki, 2005). Change in the language of discourse has serious consequences. All the projects that have been running in schools ignore language and inquiry processes, but researchers continue to make arguments against the use of local languages: indigenous languages are too limited (or primitive) to be used in scientific discourse; aboriginal communities have no scientific knowledge, they need only learn Western science; it is difficult to explain complex science in local languages; and so on. These arguments may be missing the issue. Those clinging to this view have often advised governments to suppress indigenous languages (claiming this suppresses tribalism) with disastrous consequences on concept acquisition and meaningful learning. Evidence shows that countries which have more and better scientific productivity are also those which teach in the language of their everyday life.

Fourth is the issue of teacher preparation. No educational system can rise above the quality of its teaching force. Tanzania, like a few other countries, is one of those which have neglected the training of quality teachers, especially in language, science and mathematics. And this applies specifically to primary education where in 1977 following UPE, they ran a crash training program which recruited primary school failures to fill primary school teaching posts and gave them a three month training course in college followed by on-the-job, and on-task visits. They were academically weak and poorly trained. The program might have worked if these were the bright cohort. They have since become a significant problem in many primary schools to date. As of the year 2000, there were 62,000 such teachers in primary schools, (see Ministry of Education & Culture, 2000) failing to upgrade to O-level. They faced the threat of being phased out but they pleaded with the authorities to be retained. They have resisted and most are now nearing retirement. They are weak in content knowledge, and contributed heavily to the lowering of standards in primary science, which has been passed on to secondary and tertiary teaching. Due to these policies, Tanzania shies away from encouraging a strong professional teacher association that fights for status, and enforces quality assurance in training and recruitment, unlike its encouragement for the legal, medical, nursing and accounting professions.

In 2005, the Government repeated the same UPE blunder with respect to secondary school teachers following SEDP expansion. Over 3,000 secondary school leavers with low A-level passes have been recruited and given crash training of one month in duration, and hired to teach, hoping they will improve through their experience in schools. Schools have no peer or mentor support systems for them yet as good teachers are leaving the system every day. How can untrained, or poorly trained teachers, mentor new members of the profession? This problem will plague the secondary sector for years until a watchdog for professional standards is created.

Finally the issue of work in school as part of ESR is dying out and schools are increasingly going academic. There arises the problem of work skills, and preparation for the world of work after school. This challenge faces many countries today as school graduates struggle to find employment, and are unable to enter into self employment. A recent curriculum review to provide opportunities for acquiring work skills is not proving successful, due to lack of a teacher training program. However, Tanzania had started a path to address it through education for self-reliance, 40 years ago. Again, like the arrow, once sped from the bow, and the word, spoken in haste, this missed opportunity cannot be retrieved. Or, may be it can. To paraphrase Mwalimu Julius Nyerere, ‘To plan is to choose’.

NOTE

[i] Source: Government Projections, Ministry of Planning and Development, 2005.

REFERENCES AND BIBLIOGRAPHY

Alba, X.X., Dell G., & Sandberg, J. (2006). Unveiling professional development: A critical review of stage models. *Review of Educational Research*, 76(3), 383-412.

Bevan, R. (1992). Review of English Language Support Project in Tanzania. *International Journal of Applied Linguistics*, *2*(1), 120-125.

Brock-Utne, X.X. (2002). *Language, democracy and education in Africa.* Nordiska Afrikainstituet. Uppsala, Sweden.

Cameroon, J., & Dodd, W.A. (1966). *Society, schools and progress in Tanganyika.* Oxford: Pergamon.

Chonjo, P.N, O-saki, K.M., Possi, M., & Mrutu, M. (1995). *Science education in secondary schools: Situational analysis.* Dar es Salaam, Tanzania: Ministry of Education and Culture & GTZ.

Dreyfus, H.L., & Dreyfus, S.E. (1986). *Mind over machine: The power of intuition and expertise in the era of the computer.* New York: Free Press.

Kallagho, G.L. (2003). *An evaluative study of SESS Project in Nine schools.* Unpublished MEd dissertation, University of Dar es Salaam, Dar es Salaam, Tanzania.

Kibga E (2004). *The role of practical assessment in teaching and learning of 'O' level physics in Tanzania.* Unpublished MEd dissertation, University of Dar es Salaam, Dar es Salaam, Tanzania.

Kitta, S. (2004). *Enhancing mathematics teachers' pedagogical knowledge and skills in Tanzania.* Unpublished PhD thesis, University of Twente.

Knamiller, G., O-saki, K.M, with Kuonga, H. (1995). Tanzania teachers' understanding of the science embedded in traditional technologies: A study to inform teacher education. *Research in Science and Technological Education*, *13*(1), 67-76.

Koda, G.M. (2006). *The effectiveness of teacher resource centres. Case study of TRCs in Kilimanjaro and Mbeya in Tanzania.* Unpublished PhD thesis, University of Dar es Salaam, Dar es Salaam, Tanzania.

Komba, D. (2006). *Reflections on Mwalimu Nyerere's Philosophy of Education for Self Reliance*: Paper presented at the International Seminar to Commemorate the 7th Anniversary of Mwalimu Nyerere. Dar es Salaam. Faculty of Education. University of Dar es Salaam, Dar es Salaam, Tanzania.

Mafumiko, F.M.S. (2006). *Micro-scale experimentation as a catalyst for improving the chemistry curriculum in Tanzania.* Unpublished PhD thesis, University of Twente, the Netherlands.

McKenney, Susan (2001). *Computer-based support for science education materials developers in Africa: Exploring potentials.* Unpublished PhD thesis, University of Twente, the Netherlands.

Ministry of Education and Culture, Tanzania. (2000). *Teacher education master plan.* Dar es Salaam, Tanzania: MOEC.

Ministry of Planning and Development. (2005). *Government projections.* Dar es Salaam, Tanzania: Government Publishers.

Nyerere, J.K. (1967). *Education for self-reliance.* Dar es Salaam, Tanzania: Government Publishers.

Odhiambo, T R (1981). Understanding science: The influence of African traditional view of nature. In P.D.G. Gilbert & M.N. Lovegrove. (Eds.), *Science education in Africa.* Heinemann.

O-saki, K.M. (1991). Factors influencing the sue of environment in science teaching: A study of biology teaching in Tanzania. Unpublished PhD thesis, University of Alberta, Canada.

O-saki, K.M. (1996). The changing form and content of the curriculum in Tanzania: 1961-1994. *Papers in Education and development*, *17*, 8-19.

O-saki, K.M. (2000a). *Quality education in Tanzania: A focus on standards in the secondary education sector.* Paper presented at the Annual Headmasters conference. Arusha, Tanzania.

O-saki, K.M. (2000b). *The science education in secondary schools. Internal project evaluation.* Dar es Salaam, Tanzania: Ministry of Education and GTZ.

O-saki, K.M. (2002). Science and technology education in national development. In S. Nyirenda, & A.G Ishumi, (2002). *Philosophy of education: An introduction to concepts, principles and practice* (pp. 171-196). Dar es Salaam, Tanzania: Dar es Salaam University Press.

O-saki, K.M. (2005). Reflections on the medium of instruction and interaction: A focus on science teaching and learning. In B. Brock-Utne, Z. Desai & M. Qorro, *LOITASA (Language of Instruction in Tanzania and South Africa) Research in progress*. Dar es Salaam, Tanzania: KAD Associates.

O-saki, K.M, van den Akker, J., Uiso, C., & Ottevanger, W. (2002). *Science education research and teacher development in Tanzania.* Amsterdam: Vrije Universiteit, International Cooperation Centre.

O-saki, K.M., Hosea, K., & Ottevanger, W. (2004). *Reforming science and mathematics education in Tanzania: Obstacles and opportunities*. Dar es Salaam, Tanzania: TEAMS project.

Ottevanger, W.J.W., Feiter, L., O-saki, K.M., & van de Akker, J. (2005). The TEAMS project in Tanzania: From intervention to capacity building. *Journal of International Cooperation in Education*, *8*(1), 111-128.

Qorro, M. (2004). Popularising Kiswahili as the language of instruction through the media in Tanzania. In B. Brock-Utne, Z. Desai & M. Qorro (Eds.), *Researching the language of instruction in Tanzania and South Africa: African minds.* Cape Town, South Africa: One-World Books.

Rubagumya, C.M. (1990). *Language Education in Africa: A Tanzanian perspective*. Clevedon, UK: Multilingual Matters.

Schön, D. (1987). *Educating the reflective practitioner: Towards a new design for teaching and learning in the professions*. San Francisco: Jossey-Bass.

Tanzania Government. (1967). *The Arusha Declaration and TANU's policy of socialism and self reliance*. Dar es Salaam, Tanzania: Government Printer.

Tanzania Government. (2005). *Secondary education development programme (SEDP).* Dar es Salaam, Tanzania: Ministry of Education and Culture.

Tanzania Government. (2006). *Hotuba ya Waziri wa elimu na mafunzo ya ufundi kuhusu makadirio ya matumizi ya fedha kwa mwaka 2006/7.* Dar es Salaam, Tanzania: Ministry of Education and Vocational Training.

Tanzania Government. (2006). *Basic education statistics (BEST), 2002-2006. National data*. Dar es Salaam, Tanzania: Ministry of Education and Vocational Training.

Tilya, F.N. (2003). *Design specifications for microcomputer based laboratory in the teaching of physics.* Unpublished PhD thesis, University of Twente, The Netherlands.

AFFILIATION

Kalafunja M. O-saki
University of Dar es Salaam
Tanzania

AKINSOLA AND OLAOYE

A GLIMPSE INTO MATHEMATICS CURRICULUM DEVELOPMENT AND IMPLEMENTATION IN NIGERIAN PRIMARY AND SECONDARY SCHOOLS

HISTORICAL ANTECEDENTS

The development of mathematics education as a discipline is affected by many factors. The national policy of the system of instruction plays a major part as well as the history of the country and the academic world (Furinghetti, 2006). An historical account of mathematics education and mathematics curriculum development in Nigeria is sprinkled throughout the publications of mathematics education and research (Fajemidagba, 1991). According to Fajemidagba (1991), In Nigeria the formal teaching of mathematics started with arithmetic, a component of mathematics, at the primary and post primary schools. Evans Brothers Limited's early in-roads into mathematics education in Nigeria came from the inspiration of the colonial government at the time. A study of Nigerian's educational development in the 1940s, inspired Evans UK to send an emissary to Nigeria in 1945/46 in the person of Dr D.H. Larcombe who later returned to recommend the publication of a primary arithmetic book (Akinsola, 1999). Arithmetic was compulsory for every primary school student, and had to be passed before a student could obtain the Primary School Leaving Certificate. The same condition held for the teacher training colleges - either Grade 2 or Grade 3. At the secondary school level, mathematics was taught in compartments: algebra, arithmetic, and geometry with trigonometry. The schools were largely guided by the examination syllabus at the Cambridge local syndicate, the West African Examinations Councils (WAEC), or the University London General Certificate of Education Ordinary or Advanced Level. Later in 1968, the WAEC published three alternative syllabi for mathematics at the Ordinary level: alternatives A, B and C. Alternative A contained arithmetic process, algebra, absolute geometry and trigonometry. Alternative B contained the same items as Alternative A, but with the additional topic of coordinate geometry. Alternatives A and B were for average mathematically capable students studying mathematics. The Alternative C syllabus contained topics in commercial subjects for commercial students (Badmus, 1997). By interpretation, the mathematics curriculum did not cater for the interest towards science and technology of the mathematically inclined students. These examination syllabi were invariably turned into teaching syllabi by mathematics teachers.

The major cause of the changes in school mathematics curricula and programs was the launching of the Sputnik, the first earth satellite in space, in November 1957 by the Russians. The event, according to Griffiths and Howson (1974), had

Richard K. Coll, Neil Taylor (eds.), Science Education in Context: An International Examination of the Influence of Context on Science Curricula Development and Implementation, 251–260.

an enormous effect on American complacency about their superiority in engineering capability compared with the Russians. The result of that doubt was a series of hot debates and arguments on the suitability or otherwise of the school mathematics and science curricula. These occurrences were regarded as the 'issues and forces' behind the reformation of school mathematics curricula contents, which also affected various reforms in the school mathematics program in Nigeria. This was the 'revision syndrome' which swept the world (Fajemidagba, 1991).

Elementary mathematics was introduced to primary schools in Nigeria in order to replace arithmetic and this was taken on board by teacher training colleges. The attempt was to produce a primary school teacher who could teach elementary mathematics, and in-service programs were organized for practicing teachers for the purpose of retaining them. As a continuation of the innovation, entrance examinations for secondary schools, which used to contain arithmetic and English, were changed to mathematics and general knowledge, which included some mathematics concepts and English. The national Common Entrance Examination took a lead in the changes. Moreover, the trend forced nearly all primary schools in Nigeria to commence the teaching of mathematics.

The introduction of 'modern mathematics' to secondary school students marked the beginning of the teaching of mathematics as a composite or integrated body of knowledge, rather than as a subject compartmentalized into arithmetic, algebra and geometry with trigonometry, that were separately taught to students previously. The introduction of modern mathematics was the aftermath of the wave of changes in school mathematics that traversed industrialized nations. Ohuche (1978) reported the outcomes of an international conference on Science in the Advancement of New States in Retrovoth, Israel in July 1960. This conference contained some recommendations for innovations in school mathematics for African states. The recommendations led to the establishment of the African Mathematics Programme (AMP) known as 'Entebbe Mathematics', because the program was developed in Entebbe in Uganda. The program was eventually called the African Mathematics Programme because its initiators had in mind mathematical content which would centre on the needs of Africans, and which would correlate with African cultural values and traditions without any prejudices against modern education.

The Entebbe Mathematics program started in 1962 with three macro-goals:

- to prepare and publish instructional materials in mathematics for use in schools, teachers' colleges and in-service institutions in African States;
- to trial - test the developed materials in schools and teachers' colleges; and
- to train teachers for the proper use of the developed instructional materials. (Osibodu, 1988)

Thus, the goals seem to parallel the Research, Development and Dissemination (RD&D) paradigm. That is, those involved in the project were assigned to 'fishing out' local materials that would meet the mathematics needs of African children, then to develop these materials to meet the mathematical needs of the African child.

Mr. Hugh P. Bradely was the director of the AMP and the Education Development Center; Newton, Massachusetts housed its administration. This was the case because the US government provided the bulk of the funding for the program. African and American mathematicians were employed at various workshops to develop mathematics courses and materials that would go into the program. The mathematicians were engaged as consultants. Some of them lectured at various institutes organized for the training of African mathematics teachers on how to use the mathematics materials. Eleven training institutes were established in Ethiopia, Ghana, Liberia, Malawi, Nigeria, Uganda, Sierra Leone and Tanzania, and a four-week meeting was scheduled for the development of textbooks. Three textbooks were produced for use in secondary schools, Advanced Mathematics 1, 2 and 3. These books were adopted for mathematics teaching and learning in African countries that accepted the African Mathematics Programme. In addition to the secondary school textbooks, a series of six textbooks were developed for primary schools. The mathematics concepts treated in the books include the structure of arithmetic, foundations of geometry, measurement, functions and probability, the number line and fractions.

Nigeria, according to Ohuche, (1978) participated actively in every facet of the African Mathematics Programme; however, the only successfully implemented project under the auspices of the AMP was the Lagos State Primary School Mathematics Experiment. The project started in January 1964 under the directorship of Professor Alele-Williams. Other states in Nigeria did not accept the modernization of the mathematics curriculum, dubbed 'modern mathematics' at the primary school level. According to Williams, (1974) the Lagos State Ministry of Education accepted the introduction of innovations into the primary school mathematics curriculum. Thus, the original Entebbe primary mathematics textbooks were adopted for use in Lagos state primary schools.

As a part of the project, primary school teachers in Lagos state were retrained through in-service activities. This was in tune with the objectives of the African Mathematics Programme. The actual use of the mathematics materials did not commence until 1971. However, traditional textbooks (e.g., the Larcombe Arithmetic Series) were allowed to compete with the Entebbe Primary Mathematics series, and many teachers did not understand the objectives and the subject-matter content of the series. This was probably a reflection of the inadequacies in the mathematical training received by primary school teachers. The majority of the primary school teachers studied arithmetic in the course of their training, so they could not be described as specialists in mathematics teaching at the primary school level. It is of interest to note that the funding agencies were the United States Agency for International Development (USAID) and the Ford Foundation.

In 1970, the West African Regional Mathematics Programme (WARMP) was formed by West African English speaking countries in order to cater for the mathematical needs of the Anglophone West African Countries. It was an off-shoot of the AMP. Unfortunately, Nigeria declined to participate in the Programme (Ohuche, 1978). Rather, Nigeria established the Nigerian Educational Research Council (now Nigeria Educational Research and Development Council (NERDC)) in 1969. This body was given the responsibility of charting a new course in the modernization of the school mathematics curriculum in Nigeria. Before dealing with the activities of the NERDC, a brief historical account of the Joint School Project is given. The textbooks developed from the project were recommended for the teaching of modern mathematics in Nigeria.

Another dimension to the modernization of the mathematics curriculum was the introduction of Joint School Project (JSP) textbooks into Nigeria. The JSP was started in February 1964 by eight teachers from three secondary schools in Ghana. Eight teachers worked as a subcommittee of the Mathematical Association of Ghana, with the initial work on the project commencing at Achimota School, Ghana. The organizational agencies were the University of Ghana and the Ministry of Education, Accra, Ghana. The funding agencies were the Nuffield Foundation, the Center for Curriculum Renewal and Development Overseas, London, the Guinness Award Scheme, and the World Confederation of Organization of the Teaching Profession. The macro goals of the project were:

- to produce a 'new mathematics' course for secondary schools on practical work to make up for a lack of opportunity for such experience in students' cultural and educational background;
- to make the course relevant to the environment, future needs of students, and to encourage learning by understanding rather than by rote;
- to give an intuitive development of the course, since experience had shown that an over-emphasis on logic kills all interest and hinders progress; and
- to modernize mathematics topics in order to achieve the objectives above.

The JSP course was at two levels: basic, for those students who would do no mathematics after the Ordinary level; and special, for those who would specialize in science subjects after the Ordinary level. The project further differed from similar projects because emphasis was placed on how mathematics arises naturally in the environment, and how it can be applied in various situations rather than placing emphasis on the logical structure of mathematics itself. The originators of the project felt that this approach would prove to be much more valuable for the average student in Ghana. A set of textbooks for both students and teachers was produced by the originators of projects. Several secondary schools in Nigeria adopted the textbooks for mathematics teaching and learning, and the WAEC started examining students on a syllabus based on the project in June, 1969 (for further details of JSP, see Lockard, 1970).

The modern mathematics program, however, did not succeed in Nigeria. Catalogues of criticisms were levied against the programs. The contents of the modern mathematics curriculum were meant for potential mathematicians, not for the consumers of mathematical ideas who form the bulk of the student population.

Also, parents found the content difficult to understand, and thus could not help their children. Finally, parents and teachers pointed out that the textbooks were too 'wordy' for lower secondary school students.

It is very difficult to really pin down a logical argument against the modernization of mathematics in Nigeria. It is well understood and accepted that there were a small number of mathematics teachers at the secondary school level and majority of the teachers were not adequately trained to teach the new mathematics. Also, it could be conjectured that a majority of the trained mathematics teachers lacked in-depth knowledge of mathematics content, and the quantity of mathematics learned at the various training colleges might not be adequate for an in-depth understanding of the content of the new mathematics curriculum. The unification of mathematical ideas has been a serious issue for mathematics educators and mathematicians. Attempts are made to postulate a unifying theme for mathematics at the primary and secondary school levels. The new mathematics introduced either through the Entebbe Project or the JSP project seems to have used set theory as the 'unifying concept'. Other researchers in mathematics have also proposed the 'function' as the unifying theme for secondary school mathematics. By and large, no unified mathematics idea has been provided for students. Undoubtedly, researchers in mathematics education and mathematics will continue to work on the formulation of a unifying theme for mathematics teaching in schools.

The controversy surrounding the modern mathematics prompted the Federal Government of Nigeria to organize a conference, held in Benin City in December, 1977. Mathematics educators and mathematicians were invited as participants at the conference. However, it was at the conference that a ban on modern mathematics was announced by the Government. The announcement was a shock to many participants, because it was contrary to their expectations.

After the announcement, the Nigerian Education Research and Development Council (NERDC) was mandated to produce appropriate mathematics curricula for the Nigerian children. It was stipulated that the curricula should take cognizance of the needs of the nation and those of the children. As a sequel to this, a workshop was organized at the University of Ibadan in February, 1977. The workshop's focus was on the study of the problems facing the teaching of mathematics in Nigerian schools and colleges. It was at this workshop that proposals were made for the development of a set of new mathematics curricula for schools and colleges in Nigeria. This set was intended to adequately address the provisions contained in the New National Policy on Education. In the policy, it is stipulated that a 6-3-3-4 system of education would be embarked upon in Nigeria. In the system, mathematics was to be a core subject for primary school pupils, junior and senior secondary school students.

Several workshops were organized by the NERDC in order to produce a set of mathematics curricula that met the requirements of the 6-3-3-4 system of education. A mathematics curriculum was developed for primary schools, one for junior secondary schools and two (General Mathematics & Further Mathematics) for senior secondary schools. The Further Mathematics curriculum is meant for those who have the intention of studying mathematics beyond the senior secondary school level or those who intend to study mathematics-related disciplines, for example, physics, chemistry, engineering, architecture, and the like. Another mathematics curriculum was developed for the teacher training colleges.

These mathematics curricula are in operation in every school and teachers' Grade 2 in Nigeria. Attempts are being made to produce textbooks based on the curricula. The NERDC has produced a set of six textbooks for the primary schools, covering Primary 1 through Primary 6. Also the Mathematical Association of Nigeria, (MAN) as well as Science Teacher Association of Nigeria (STAN), has produced a mathematics textbook for junior Secondary Year 1 up to Senior Secondary 3. Other government agencies and publishing houses have also produced mathematics textbooks based on the new set of mathematics curricula. However, no serious evaluation has been performed on the curricula and the accompanying mathematics textbooks. This is a challenge to mathematics educators and mathematicians.

MATHEMATICS CURRICULUM TRANSITION (1970-1977)

Modern mathematics spread its tentacles faster among the primary schools though with varying times of commencement in the different parts of the country. It started in Lagos State primary schools in 1971, while other states of the federation started in 1974. The political undertones experienced during that period include an acute shortage of personnel to implement the curriculum, lack of suitable textbooks, multiplicity of courses and poor level of information dissemination by existing educational bodies. The evidence was glaring that the northern states had not trained teachers to handle the curriculum. In 1973, a modern mathematics curriculum was drawn up the secondary schools intended to make mathematics more meaningful, and it used set theory to make mathematics concepts meaningful - but few teachers could do this. This was the political undertone to the implementation of modern mathematics curriculum. The efforts of the NERDC, Mathematical Association of Nigeria (MAN), Institutes of Education of Universities and West African Examination Council made the nation realize the flaws in the scope and contents of modern mathematics, which some educators believed were the handiwork of some collaborator with colonial powers to re-enslave Nigeria via her school system. The poor mathematics results in the West African School Certificate Examination (WASCE) in 1974 were part of the evidence that there was problem with modern mathematics. As a result of consistent failures in mathematics, especially in modern mathematics, there was a general tension among educationists, parents and other stakeholders as to the desirability of continuing with the then mathematics syllabus. This made different mathematicians like Professors Iya Abubakar, Chike Obi, Adegoke Olubunmo, James Ezeilo and their contemporaries call for the cancellation of modern mathematics. The activities of some those mathematicians led to the formation of a different mathematics movement to protect the national curriculum from further degenerating into backwardness compared with other nations. These movements include African Mathematics Union's Commission on Women in Mathematics in Africa (AMUCWMA) founded by Grace Alele Williams, and War Against Poor Achievement in Mathematics (WARPAM). It was obvious that the non-appropriation of the modern mathematics curriculum also slowed down the pace of science and technology. This affected the teacher training curriculum that was prepared to cater for the teeming pupils of the primary schools. Furthermore, the government, nursing an ambition for universal primary education (UPE) to start in 1976, saw the need to arrest students' mass failure in mathematics in order to put the nation on the threshold of science and technology.

MATHEMATICS CURRICULUM TRANSITION (1977 TO 2007)

The 1977 Benin conference on modern mathematics marked a turning point for the mathematics curriculum. While the participants were deliberating on the way out of correcting the problems engendered by modern mathematics, the then Federal Commissioner of Education, Dr A.A. Alli, announced that the Government was no longer interested in the continuation of the subject. This singular pronouncement shocked the stakeholders. At the conference, NERDC was mandated to come up with a proposal to prepare more relevant mathematics curricula for all levels on sound footing. This body organized a series of workshops (among which was a national critique workshop on mathematics curriculum at Onitsha in March 1978 under the eminent mathematician, Professor J.O.C. Ezeilo) which produced a primary school mathematics curriculum, primary teacher education mathematics curriculum and secondary school mathematics curriculum. Table 1 shows the nature of mathematics curriculum in secondary schools, testifying to the scope and contents of mathematics at that level.

Unlike the old secondary school mathematics curriculum that comprised arithmetic processes, algebra, geometry and trigonometry, the new integrated mathematics curriculum split the contents areas of the curriculum into major topics and aligned it with the group of students at which these topics were aimed. This clarified the confusion which the old curriculum brought about as there had been no such specification to help the then teachers understand how to handle the topics, even when few of them know the topics. Also, there was no other mathematics specifically prepared for the talented students who would pursue mathematics and related subjects further in life. This might be one of the reasons for the proliferation of liberal arts courses like classics, history, and law at the expense of science and technology. The new mathematics curriculum provided separate mathematics, and further mathematics curricula as shown in Table 2, in the upper three years of the secondary school levels for mathematically-inclined students who might wish to pursue careers in the areas of science and technology. More scientific and technological mathematical topics were introduced to ensure that the national goal of self reliance in the area of science and technology that had might be realized.

Table 1. Nature of mathematics curriculum in secondary schools

S/N	*Major Topics*	*Junior School Topics*	*Senior School Topics*
1	Numbers and Numeration	Number systems, functions, decimals, factorization, proportion and approximation	Indices and logarithms, exponents, set theory and progression
2	Algebraic processes	Mathematical statements, simple equations and Inequalities variables	Quadratic equations, simultaneous equations, and Inequalities
3	Geometry and Measurement	Basic properties of 2-and 3-dimensional-shapes, properties of angles, scales and the elementary trigonometry-ratios	Geometrical constructions, basic Euclid theorems, Advanced measurement in 2-and 3-dimensional shapes
4	Everyday Statistics	Data collection techniques and graphical representation, frequency tables, measures of central tendency (mean, median and mode), elementary ideas of probability	Construction and interpretation of statistical representation, histograms, bar graphs and charts, group data analysis, measures of central tendency and dispersion in a continuous case and data interpretation
5	Trigonometry		Further treatment of angles and compound angles, equation involving trigonometry identity
6	Probability		Experimental and theoretical probability, conditional and discrete cases

IMPLEMENTATION

The Federal Government of Nigeria introduced the 6-3-3-4 system of education in 1979. This development made it necessary to have two separate mathematics curricula for use in secondary schools (junior and senior). The implementation of the senior secondary mathematics curriculum was scheduled to be launched with the first graduates of the junior secondary school students in 1985/86 school year. Unfortunately, the implementation of the new system of education could not commence as scheduled in all states during the second republic (1979-1993). Only a few states took the giant step to implement the new policy even though education is a part of the nation's constitution. However, with the coming of military administration at the end of 1983, considerable progress was made in the implementation of the new policy and all the states of the federation embraced the policy (Jahun & Korau, 1991). This new dispensation was made possible simply because the military run a unitary kind of government.

The earlier curriculum for school mathematics was developed without much involvement of the classroom teachers. This is contrary to the suggestion of many curriculum specialists (Hawes, 1979). Proper implementation of the mathematics curriculum at any level of education depends largely on the classroom teachers attitudes, ingenuity, creativity and devotion to the teaching of mathematics (Gould 1960). In a study conducted by Jahun and Korau (1991) on mathematics teachers perception of the new curriculum, it was found that there was a sharp difference between the curriculum innovations and the way the mathematics teacher perceived the mathematics curriculum. Probably this was what the organizer of th National Curriculum Review Conference of September, 1991 had in mind when various stakeholders (including teachers and teacher unions) were invited to participate in a review of the various school subjects, including mathematics. The outcomes of this review led to new mathematics curricula which as of 2007 are in operation at the primary, junior and senior secondary schools.

Table 2. Nature of further mathematics curriculum in the secondary schools

S/N	*Major Topics*	*SS I*	*SS II*	*SS III*
1	Pure Mathematics	Identification and Measurement in 2 and 3 dimensions, Construction, Formal Geometric proofs, Trigonometric Ratios, Indices and logarithms, Calculating and Processing Devices, Algebraic Equations, Polynomials, Partial Fractions, Linear Inequalities and Logic	2-dimensional Rectangular Cartesian Coordinates, Differentiation, Logic, Sets, Mapping, Operations, Calculating and Processing Devices, Geometry and Trigonometry	Mensuration in 3 dimensions, Analysis, Functions, Calculating and Processing Devices, Matrices and Determinants
2	Mechanics	Vectors and Dynamics	Vector Geometry, Dynamics	Statistics
3	Statistics	Measures of Location and Dispersion	Correlation, probability and, Permutations and Combinations	Probability Dimensions, Scatter Diagrams

Source: National Curriculum for SSS by Federal Ministry of Education (1985)

REFERENCES AND BIBLIOGRAPHY

Akinsola, M.K. (1999). Mathematics curriculum in the primary and secondary schools. In A. Dada, (Ed.), *The teacher and the curriculum* (pp. 124-134). Ibadan, Nigeria: Tejama General Enterprises.

Badmus, G.A. (1997). *Mathematics education in Nigeria.* Lagos, Nigeria: Nigerian Academy of Education.

Fajemidagba, O. (1991).Trends in mathematics education in Nigeria: Issues and problems. *Abacus: The Journal of Mathematical Association of Nigeria, 21*(1), 139-153.

Furinghetti, F. (2006). Not out of the blue: Historical roots of mathematics education in Italy. *The Montana Mathematics Enthusiast, 3*(1), 99-103.

Gould, S.B. (1960). Curriculum implementation. In E.C. Short & G.D. Marconnit (Eds.), *Contemporary thought on public school curriculum* (pp. 124-138). Dubuque, WMC: Brown.

Griffiths, H.B., & Howson, A.G. (1974). *Mathematics society and curricula.* London: Cambridge University Press.

Hawes, H. (1979). *Curriculum and reality in African primary schools.* Essex, UK: Longman.

Jahun, I.U., & Korau, Y. (1991). Teachers' perceptions of the new senior secondary school mathematics curriculum. *ABACUS: The Journal of Mathematical Association of Nigeria, 21*(1), 58-69

Lockard, J.D. (1970). *Seventh report of the international clearing house on science and mathematics curricular developments.* A Joint Project of American Association for the Advancement of Science and Science Teaching Centre, University of Maryland.

Ohuche, R.O. (1978). Change in mathematics education since the late 1950s: Ideas and realization, Nigeria. *Educational Studies in Mathematics, 9*, 271-281.

Osibodu, B.M. (1988). Mathematics at Primary Level. *ABACUS: The Journal of Mathematical Association of Nigeria, 18*(1), 13-20

Williams, G.A. (1974). Dynamics of curriculum change in Mathematics: Lagos State Modern Project. *West African Journal of Education, 27*(21), 231-240.

AFFLIATIONS

Mojeed K. Akinsola
Department of Primary Education
Faculty of Education
University of Botswana
Botswana

Adetunji A. Olaoye
Department of Curriculum Studies
Lagos State University
Nigeria

DE BEER

INCLUSIVE SCIENCE EDUCATION FOR THE RAINBOW NATION: REFLECTIONS ON SCIENCE TEACHING AND THE DEVELOPMENT AND IMPLEMENTATION OF THE NATIONAL CURRICULUM STATEMENT IN SOUTH AFRICA

INTRODUCTION

The following narratives contextualize some of the educational changes in post-apartheid South Africa over the last two decades.

A Township School in Pietersburg, 1990

It is the year 1990 in an *Apartheid*-South Africa. Caiphus is a biology teacher in a so-called township school in Pietersburg, in the northern part of the country, in the Transvaal province. The school he is teaching in comes under the Department of Education and Training, as do all the other schools for black learners in the country. Unfortunately there is a racial divide in the country. On his way to school Caiphus passes a school for white learners. The school is an inviting place - beautiful buildings and gardens, well-equipped science laboratories, and an established culture of learning. The learners are taught in their mother tongue, Afrikaans. The school is under the governance of the Transvaal Education Department. When he started teaching, every day he noticed how the environment changes as he travels to his school. Gracious houses make way for squatter camps, where people live in tin houses. Nowadays, he does not notice it anymore. The informal shacks have become part of the scenery. It is in these houses, often without electricity, where his learners need to do homework in the evening by candlelight.

His science classroom has 30 school desks and chairs. Since some of his classes consist of more than 40 learners, some learners have to sit on the floor. The simple classroom has a chalkboard in the front of the class. Next to the chalkboard stands his desk. The only other piece of furniture in the classroom is a steel cabinet, in which he keeps books and stationary, and 'Mr. Hotpants', which is used for corporal punishment. He does not have running water in the classroom.

The bell rings for his first biology class of the day. Caiphus has to teach about gymnosperms today. Although pine trees are invaders in the country, the syllabus prescribes a discussion of these invader plants, responsible for the extinction of

Richard K. Coll, Neil Taylor (eds.), Science Education in Context: An International Examination of the Influence of Context on Science Curricula Development and Implementation, 261–270.

some Fynbos species in the Cape Floristic region. He starts his class by explaining the morphology of the tree. He explains that *Pinus* sp. has *cylindrical* leaves, and writes it on the chalkboard. He notices that his learners, who all speak African languages, have difficulty in understanding this English word, and he code-switches to Sesotho. After explaining *cylindrical* in Sesotho, he decides to draw the leaves on the chalkboard. He never notices the pine tree growing in the schoolyard.

Any observer of his class would immediately categorize his teaching method as 'chalk and talk' - highly expository, with very little learner engagement. At the end of the lesson he makes provision for assessment, and asks a few questions. *What is cylindrical? Why is the pine tree a gymnosperm?* An unfortunate boy, who daydreamed, has a painful encounter with Mr. Hotpants, when he cannot answer a question. Caiphus gives the learners homework: they have to make drawings of leaf fascicles and pine cones in their workbooks.

The Same School in Polokwane, 2006

Sixteen years later Caiphus is still teaching in the same school. It is 12 years into a new democratic dispensation in South Africa. This northern region of the old Transvaal now forms the Limpopo Province, and Pietersburg has been renamed Polokwane. A National Curriculum Statement (NCS) has been published in the Government Gazette. All schools now follow the NCS curriculum, which is governed by a single Department of Education. The subject biology has been renamed 'life sciences', and the new Grade 10 curriculum was implemented this year, to be followed by the new Grade 11 curriculum in 2007, and Grade 12 in 2008. Caiphus attended a workshop on outcomes-based education (OBE), and the new NCS in 2005, but still feels very insecure with the new curriculum. He finds it hard to understand how the defined learning outcomes now navigate his teaching, and not content. The emphasis on a more learner-centered approach, group work and portfolio assessment are also issues that he struggles to come to grips with.

He is introducing a new topic to his Grade 10 learners today, as part of the knowledge area 'diversity, change and continuity' - plants in our daily lives. The textbook they are using emphasizes the inclusion of indigenous knowledge (IK) in this section of the work. The teacher's guide that he received with their purchase of learners' textbooks, suggests that learners do small group research into the cultural use of plants - as food and beverages, medicine, cosmetics and in weaving and basketry. The guide also says that learning outcome 3, 'the role of science in society', should be emphasized. Caiphus finds it hard to think of methods by which he will assess this in the examination. "We are going to look at the importance of plants in our daily lives", he informs his learners. "Let us look at the pine tree. Pine trees have cylindrical leaves …", and he writes *cylindrical* on the chalkboard.

After this class he explains that he is "*an old horse at this*" - and that, as a star teacher, he assists in the professional development of novice teachers. Though he can refer to the underpinning principles in the NCS, it seems as if he is merely paying lip service to OBE. Very little has changed regarding his teaching strategies. All the equipment he has received from the Department of Education remains unopened in the library, neatly stored.

EDUCATIONAL CHALLENGES FOR SOUTH AFRICA IN A POST-APARTHEID REGIME

Prime Minister Hendrik Verwoerd, considered by many as the architect of apartheid, asked when he was Minister of Native Affairs in the previous apartheid regime: "What is the use of teaching a Bantu child mathematics when it [sic] cannot use it in practice?" (Fiske & Ladd, 2004, p. 42). Educational and curriculum reform today is taking place in a context of radical sociopolitical change in a democratic South Africa. However, it is naïve to think that the legacy of problems created when apartheid was introduced in education with the 1954 Bantu Education Act, will, after 40 years of segregation, disappear overnight. South Africa still faces major challenges regarding science education. If one looks at the statistics, they are daunting. South Africa's performance in the Third International Mathematics and Science Study (TIMSS), and statistics on the number of scientists compared with other countries, leave much to be desired. For example, recent statistics show that Japan has 71 scientists per 1000 people and the USA has 22, only 3 out of every 1000 people in South Africa are scientists. The situation is no better if we compare South Africa with other developing countries. Brazil, for example, has 11 scientists per 1000 people. In order to be a global player in the world economy, South Africa needs good scientists, and the key to achieving this is the drastic improvement of science education in schools. According to a research report released by the Centre for Development and Enterprise (CDE): *From Laggard to World Class: Reforming Maths and Science Education in South Africa's Schools*, the failure to improve mathematics and science education is one of the most significant obstacles to African advancement in South Africa. Research by the CDE suggests that from 1991 to 2003, enrolment in higher grade mathematics, the essential requirement for entry into tertiary education, plummeted from 53,631 to 35,959 (Science in Africa, 2004). So, despite South Africa's 'economic muscle', it seems it is performing poorly in science education, and is regularly outclassed by its Tunisian and Moroccan counterparts (Science in Africa, 2003). A national newspaper proclaimed a few years ago that, "SA pupils are the dunces of Africa" (Sunday Times, 2001, p. 1).

So what are the challenges that South Africa faces in science education in a post-apartheid regime? They are many, as we shall see.

Under-resourced Science Classrooms

Many science classrooms in South Africa are under-resourced. However, it also seems clear that even teachers who have scientific equipment do not necessarily use it (Muwanga-Zake, 2004). In a project in the Eastern Cape, where 21 schools were equipped with microscience kits, only five schools have attempted to use the science equipment. In 16 schools the equipment was never even opened. Research suggests that many teachers, even well qualified teachers, do not know how to conduct certain experiments. For example, cases have been documented where teachers were given beam balances to use in their science laboratories, but these were never used because the teachers did not know how to assemble them (Muwanga-Zake, 2004).

Under-qualified Teachers

Research suggests that poor teacher education results in teachers relying heavily upon textbooks (Muwanga-Zake, 2004). Institutions of higher learning thus should critically assess their programs to ensure that student teachers receive training that will foster creative teaching and improvisation, for example, implementing 'science-on-a-shoestring' approaches if they end up teaching in a poorly equipped school.

Science Education Not a Popular Career Choice and Has High Attrition

There is a lack of interest amongst university students in science teaching as a career. There is also a very high attrition rate amongst science teachers in South Africa. The current HIV/AIDS pandemic in South Africa tragically exacerbates the natural rate of attrition. Recent, conservative estimates suggest that 12.7% of South Africa's teachers are HIV positive, and for those aged 25–34 years this rises to 21.4% (Blaine, 2005).

Chalk and Talk

Science in South African classrooms is generally presented in a very formal way. Even where teachers do practical work, this often consists merely of verification activities - 'hands-on' not 'minds-on' activities, with very little realization of 'science as enquiry' (De Beer, 1993). Once again, one can question the training that many of these teachers received.

An Insufficient Culture of Learning

A culture of learning does not exist in many schools, especially in rural areas. Violence is common in South African schools, and in 2006 in less than one month violence claimed the lives of three learners in three separate incidents. This led Education Minister Naledi Pandor to require a national code of conduct to be developed for all schools (Allafrica.com, 2006).

Language Barriers

In countries where the language of education is a former colonial language, such as in South Africa, the scholastic performance of learners is inevitably influenced by their competence in the medium of instruction (Van der Walt, Mabule & DeBeer, 2001). Learners in South Africa are taught in English, yet, for a majority of learners, English is a second or even third language. Research suggests that many textbooks used in South African schools, do not consider language barriers that learners might experience (Muwanga-Zake, 2004). Whereas some researchers advocate that the students' first language should be let in the backdoor by means of code switching (Van der Walt et al., 2001), other researchers and educators warn against this. Responsible code-switching might be effective as a didactic tool in science classes in South Africa, but teachers should receive training in responsible code-switching.

Gender and Race Issues

Many learners, especially black learners and girls, were marginalized in the past, and many of these learners are not interested in pursuing careers in the natural sciences. Howie (2003) highlights the problem, comparing a career in the sciences for many African children with climbing Mount Everest. The stereotype of a scientist being a white male looking like Einstein thus needs to be changed.

LIGHT IN THE TUNNEL: THE NEW NATIONAL CURRICULUM

The Department of Education and various NGOs are working to address the above-mentioned problems. The Centre for the Advancement of Science and Mathematics Education (CASME) has introduced a number of initiatives such as Dinaledi ('Stars' in Sesotho) targeting better science education in schools, by exposing teachers to innovative teaching methods and hands-on practical experiences in key areas of the curriculum. Many of the problems of the past, such as an inflexible and often irrelevant curriculum, have been addressed with the introduction of the new National Curriculum.

Basic Principles Underpinning the South African National Curriculum Statement

The new National Curriculum Statement is based on the following nine basic requirements (Sanders & Nduna, 2006):

- Education should be outcomes-based;
- Teachers should be facilitators of learning;
- Learning needs to be learner-centered;
- Learning should be activity-based;
- The curriculum should be relevant;
- Skills development should be a focal part of the curriculum;

- Group work should be promoted;
- Continuous assessment should be used to foster learning; and
- Learning should be integrated across learning areas.

Three learning outcomes (LOs) were identified for the natural sciences (Grades 7-9), life sciences (Grades 10-12), and physical sciences (Grades 10-12):

LO1 Use process skills, critical thinking, scientific reasoning and strategies to investigate and solve problems in a variety of scientific, technological, environmental and everyday contexts;

LO2 State, explain, interpret and evaluate scientific and technological knowledge and apply it in everyday contexts; and

LO3 Identify and critically evaluate scientific knowledge claims and the impact of this knowledge on the quality of socio-economic, environmental and human development.

Whereas in the past science teaching to a large extent focused exclusively on the cognitive domain, and in many classrooms on the lower levels of Bloom's taxonomy, there is now more emphasis on the psychomotor and affective domains. There is also more emphasis on teachers utilizing more heuristic teaching strategies: a huge problem for many teachers that are used to chalk and talk approaches, and who fear loss of control with self-discovery methods (De Beer, 1993).

Indigenous Knowledge

One particularly positive change introduced in the NCS is incorporation of indigenous knowledge systems in the curriculum. This is a legacy of knowledge and skills unique to a particular indigenous culture and society, involving a 'capital of wisdom' which has developed and been passed on over generations. It is the basis for local decision-making in areas like agriculture, health, and natural resource management. Indigenous knowledge is thus local knowledge, derived from interactions between people and their environment. Indigenous knowledge is an excellent starting point when introducing science concepts to be taught. This concurs with Ausubel's (1968) oft-quoted statement about the importance of prior knowledge when teaching: "The most important single factor influencing learning is what the learner already knows. Ascertain this and teach him accordingly" (pp. 18).

Indigenous knowledge and Western science can blend beautifully, as the following example illustrates. Researchers report that *Sutherlandia frutescens*, the so-called cancer bush - indigenous to South Africa - has potent medicinal qualities known by Khoi, San and Zulu healers for many years. People suffering from cancer responded well to the use of extracts made from this plant as a treatment. This led scientists to hypothesize that *Sutherlandia frutescens* as well as being a useful treatment for cancer might be helpful in treatment of AIDS, since it has such strong antimicrobial activity. Recent research has shown that the shrub contains an amino acid that fights depression; *pinitol*, which helps patients to pick up weight; and *canavanine*, which is successful in treating retroviruses (De Beer, 2005). Although *Sutherlandia frutescens* extracts do not cure AIDS, they definitely help people with AIDS enjoy a better quality of life. Drawing on an ethnobotanical approach to conventional plant studies, thus allows teachers to make their teaching much more relevant for South African learners.

However, a note of caution is appropriate here. Teachers may experience difficulties when trying to incorporate indigenous knowledge in science lessons; another example serves to illustrate this point. Some people believe that vultures can see the future, and if a person eats the brain of a vulture, this allows them to, say, win the lottery. Given that vultures are endangered in South Africa, incorporation of such theories could obviously have detrimental results. One weakness of the NCS is that it is rather vague on what indigenous knowledge should be incorporated in classroom teaching.

THE MISSING LINK: WELL-PLANNED PROFESSIONAL DEVELOPMENT OF SCIENCE TEACHERS

As Rogan (2004) points out, the past decade has seen a renewed interest in the development of science curricula in many parts of the world. In most cases these curricula are well designed, and their aims commendable. However, very often the policy makers become so focused on the 'what' of the desired educational change, that they marginalize the 'how'. Verspoor (1989, p. 133) notes that "Programs tend to emphasize adoption and neglect implementation … In many cases low outcomes result from poor implementation of what was essentially a good idea".

Very often the missing link is a well-planned professional development program, to complement new curricula and policies (Rogan, 2004). Rogan refers to the construct 'profile of implementation' as an attempt to understand and express the extent to which the ideals of a curriculum are being put into practice. Inherent in the notion of a profile of implementation is the recognition that the implementation of a new curriculum is not an all or nothing proposition. The profile of implementation proposed by Rogan provides levels of implementation, or a map for implementing a new curriculum. It enables curriculum planners at school level to determine where they are - what their strengths and weaknesses are. They can then work towards meaningful change.

Curriculum implementation strategies are effective when they proceed just in advance of current practice. Rogan, building on Vygotsky's notion of zones of proximal development, suggests that curriculum implementation strategies should be within a 'zone of feasible innovation'. For example, a teacher whose practices are limited to those described by level 1 of the profile of implementation will be unlikely to be able to immediately employ practices described by level 4. In essence, the Department of Education should set realistic goals during the implementation of the new curriculum.

RECOMMENDATIONS

With the introduction of the South African National Curriculum Statement, an exciting era has dawned for science education in the nation. However, overcoming problems embedded in the education system by the previous political regimes requires innovative approaches, especially in terms of staff development. Teachers clearly need support when implementing what is an exciting, but challenging new curriculum. The following recommendations may serve to aid effective implementation of the NCS:

- Training of teachers at universities should be more practical in nature. For instance, prospective teachers should be taught how to conduct experiments and use equipment. Outcomes-based education was too hastily passed on to teachers, and student teachers need more effective training in the principles of outcomes-based-education;
- In-service training of teachers in outcomes-based education principles and in the National Curriculum Statement is essential. However, Rogan's Profile of Implementation warns us to be pragmatic in setting goals for the professional development of teachers;
- The NCS emphasizes a more 'concept rich but content light' approach, with skills development receiving much attention. However, the new textbooks published (especially for further education and training (FET), Grades 10-12), suggest that little has changed in practice. The new curriculum is just as content-heavy as the previous curriculum. There is also still a strong focus on summative, end-of-matriculation (Grade 12) examinations, despite claims that formative and continuous assessment is to guide teaching and learning. The Department of Education should look at these issues when revising the NCS, and should ensure that we do not only pay lip-service to the principles stated in the NCS;
- Initiatives such as the Dinaledi Programme (where good science schools are nurtured) should be implemented on a much broader scale; and
- The Government and the Department of Education should seek to make the teaching profession more attractive, in terms of remuneration, support for in-service professional development, and career advancement. Many novice teachers experience something of a 'reality shock' when they first start teaching, and this often leads to a high attrition rate. Better induction of teachers is needed, and novice teachers should also have mentors to assist them in their professional development. In communities of practice an identified mentor could assist the novice teacher in adapting better. However, research suggests that, in many schools, one should actually speak of 'pseudo communities of practice' (Whitelaw, 2006) because of the 'institutionalized bullying of young teachers' (Jarzabkowski, 2002).

Table 1. Example of a profile of implementation for science education (Rogan, 2004)

Level	*Classroom Interaction*	*Science Practical Work*	*Science in Society*	*Assessment*
1	Teacher presents organized lessons; uses textbook effectively. Learners stay engaged, respond to questions	Teacher uses classroom demonstrations to help develop concepts.	Teacher uses examples from everyday life to illustrate scientific concepts.	Written tests are given. Most questions of recall type. Tests marked and returned promptly.
2	Textbooks used along with other resources; engages learners with questions that encourage deeper thinking. Learners engaged in meaningful group work.	Teacher uses demonstrations to promote a limited form of inquiry. Learners participate in cookery-book practical work.	Teacher bases lesson on specific problem faced by community.	Written tests include 50% questions on higher cognitive levels.
3	Teacher structures learning activities along 'good practice' lines. Learners engage in minds-on learning activities	Practical work designed to encourage learner discovery of information (guided discovery)	Learners actively investigate the application of science & technology in their own environment, e.g. by surveys	Written tests include 'guided discovery' type activities; include other forms of assessment such as reports.
4	Learners take major responsibility for own learning; undertake long-term investigations & projects. Teacher facilitates learning.	Learners design & do open investigations.	Learners actively undertake a project in their local community; explore long term effects of community projects.	Open investigations & community based projects included in final assessment. Learners create portfolios.

REFERENCES AND BIBLIOGRAPHY

All-Africa.com. (2006). Retrieved 13 May 2007, from http://allafrica.com

Ausubel, D.P. (1968). *Educational psychology: A cognitive view*. New York: Holt, Rinehart & Winston.

Blaine, S. (2005). Shortage of teachers looming. *Business Day*, Johannesburg. Retrieved 10 September 2007, from http://allafrica.com/stories/printable/200512070232.html

Centre for Development & Enterprise. (2004). *From laggard to world class: Reforming maths and science education in South African schools*. Retrieved 13 May 2007, from http://www.scienceinafrica.co.za

De Beer, J.J.J. (1993). 'n Evaluerende ondersoek na die waarde van praktiese werk in Biologie-onderrig op senior sekondere skoolvlak. Pretoria: Ongepubliseerde PhD proefskrif.

De Beer, J.J.J., Bowie, M., Whitlock, C.E., & Cassim, F. (2005). *Life sciences grade 10*. Nasou-Via Afrika.

Department of Education. (2003). *National curriculum statement*. Gazette No 25545. Pretoria: Seriti Printing.

Fiske, E.B., & Ladd, H.F. (2004). *Elusive equity: Education reform in post-apartheid South Africa*. Washington: Brookings Institution Press.

Howie, S. (2003). Why don't kids learn maths and science successfully? Retrieved 13 May 2007, from http://www.scienceinafrica.co.za/2003/june/maths.htm

Jarzabkowski, L.M. (2002). The social dimensions of teacher collegiality. *Journal of Educational Enquiry*, *3*(2), 1-20.

Muwanga-Zake, J.W.F. (2004). *Is science education in a crisis? Some of the problems in South Africa*. Retrieved 13 May 2007, from http://www.scienceinafrica.co.za/scicrisis.htm

Rogan, J. (2004). Professional development: Implications for developing countries. In K. O-saki, K. Hosea & W. Ottevanger (Eds.), *Reforming science and mathematics education in sub-Saharan Africa* (pp. 155-170). Dar-es-Salaam, Tanzania: TEAMS Project.

Science in Africa on-line magazine. (2004). Retrieved 13 May 2007, from http://www.scienceinafrica.co.za

Sanders, M., & Nduna B. (2006). Understanding the requirements of the new South African curriculum: Implications for the teachers of science. In H. Van Rooyen & J. De Beer (Eds.). *Science teaching in the OBE classroom* (pp. 178-186). Johannesburg: Macmillan.

SA pupils are the dunces of Africa. (2000, July 16). *Sunday Times*, p. 1.

Van der Walt, C., Mabule, D.R., & De Beer, J.J.J. (2001). Letting the L1 in by the back door: Code switching and translation in science, mathematics and biology classes. *Journal for Language Teaching*, *35*(2&3), 170-181.

Verspoor, A. (1989). *Pathways to change: Improving the quality of education in developing countries*. Washington DC: The World Bank.

Whitelaw, S. (2006). *Novice teachers in a social context: Enculturation in a pseudocommunity of practitioners*. Unpublished EdD thesis, University of Johannesburg, South Africa.

AFFILIATION

Josef de Beer
Department of Mathematics, Science, Technology & Computer Education
Faculty of Education
University of Johannesburg
South Africa

PRINSLOO

A CRITICAL SENSE OF LOCATION: ANALYZING THE IMPACTS OF STUDENT AND DISCOURSE CONTEXTS IN SCIENCE EDUCATION

INTRODUCTION

Within the unique context of open and distance learning (ODL), the University of South Africa (Unisa) provides opportunities for students that would otherwise be excluded from higher education. Unisa is one of the largest distance education institutions in the world with over 250,000 students, nationally and abroad. As the only dedicated public distance education institution in South Africa, Unisa is seen to play a pivotal role in addressing the disparities in mathematics and science competencies in South Africa and on the African continent.

The appalling retention and through-put rates of students in the natural and decision sciences at Unisa are no exception to the national and international situation. This is the exact opposite of what is required to meet the urgent need for more qualified and competent graduates in the natural and decision sciences. Most educators blame this on the massification of higher education, the poor quality of students entering higher education, and the lack of sufficient numerical and language proficiency of students entering higher education. Distance education in particular requires competencies and autonomy from students that many, especially younger, students lack.

In the light of the possible geographical reach of ODL as well as the number of students involved, it is critical to reflect on strategies to improve student retention and success in the sciences. At present these strategies mostly involved are allocated to the more effective use of technology and increasing tutorial support, as well as redesign of learning materials and experiences. Though these strategies are laudable, the reasons for student failure may *also* lie in students and educators not having a 'critical sense of location'. This chapter explores the possible impacts of student and discourse contexts on the teaching of science.

I will explore 'context' in the sense of *locating* students with regard to their personal circumstances, their histories of learning and personal ways of knowing, their support networks (or lack thereof) as well as their expectations of their studies in the fields of science. I will also explore the impact of students being induced into a *new* location, namely the context of 'science as subculture' (Aikenhead, 1996). These contexts and new locations are furthermore multi-layered and multi-textured. This chapter suggests that having a 'critical sense of location' may impact positively on students' as well as educators' experiences in science education.

Richard K. Coll, Neil Taylor (eds.), Science Education in Context: An International Examination of the Influence of Context on Science Curricula Development and Implementation, 271–279.

This chapter will explore the importance of context; discuss the impacts of different contexts; identify teaching science as relocation and enculturation; and examine the need for critical *self*-location

THE IMPORTANCE OF CONTEXT

The legacy of apartheid in education, and specifically its effect on science education, will haunt educators and policy-makers for many years to come. Kahn (2004) postulates that the consequence of the separate and unequal education policies of the apartheid regime, "was and still is catastrophic" (p. 149). Blankley and Arnold (2001) report on a nationally representative survey that was undertaken by the Human Sciences Research Council (HSRC) in 1999 in which it was found that over "30% of adults had never studied Mathematics at school, 50% had never studied Biological Science, and 55% had never studied Physical or Chemical Science" (p. 65). The research also indicated that "men had passed more Science and Mathematics courses than women. The gender differential was least pronounced in the Biological Sciences, with about 18% of both sexes having passed matriculation Biology and 4-5% of both sexes having passed courses at tertiary level" (p. 66). The research further indicates that "71% of the people in the highest income bracket had passed at least matriculation Mathematics" and that "those in the lowest income bracket had not passed Mathematics nor Biology, Physical Science or Chemical Science" (p. 66).

Prior to 1994, the education of different races in South Africa was administered by different departments and in many cases the curricula in specific subjects also differed. Subjects such as mathematics and physical and chemical science were mostly deemed inappropriate for people of color. In research reported by Blankley and Arnold (2001), it was noted that more than half of Africans, colored, and Indians in the study, had not studied physical or chemical sciences at all, and, for colored people, the proportion climbed to two-thirds.

In responding to this legacy, the teaching of natural and decision sciences today often focuses on a *deficit* model of addressing students' perceived, and often real, lack in numerical and language competencies. A deficit model may *not* be the most effective or appropriate strategy in addressing the legacy of apartheid and the critical need for more scientists. Taking context seriously, on the other hand, may open possibilities for engagement and more effective teaching.

THE IMPACT OF CONTEXT

Tessmer and Richey (1997) have found a number of assumptions to underlie our understanding of context.

We are condemned to context. Learning does not take place in a vacuum. Students do not leave behind their personal histories when they open their study materials or enter the tutorial. "We cannot choose to be separated from or avoid the context in which we operate, all cognition and reasoning is situated" (Greeno,

1989, cited in Tessmer & Richey, 1997). Although we can try to ignore context, we are all prejudiced by it.

Context is made up of a variety of factors that either inhibit or facilitate learning. In each context there are various interacting factors that can either enhance or frustrate the learning event.

There are multiple contexts for a given learning or performance. A student is at any moment in time a participant in multiple contexts.

Student support strategies can accommodate context but cannot control it. We can analyze contexts, we can try to lessen the impact of factors that obstruct successful learning, we can try to increase the factors that support and facilitate learning *but we do not control contexts.*

Effective learning will be 'context-rich'. It will attempt to effectively address the multiple contexts of students.

We must think systemic not only systematic. The different aspects of a student-context are part of a bigger context. Somehow everything is related and everything is influenced by everything.

Tessmer and Richey (1997) continue to explore three different contexts in a teaching or training scenario. They refer to the *orienting*, *instructional* and *transfer* contexts. These three contexts are in one sense linear, in the sense that the orienting context mostly precedes the instructional context. On the other hand, although there is a sense of progression, these three different contexts overlap. I will now briefly explore each of these contexts, before we reflect on 'teaching science as enculturation' which encompasses all three contexts.

The *orienting* context refers to the location of students *prior* to the learning event - whether it is pre-registration, or before opening their study guides, or even before a tutorial. 'Location' refers to much more than just the physical circumstances of students. It refers to factors that influence the "prospective student's motivation and cognitive preparation to learn" (Tessmer & Richey, 1997, p. 90) as well as students' prior knowledge and experiences with regard to the subject area to be covered.

A South African study by Maree, Aldous, Hatting, Swanepoel and Van der Linde (2006), focused on "rural schools with only black learners in attendance" and probed five different foci namely "demographic data, learner background, learner attitude to science and mathematics, learners' perceptions of what was happening in the science and mathematics classroom, respectively and a test" (p. 232). Maree et al. (2006) found that demographic characteristics and physical circumstances of students' location play an important role in their orientation and success or failure in the sciences. Maree et al. concluded that "science and mathematics teaching can never be divorced from the socioeconomic context in which it is taught" (p. 238). Barton (1998) explored the factors influencing teaching science to urban homeless children and found that:

> Urban homeless children, like all children, learn science in and through a variety of contexts: in school, out of school, and from families, friends, and the media. As they learn science, they also learn a lot about who they are (and

can be) and what science is (and can be). Children learn about the production of knowledge, values, social identities, and power, and their interconnections. (p. 382)

In other words, pedagogy in science classrooms is also about the struggle for identities and representations.

In ODL, most students' location (their personal circumstances, their histories of learning and ways of knowing, their support networks (or lack thereof) as well as expectations of their studies in the field of science) are unknown to the teaching institution. While residential universities may allocate resources to alleviate some of the negative impacts of students' physical environments, addressing students' immediate contexts is mostly out of reach in ODL. Research by Simpson (2003) and Pizzolato, (2003, 2004), however, indicates that should students be able to assess their *own* risks and resources and establish their *own* locus of control, their chances of success increase accordingly. The following questions are examples of questions that may be critical to sensitize students to 'locate' themselves and establish their locus of control (Barton, 1998):

- Where are you located? How does your location impact on your studies?
- How does your age influence your learning?
- How do your gender and family roles/immediate context influence your learning?
- How does the fact that you are of a certain race influence your learning?
- How does your home language influence your learning?
- How do your previous qualifications assist you in your studies?
- How do your primary and secondary school experiences influence your chances of success? Were they good and happy experiences? Which of those experiences have prepared you for your studies in science?
- What skills and values have you developed through your school years that you think will assist you in your studies?
- How useful do you think this module will be one day in your professional life?
- Do you know anyone who has passed this module? Do you know any scientists in your community?
- Is your postal address reliable? Will you have regular access to your mail?
- Where will you study - do you have your own room? What arrangements can you make to have an own space to work in?
- How much time will you have per day to study?
- Do you have sufficient light at night?
- Where will you store your books/ stationary?
- How far are you from your nearest regional centre/ computer lab/ examination centre?
- How many people in your immediate family have studied further? How does this impact on your studies?

- Into which resources in your immediate community (e.g., teachers, other students) can you tap to assist you this year?
- Your peers who do not study through Unisa - how do they support your studies?
- If you are employed, how do you think your co-workers will support your studies?

The questions above are examples of questions, when reflected upon, may assist students in *locating* themselves with regard to their studies. Blankley and Arnold (2001) state that in South Africa there is a vast lack of public understanding of science and a scientific vocabulary or scientific constructs "that form the basis for the assimilation of scientific knowledge" (p. 69), going on to note that "Whatever means communicators and policy-makers choose to promote the understanding of science and technology among the general public, care and forethought are needed where levels of conceptual development regarding science appear to be extremely low" (p. 69).

The *instructional context* comprises factors and environments that are directly involved in the delivery of instruction, that is, "the immediate physical, social and symbolic resources outside the person" (Tessmer & Richey, 1997, p. 91). The instructional context normally includes the impact of seating arrangements and the duration of the class or choosing appropriate activities. In an ODL setting, the instructional context involves carefully designed and structured learning materials and experiences, whether in print or online. The instructional context is also the culmination of not only the life-history and teacher-preparation of the teacher, but also the life-histories and learning trajectories of individuals embracing socioeconomic, political and personal and tribal/collective consciousnesses. In the instructional context the student is relocated to another location, namely science as subculture. Although this relocation often results in a dislocation ending in frustration and failure, being dislocated is inherent to learning and teaching, as explored by Edwards and Usher (2000), and even necessary (Doll, 1986).

Relocating students to the subculture of science can be compared to being introduced to a totally foreign culture. One just has to reflect on the experience of arriving at a foreign airport where the last person who you could understand was the flight attendant saying goodbye. The moment you enter the airport buildings, you see that the signs, symbols, self-help computers and directions are in a different language, and often a different alphabet. Teaching science is, however, more than a merely introducing students to a different culture, teaching science has as its purpose to make students at home in science as subculture, a process that is described by the term 'enculturation'.

The *transfer context* refers to the environment where the learning will be applied. Many ODL students are employed and may have the opportunity to apply what they learn while they work, if employed within the fields in which they study. Many students are, however, unemployed or employed within fields far removed from science as subculture. As Blankley and Arnold (2001) found, there is in South Africa a very low conceptual understanding regarding science, and this includes preconceived ideas (or the lack thereof) of what scientists do in practice. While students at residential universities may get a 'picture' of what scientists look like and think like, students at ODL institutions get glimpses during compulsory practicals. The enculturation process in ODL, therefore, poses unique challenges to the design of learning materials and learning experiences.

We now turn to exploring teaching science as enculturation which encompasses the orienting, instructional and transfer contexts.

TEACHING SCIENCE AS RELOCATION AND ENCULTURATION

Enculturation refers to the learning of a specific language, of a culture different to your own, reading the signs and symbols, decoding and recoding them. Science is a unique system with unique signs and interpretation systems and schemas. Often the teacher has forgotten the strangeness of the signs and symbols and of the reasoning and has come to accept it as just the way it is.

When teachers see the teaching of science as enculturation (Hodson, 1999), they recognize that students are required to move from one culture or space, to another, a new and different culture. Aikenhead (1996) describes the need for science educators to "recognize the inherent border crossings between students' life-world subcultures and the subculture of science, and … to develop curriculum and instruction with these border crossings explicitly in mind" (p. 2). Aikenhead, therefore, talks of learning science as "*culture acquisition*" (1996, p. 4, original emphasis).

An essential aspect of enculturation is to induce the student into the habits, vocabulary and ways of seeing, thinking, knowing and being of a different culture. Successful enculturation is apparent when the student has an insider's use and knowledge of the signifiers and the signified. In enculturation, the acquisition of the new culture's language is a critical element in the success of enculturation. Different subjects and disciplines, like cultures, have distinctive vocabulary and ways of reasoning and of linking ideas.

While the role of language has been acknowledged in research that investigated the factors that impact on the success of students in mathematics, the emphasis is often on the readability of a text and not its comprehension. Reading comprehension is described by Bohlman and Pretorius (2002) as entailing both 'decoding' and 'comprehension':

> "Decoding" involves those aspects of reading activity whereby written signs and symbols are translated into language. "Comprehension" refers to an overall understanding process whereby meaning is assigned to the whole text. The interaction between decoding and comprehension in skilled readers happens rapidly and simultaneously. (p. 196)

Bohlman and Pretorius (2002) go on to explain that comprehension cannot occur unless decoding is successful. Successful decoding on the other hand does not, however, imply comprehension. Mathematical texts are also "hierarchical and cumulative, in the sense that understanding each statement or proposition is necessary for understanding subsequent statements" (pp. 196-197). Bohlman and Pretorius's research focused on several language constructs in mathematics such as 'causal and conditional relations', contrastive relations and sequencing, and the use of graphics. Some of their findings are that:

- The worse the reading ability of the student, the worse their performance in mathematics;
- The sequencing task was the most difficult, followed by the contrastive and then the conditional text-semantic relations. "These results indicate that the weaker students regularly miss vital clues that aid in constructing and keeping track of meaning in a text" (2002, p. 205);
- Students' ability to resolve anaphors impact on their success in mathematics.

> Anaphora basically involves repeated reference in discourse. In other words, a referent relating to a person, entity, event, state or idea that is introduced into a discourse is referred to again, either by means of repetition of the same linguistic term (e.g., noun or proper noun) or by means of another linguistic item (e.g., pronoun or synonym) (2002, p. 197-198).

  If a student cannot resolve an anaphor, he or she will miss the point; and
- Students who grasp conditional or contrastive relations in a text do better.

Language and not numerical skills were found to be a major influence in the success of students in a bridging course in mathematics. In a follow-up study, the same researchers report on a 22 week intervention plan to increase students' reading abilities in the mathematics bridging course (Pretorius & Bohlman, 2003). Their research found that students identified to be at risk benefited from the 22 week intervention, but not enough to prepare them sufficiently for the course materials.

It would be immoral for higher education not to address language proficiency of prospective students (Pretorius & Bohlman, 2003).

THE NEED FOR CRITICAL *SELF*-LOCATION

Although language may be a critical element in successful enculturation in science education, Aikenhead (1996) explores four types of cultural transition describing students' relocation to science as a subculture. Aikenhead (1996) refers to the model of Phelan, Davidson and Cao (1991) which proposed four types of transition where students move from one world to another. *Congruent worlds* support smooth transitions, *different worlds* require transitions to be managed, *diverse worlds* lead to hazardous transitions, and highly *discordant worlds* cause students to resist transitions which, therefore, become virtually impossible.

Congruent worlds entail students' current world view and content as being congruent with 'science as culture'. A student should be able to cross into the world of science without a problem. Diverse worlds are where students currently find themselves in a different 'world', but the transition to the world of science will be possible if 'managed'. Should a student find him or herself, however, part of diverse worlds, a transition to the world of science could be hazardous. Transitions from discordant worlds are virtually impossible.

In another attempt to describe the potential transition between students' 'current' location and science as a subculture. Costa describes five categories:

- *Potential scientists:* worlds of family and friends are congruent with worlds of both school and science;
- *Other smart kids:* worlds of family and friends are congruent with the world of school but inconsistent with the world of science;
- *"I don't know" students:* worlds of family and friends are inconsistent with worlds of both school and science;
- *Outsiders:* Worlds of family and friends are discordant with the worlds of both school and science; and
- *Inside outsiders:* worlds of family and friends are irreconcilable with the world of school, but are potentially compatible with the world of science. (cited in Aikenhead, 1996, p. 15)

Should students be encouraged to locate themselves in one of these worlds (whether in face-to-face interviews or by filling in a questionnaire), they may be able to establish their risk profile and take the necessary steps or de-register. There is a positive correlation between students' awareness of their location (their resources, risks and networks of support) and their success in crossing borders (Pizzolato, 2003, 2004).

CONCLUSION

A critical sense of location will provide science educators as well as students with an acute sense of knowing their present locations and preparing students for being dislocated and relocated to science as subculture. A critical sense of location also allows students to establish their own loci of control in order to design appropriate strategies to ensure success. When science educators develop a critical sense of location, they have an informed understanding of their students' contexts and the factors shaping students' attempts to learn the language and subculture of science.

In the ODL context where students often experience their studies as if they are entering a liminal space of studying and yet not being on a campus, it is crucial that they are appropriately oriented and guided through these dislocating and relocating experiences.

REFERENCES AND BIBLIOGRAPHY

Aikenhead, G.S. (1996). Science education: Border crossing into the subculture of science. *Studies in Science Education*, *27*, 1-52.

Barton, A.C. (1998). Teaching Science with homeless children: Pedagogy, representation, and identity. *Journal of Research in Science Teaching, 35*(4), 379-394.

Blankley, W., & Arnold, R. (2001). Public understanding of science in South Africa: Aiming for better intervention strategies. *South African Journal of Science, 97*, 65-69.

Bohlman, CA., & Pretorius, E.J. (2002). Reading skills and mathematics. *South African Journal for Higher Education, 16*(3), 196-206.

Doll, W.E. (1986). Prigogine: A new sense of order, a new curriculum. *Theory into Practice, 25*(1), 10-16.

Edwards, R., & Usher, R. (2000). *Globalisation and pedagogy: Space, place and identity*. London: Routledge.

Hodson, D. (1999). Going beyond cultural pluralism: Science education for sociopolitical action. *Science Education, 83*, 775-796.

Kahn, M. (2004). For whom the bell tolls: Disparities in performance in senior certificate mathematics and physical science. *Perspectives in Education, 22*(1), 149-156.

Maree, K., Aldous, C., Hatting, A., Swanepoel, A., & Van der Linde, M. (2006). Predictors of learner performance in mathematics and science according to a large-scale study in Mpumalanga. *South African Journal of Education, 26*(2), 229-252.

Pizzolato, J.E. (2003). Developing self-authorship: Exploring the experiences of high-risk college students. *Journal of College Student Development, 44*(6), 1-16.

Pizzolato, J.E. (2004). Coping with conflict: Self-authorship, coping and adaptation to college in first-year, high-risk students. *Journal of College Student Development, 45*(4), 425-442.

Pretorius, E.J., & Bohlman, C.A. (2003). A reading intervention programme for mathematics students. *South African Journal for Higher Education, 17*(2), 226-236.

Simpson, O. (2003). *Student retention in online, open and distance learning*. London: Kogan Page.

Tessmer, M., & Richey, R.C. (1997). The role of context in learning and instructional design. *Educational Technology Research and Development, 45*(2), 85-115.

AFFILIATION

Paul Prinsloo
Education Consultant
Institute for Curriculum and Learning Development
University of South Africa
South Africa

ASIA

KOH, TAN, AND CHEAH

SCIENCE EDUCATION IN SINGAPORE: MEETING THE CHALLENGES AHEAD

INTRODUCTION

Science education in Singapore is regarded by many as successful. Singapore students have performed well in international educational evaluation studies such as the Third International Mathematics and Science Study (TIMSS). Singapore educational materials such as textbooks and curricula documents are well regarded internationally. However, in Singapore there is a prevalent view that, in order for Singapore to maintain economic progress, the country cannot afford to become complacent in the provision of education, especially science education. There is a need to constantly make timely changes to the science curriculum to ensure that it meets future needs. Here, we report on several new developments in Singapore science education that include an increased emphasis on inquiry-based science, encouraging school-based innovations, more curricular options in the form of applied science elective modules, generating interest in science, the establishment of a specialized school to foster talent in science and mathematics, and enhancing teacher professional development especially for non-science graduate teachers teaching science.

Singapore is a small island nation with a multi-racial population of about 4.4 million people.[i] It occupies a land area of only about 700 square kilometers. The demography of the population, as of 2005, was 75.6% Chinese, 13.6% Malays and 8.7% Indians. The remaining 2.1% consist of people from the other ethnicities. Apart from Singapore's strategic location at the heart of South East Asia, her only natural resource is the 3.6 million residents. Investment in a high quality education system that provides a strong foundation in science and mathematics education is one of the key strategies that Singapore adopts to ensure her continued survival as a competitive economy in a changing global landscape. This global landscape is driven by interrelated forces of the knowledge-based and innovation-driven economy, globalization and the rapid advances in science and technology. After gaining independence in 1965, Singapore did not sever its educational ties with the UK, and the University of Cambridge Local Examination Syndicate (UCLES), first introduced in 1891, still collaborates with the Singapore Ministry of Education (MoE) to develop the school curricula and national examinations at the end of secondary (i.e., Year 10) and pre-university education (i.e., Year 12).

Pre-schooling in Singapore starts as early as three years old when the child enters nursery level followed by the K1 and K2 levels, each level taking

Richard K. Coll, Neil Taylor (eds.), Science Education in Context: An International Examination of the Influence of Context on Science Curricula Development and Implementation, 283–290.

approximately one year. Formal education begins at six years of age and spans at least 10 years of general education. This consists of six years of primary (elementary) schooling and four to five years of secondary (middle) schooling. The school system consists of about 170 primary schools, 160 secondary schools, and 17 junior colleges (high schools) respectively. The typical roll of a primary school, secondary school and a junior college are 1,700, 1,300 and 1,500 respectively. The total enrolment from primary 1 to pre-university 2 levels (from Year 1 to Year 12) is about 530,000 students. The number of teachers in the system is about 26,000. Ratios of students to teaching staff at primary, secondary and pre-university (high school) are 24, 19 and 11, respectively. Education is compulsory for the first six years of schooling.

OVERVIEW OF SCIENCE EDUCATION IN SINGAPORE

In order to provide students with a strong foundation in the sciences so as to prepare them for future challenges, science is a compulsory subject taken by all students from Years 3 to 10. Many students continue to do at least one science subject in Years 11 and 12. The framework for Singapore science education is shown in Figure 1.

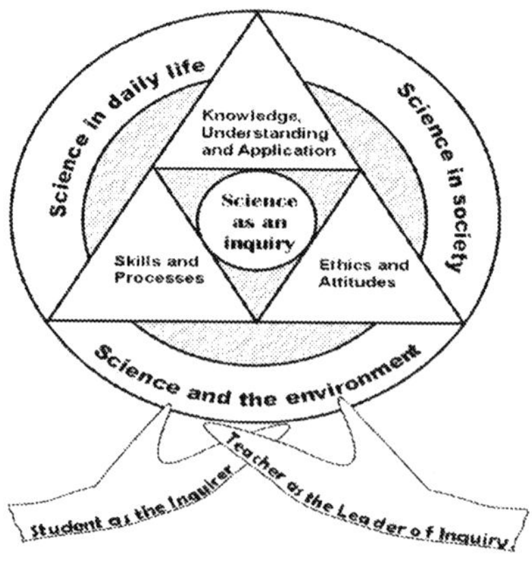

Figure 1. Framework for Singapore science education (reproduced with permission of the Curriculum and Development Planning Division)[ii]

In Years 1 and 2, the emphasis in the curriculum is on the development of students' literacy and numeracy skills, that is, Singapore students focus on the learning of English language, their mother tongue and mathematics, so they do not learn science. In the primary (Years 3 to 6) and lower secondary (Years 7 to 8) levels, science is taught as an integrated subject and organized into themes, five in primary science (Diversity, Cycles, Systems, Energy, and Interactions) and six in lower secondary science (Diversity, Cycles, Energy, Interactions, Models &

Systems, and Measurement) (Curriculum Planning & Development Division, 2004a, 2004b). The science curricula from Years 3 to 8 encompasses a core body of concepts in both the life and physical sciences to provide a broad understanding of the environment, and the interaction between science, technology and the environment. It also focuses on the development of skills and attitudes necessary for scientific inquiry and problem-solving. The curricular time for science starts from one and a half hours per week in Year 3 and increases progressively to three and a half hours per week in Years 7 and 8. The Year 9 to 12 science curricula (upper secondary and pre-university levels) build upon the earlier foundation to allow students to develop in-depth knowledge, skills and attitudes in at least one of the three sciences, namely, biology, chemistry and physics. In addition to providing students at Years 9 to 12 with the ability to understand and make informed decisions on science and technology in society, these studies also provide them with a strong foundation for tertiary studies in science, engineering and technology. The curricular time per science subject per week ranges from three hours in Years 9 and 10, to five hours in Years 11 and 12. The actual curricular time spent per week on a science subject depends on the number of science subjects taken. A student in a science-based course can take up to three science subjects, that is, biology, chemistry and physics.

RECENT DEVELOPMENTS IN SINGAPORE SCIENCE EDUCATION

Although Singapore achieved considerable success in the Third International Mathematics and Science Study (TIMSS) (Toh, Riley, Atputhasamy & Subramaniam, 2006), past success is no longer seen as a predictor of future success. In this complex and uncertain world, Singapore needs to constantly make timely changes to its science education programs to anticipate future challenges. Some recent developments in Singapore science education include:

- Greater push towards authentic and inquiry-based science learning in curriculum;
- School-based innovations;
- More curricular options in terms of applied science elective modules;
- Generating interest in science; and
- Establishing a specialized school to develop talents in mathematics and the sciences.

Although an inquiry-based science curriculum is not new to Singapore schools in terms of actual practices, there have been systemic efforts at the policy level in recent years to place greater emphasis in curriculum design on authentic and inquiry-based science learning through explicit integration of exemplars into the science curricula, especially in terms of authentic learning of science at all levels. In the science curriculum guides prepared by the MoE for teachers, possible authentic and inquiry-based activities are mapped onto the syllabuses explicitly to encourage teachers to use them in their lessons when they teach. For example, in

Years 7-8 Science, the topic on 'energy forms and uses' (Curriculum Planning & Development Division, 2007) includes a suggested activity where students would first be asked to gather information on the last time that Singapore had a major power failure and the consequences that this failure had on the economy and daily living. The students would then be brought on a visit to a local power station to learn how Singapore ensures a secure and reliable supply of power to meet Singapore's energy needs. Finally, it is suggested that the students could communicate their findings and thoughts on Singapore's energy needs on a discussion platform such as an online discussion board or a blog. Textbook authors are also encouraged to increase the number of authentic and inquiry-based examples and activities in their science textbooks and suggest activities that teachers could use to engage their students in inquiry-based science. For example, in a Year 4 science textbook, one suggested activity that students could do in order to understand the effects of pollutants was to investigate what would happen to animals and water plants if the water in the school pond were to become polluted (Teo-Gwan & Kwa, 2002). Such 'disasters' do happen during the school vacations when there are no students to take charge of the pond. School-based science practical assessment has also been implemented in Years 10 and 12 to replace the one-off national end of year practical examinations in order that inquiry-based science can be included in the assessment.

All Singapore schools are well equipped with the necessary science laboratories that have the appropriate equipment such as dataloggers (Tan, Hedberg, Koh & Seah, 2006) and materials to conduct various demonstrations and experiments to deliver their respective science programs. Many primary schools also are equipped with ponds and gardens for ecology studies. Students deemed to have talent in science are provided with opportunities to work in science and technology centers established in selected schools in partnership with industry and institutes of higher learning. These centers provide sophisticated laboratory and research equipment for students to use so that they can do advanced inquiry-based science projects under the guidance of their teachers and university academics. To date, there are 12 centers that focus in areas that range from life sciences to photonics and nanotechnology.

School leaders and teachers are key agents in bringing about pedagogical innovations in science teaching and learning in the classrooms. To encourage them to implement innovative pedagogical practices in schools, the MoE has launched various initiatives such as Teach Less Learn More (TLLM) and Information and Communication Technology (ICT) Masterplan. These initiatives provide incentives to schools in the form of additional funding, expert manpower support and/or curricular exemplars to support them in the implementation of their innovative programs from Years 1 to 12 levels that include the teaching and learning of science. Successful models of pedagogical practices and curricular resources arising from these schools are shared with other schools so that other teachers can adapt and apply the practices and resources in their respective schools. For example, an integrated program school[iii] (Years 9 to 12) has developed a PBL framework called THINK which stands for Trigger, Harness, Investigate, Network and Know. Based on this framework, students learned essential scientific principles by solving authentic problems with the help of knowledge building software called Knowledge Constructor on their Tablet PCs. A secondary school (Years 7 to 10) is collaborating with a polytechnic to develop an interactive electronics teaching module with rich media content for the learning of electronics in physics. Elementary schools are not left behind in the use of innovative practices to enhance science teaching and learning. For example, one elementary school is leveraging on ICT to support inquiry-based science learning along a continuum of strategies that ranges from 'structured inquiry' to 'guided inquiry' to 'student-initiated inquiry'. At that school, all students make extensive use of concept mapping to express their understanding of the science they have learned and are provided with ample opportunities to make presentations to their peers and teachers on their learning. Another primary school is collaborating with university researchers and the MoE to develop mixed reality applications to help students visualize the solar system and the movement of the planets. The use of ICT seems to motivate and engage the 'digital natives' (Prensky, 2001), students who grew up playing video games and using computers and the Internet, and provides visualization and simulations necessary to elaborate on abstract concepts, thus increasing students' conceptual understanding of these concepts.

The MoE recognizes that more time and manpower is necessary to allow for classroom innovations and inquiry-based learning. Mr. Tharman Shanmugaratnam, the Minister of Education, announced at the MoE Workplan Seminar 2006 that the MoE was considering judicious reduction in the curriculum of all subjects and providing schools with more teachers to reduce teacher workload so that they had time to develop, implement and revise their innovative programs. There will also be a greater two-way flow of expertise between the MoE and schools intended to facilitate sharing between specialists in the MoE and school teachers to support the schools' endeavors.

To encourage students to discover and develop a range of talents, schools in Singapore will be providing more subject options in the science curricula. Secondary schools will be offering, from 2007, Advanced Elective Modules (AEMS) developed by the local polytechnics in applied areas such as creative game design and food innovation. Although the AEMS will not be examined at the Year 10 national examination, students offering these subjects would have the opportunity to experience practice-oriented learning in diverse fields that interest them. From 2008, examinable subjects called Applied Graded Subjects (AGS) at General Certificate of Education Ordinary Level (Year 10 national examination) will also be developed in collaboration with the local polytechnics in applied areas such as creative 3D animation, fundamentals of electronics, and cell and molecular biology (MoE, 2006). The AGS would give students the opportunity to pursue applied subjects in greater depth than would be allowed in the current science curricula. These two categories of science-based applied subjects will complement the broad-based education in secondary schools by providing a greater diversity of curricula choices to cater to students' interests and talents.

To generate interest in the sciences among students, schools provide opportunities to participate in a variety of programs and activities that cater to students' special interests and aptitude. These include mentorship programs where students are offered the chance to work on research projects in real-life conditions under the guidance of mentors in the universities and research institutes. They start on the projects during their June vacation and meet their mentors regularly until September to complete their project. On completion of the project, they are expected to write a scientific paper to be presented at an annual student research conference organized by the universities. Some of the better research work done by the students has been published in international refereed journals. There are also science-based competitions and science fairs, attachment to industries, talks by distinguished international scientists and Nobel Laureates, overseas science enrichment programs and International Science Olympiads to excite the students in science. Students are also provided with out-of-school experiences to complement the lessons taught in schools by the students' visits to informal science-based institutions such as the Singapore Science Centre, Singapore Zoological Garden, Jurong Bird Park and Raffles Museum of Biodiversity Research as well as places such as the Singapore Botanical Garden, Bukit Timah Nature Reserve and Sungei Buloh Nature Park (Subramanium & Toh, 2004).

The National University of Singapore (NUS) High School of Mathematics and Science[iv] was established in 2005 to cater for Years 7 to 12 students identified as having talent in mathematics and science. Being a specialized school, it has the flexibility to determine its curriculum; it has an enriched and accelerated curriculum to cater to the students' interests and abilities, and emphasizes research to arouse students' enthusiasm and curiosity in mathematics and science. The school adopts various student-centered pedagogies that include peer tutoring, and independent and exploratory learning to engage their students in science. The students at the school are able to accelerate their studies and do advanced modules at the first-year undergraduate level that are tailored to their interests and talents. Being a school supported by the university, students are mentored one-on-one by university faculty members in subjects where they show exceptional talent. NUS high school is the only school in Singapore that awards its own NUS high school diploma to its graduates.

TEACHER EDUCATION AND PROFESSIONAL DEVELOPMENT

Teachers need continuous professional development to help them meet the changing demands of society and science education in Singapore. The Ministry of Education launched the Professional Development Continuum Models (PDCM) in 2004 to provide teachers with several models of professional development and opportunities for advanced certification. Non-graduates can enroll in structured in-service modules which lead to advanced diploma and degree certification, while graduates can enroll in modules in advanced diploma, masters and doctoral programs, and obtain the relevant certification when they complete the program requirements. For example, primary science teachers who are non-graduates, or graduates in disciplines other than science, may not have sufficient content knowledge and process skills to teach science effectively and confidently, especially in inquiry-based lessons. Thus, a primary science in-service program comprising seven modules, which can lead to an advanced diploma, has been designed for the teachers to equip them with the knowledge and understanding of the primary science syllabus, and to guide them in the use of various pedagogical approaches to teach the topics in the syllabus. The teachers have to take one content module in biological sciences and one in physical sciences to deal with the theoretical and laboratory aspects of the topics. The remaining five modules focus on student-centered and inquiry-based approaches in the teaching of the five themes in primary science (Diversity, Cycles, Systems, Energy, & Interactions) to help teachers gain insights on how to integrate content and process skills in their lessons to make them engaging and enjoyable for the students.

In addition to academic pursuits, the MoE also supports programs for teacher-attachments to industry. Science teachers can be attached to firms in the energy, ICT, engineering and manufacture, and health sectors, or research institutes and universities. This exposes the teachers to actual scientific workplace environment and interaction with the people in such places, thus broadening their outlook and experience. The attachment may also open teachers' eyes to an industry need for employees who can generate ideas and/or utilize those postulated by others, design and conduct experiments, and evaluate data and outcomes of the experiments (Duggan & Gott, 2002). Teachers need to realize that scientific habits of mind are more valued at the workplace than content knowledge, and that they should place more emphasis on problem-solving and procedural understanding in addition to conceptual understanding in their lessons.

CONCLUSION

The Ministry of Education has already, and will continue to implement policies and programs to help Singapore face the challenges in the future resulting from globalization, and scientific and technological advances. However, education is essentially a human enterprise. It is about people. To achieve success for the recent and future changes to the Singapore education system, school leaders and teachers must be engaged and be convinced of the importance of the changes made. They must also have the necessary professional expertise to effectively implement the policies and programs. Teachers with the necessary conviction and professional expertise to bring about pedagogical innovations in their classroom practices have become crucial agents of change as the schools are given more autonomy to shape their own direction and take more responsibility for their decisions. The ability of schools to achieve a balance of 'local' decision-making with central support will determine the success of the Singapore education system in meeting the challenges ahead.

NOTES

i Basic statistical information on Singapore is found at SINGSTAT Online, Singapore Department of Statistics (http://www.singstat.gov.sg/)

ii Curriculum Planning & Development Division. (2007)

iii Students at the integrated program schools do not take the national examination at the end of Year 10 but proceed on to take the national examination only at the end of Year 12. This frees up curricular time (up to 6 months) for the integrated program schools to put in place innovative educational programs for their students.

iv http://www.highsch.nus.edu.sg/

REFERENCES AND BIBLIOGRAPHY

Curriculum Planning and Development Division. (2004a). *Primary science syllabus*. Singapore: Ministry of Education.

Curriculum Planning and Development Division. (2004b). *Science syllabus: Lower secondary*. Singapore: Ministry of Education.

Curriculum Planning and Development Division. (2007). *Draft science syllabus: Lower secondary special/express/normal (academic) 2008*. Singapore: Ministry of Education.

Duggan, S., & Gott, R. (2002). What sort of science education do we really need? *International Journal of Science Education*, *24*(7), 661-679.

Hoe, W.M., & Tong, I. (2006, May). *Authentic assessment through innovative project work in Jurong secondary school*. Paper presented at 32nd annual conference of the International Association for Educational Assessment. Singapore.

Koh, K.H., Lee, A.N., Gong, W., & Wong, H.M. (2006, May). *Development of the Singapore prototype classroom assessment tasks: Innovative tools for improving student learning and performance*. Paper presented at 32nd annual conference of the International Association for Educational Assessment. Singapore.

Ministry of Education. (2002). *Report of the Junior College/Upper Secondary Review Committee*. Retrieved 10 July 2006, from : http://www.moe.gov.sg/jcreview/JC_Upp_Sec_Review_Report.pdf

Ministry of Education. (2006). *Report of the Polytechnic-School Review Committee*. Retrieved 10 July 2006, from www.moe.gov.sg/corporate/pdf/psrc-report.pdf

Prensky, M. (2001). Digital natives: Digital immigrants. *On the Horizon*, *9*(5), 1-6.

Subramanium, R., & Toh, K.A. (2004). Enrichment learning outside the primary science curriculum. In K.C. Yap, N.K. Goh, K.A. Toh & H.K Bak (Eds.), *Teaching primary science* (pp. 295-209). Singapore: Pearson.

Tan, K.C.D., Hedberg, J.G., Koh, T.S., & Seah, W.C. (2006). Datalogging in Singapore schools: Supporting effective implementations. *Research in Science and Technological Education*, *24*(1), 111-127.

Teo-Gwan, W.L. & Kwa, S.H. (2002). *My Pals are here: Science 4A*. Singapore: Federal Publications.

Toh, K.A., Riley II, J.P., Atputhasamy, L., & Subramaniam, R. (2006). School science achievement in Japan and Singapore: A tale of two cities. *Educational Research for Policy and Practice*, *5*, 1-13.

AFFILIATIONS

Thiam Seng Koh, Kim Chwee Daniel Tan, Horn Mun Cheah
National Institute of Education
Nanyang Technological University
Singapore

DAHSAH AND FAIKHAMTA

SCIENCE EDUCATION IN THAILAND: SCIENCE CURRICULUM REFORM IN TRANSITION

INTRODUCTION

A Thai science curriculum was developed in 1999 to meet national needs and help Thailand cope with the effects of globalization. The intention of this curriculum is that students develop lifelong learning skills with a balance of knowledge, skills, and attitudes. School-based curricula, and constructivist-based teaching and learning were introduced, but these ideas are new for most teachers, students and the community. Examination of the implementation of the science curriculum seems to suggest that science teaching is in something of a state of a transition with regards to pedagogy, gradually shifting to a constructivist-based teaching approach from the traditional didactic teaching that formed the historical landscape of teaching in the country. Here we discuss this transition, including problems and issues associated with curriculum implementation.

Thailand is situated in the heart of the Southeast Asian mainland, covering an area of 513,115 square kilometers, bordered by Myanmar, Laos, Cambodia and Malaysia. Thailand is a country proud of its own unique society, and unusually for the region, was never colonized by a European power. It is a constitutional monarchy and the King is highly respected by all Thais. The main religion in Thailand is Buddhism, but there are also significant numbers of Muslims, Sikhs, Hindus, Christians and other religions. Buddhism plays a central part of community life and has had a large effect on the culture and attitude of Thai people. The general characteristics of Thai people are that they are peaceful, generous, have concern for others, adhere to the religious moral teachings, and uphold the institution of the monarchy. Despite this uniqueness, modern and Western technologies and beliefs have affected the way Thai people live and their social activities. The advance of modern technology, in particular, has caused tremendous changes in the primary structure of Thai society, and the country is moving away from dependence on agriculture to become a semi-industrial and information-based economy.

In 1997, there was a major economic crisis in the whole Southeast Asian region, and later analysis of this crisis suggested to Thai authorities that there were major deficiencies in many areas of the country's economic development, especially in the areas of science and technology (Office of the National Education Commission [ONEC], 2001). The so-called Asia economic crisis thus induced a growing awareness that the Thai education system needed to equip Thais with the skills to

Richard K. Coll, Neil Taylor (eds.), Science Education in Context: An International Examination of the Influence of Context on Science Curricula Development and Implementation, 291–300.

engage in an information-based society, in order to cope with the economic and technological features of globalization. As a result, The 1997 Constitution of the Kingdom of Thailand (Office of the Council of State, 2005) and The Ninth National Economic and Social Development Plan (National Economic & Social Development Board [NESDB], 2002) were developed. These documents emphasize the development of human resources in science and technology in order to improve Thai people's quality of life, and to develop a nation capable of competing with others in an age of globalization. In this development, education is seen as a key enabler of the development of science and technology literacy for all Thai citizens.

THAI SCIENCE EDUCATION REFORM

Reform of science education in Thailand began in 1999, and the first step was to align science education with the 1997 Constitution, and the 1999 National Education Act. Under the 1997 Constitution the State is required to help Thais attain knowledge and morality; to issue laws relating to national education; to improve education so as to be attuned to economic and social change; to create and strengthen knowledge and inculcate sound awareness of politics and a democratic system of government under a constitutional monarchy; to promote research in various disciplines; to accelerate the application of science and technology for national development; to promote the teaching profession; and to encourage the revival of the wisdom, art and culture of the nation.

Subsequent to the promulgation of the Constitution, the National Education Act became the master law that provides the framework and guidelines for the provision and development of Thai education. It emphasizes lifelong learning with a balance of knowledge, skills and attitudes. The aim is for all learners to be able to adjust to world trends and events, and to develop desirable characteristics including virtue, competency, happiness, and self-reliance (ONEC, 1999). The educational guidelines of this Act suggest that the education system needed reform with respect to approaches to learning, curriculum, assessment and evaluation methods, instructional media, and learning resources (Fry, 2002; Institute for the Promotion of Teaching Science & Technology [IPST], 2002; ONEC, 1999; Pillay, 2002).

As noted above, the Act stipulates that a learner-centered teaching approach is to be used in classrooms, and that this approach is to be based on constructivist views of learning which posit that students actively and purposefully construct their own knowledge in order to make sense of scientific phenomena by using and building on their existing knowledge and new information (see, e.g., Bell, 1993; Tobin, Tippins & Gallard, 1994). Students also should engage in cooperative learning activities which will provide them with opportunities to co-construct meanings through interactions and exchange ideas with peers and teachers (IPST, 2002). Based on this constructivist view of learning, the use of a variety of instructional media, technology and learning resources is seen to be necessary to enhance student learning of science. The learning resources can vary, and need not be confined to the classroom but can include, for example, the use of public libraries, museums, zoological gardens, botanical gardens, and science and technology parks (Fry, 2002; Pillay, 2002). Assessment and evaluation methods also are seen as crucial factors to promote student learning and skill development. To assess student performance, a variety of assessment and evaluation methods also are required. Things such as observation of student development, learning behavior, participation in activities, self-evaluation, and classroom topic tests, are recommended.

In order to provide students with opportunities and activities to learn science in line with the education reform guidelines, science teachers need to use teaching approaches associated with constructivist views of teaching and learning. In doing so, it is expected that science teachers need to shift from being a teacher-centered 'director' towards a learner–centered 'facilitator' of learning (Office of the Education Council [OEC], 2004). The teacher as facilitator provides students with activities to shift them from any alternative conceptions towards the scientific conceptions. The facilitator also has roles to play in discovering what the students are thinking; helping students clarify and reflect on their own ideas; challenging students' ideas; helping students change their ideas; developing school-based science curriculum, planning lessons, developing instructional media; assessing and evaluating student learning; and so on.

A NEW SCIENCE CURRICULUM FRAMEWORK

Based on the 1999 National Education Act, the Basic Education Core Curriculum was promulgated in 2001, and this changed the Thai education system, teaching, student learning, and community participation. The new Thai education system comprises four levels: pre-school, primary, secondary, and higher education. The basic education curriculum has a 12-year core curriculum with four grade levels - each level comprising three years. First is first-level primary education (Grades 1-3), then is the second-level of primary education (Grades 4-6), next is the first-level of secondary education (Grades 7-9), and the final level is the second-level of secondary education (Grades 10-12). Each grade level has the same goals and objectives (but with different emphasis), and the same 'substance'. The substance consists of a body of knowledge, skills or learning processes, values or virtues, morality and right behavior. This substance is assembled into eight subject groups of learning: Thai Language, Mathematics, Science, Social Studies, Religion and Culture, Health and Physical Education, Art, Career and Technology, and Foreign Languages.

A Science Curriculum Framework was subsequently developed by the Institute for the Promotion of Teaching Science and Technology (IPST). This is a broad framework that provides objectives, curricular strands, standards for curricular contents and learning outcomes, and assessment and evaluation methods for teaching and learning for science subjects (IPST, 2002). Specifically, all learners are expected to be scientifically literate persons who:

- Understand principles and theories of basic science;
- Understand the boundaries, nature and limitations of science;
- Use skills to inquire about and explore science and technology;
- Develop thinking processes and imagination, ability to solve problems, and management, communicative and decision-making skills;
- Realize the influence and effects of the relationships between science, technology, people and the environment;
- Use knowledge of science and technology to advance society and everyday life; and
- Have scientific minds, morality, and values for using science and technology creatively.

The Science Curriculum Framework provides eight strands, covering the 12 years of basic education (i.e., Grades 1-12). The curricular strands consist of Living Things and Living Processes; Life and the Environment; Substances and Their Properties; Force and Motion; Energy; Changing Process of the Earth; Astronomy and Space; and the Nature of Science. At the first to third level (primary Grades 1-6 and secondary Grades 1-3), general science courses are offered as core-compulsory and electives. At the fourth level, students are divided into science and non-science streams (IPST, 2002). Science stream students are for those who intend to pursue higher education in pure science, applied science, technology or science-related areas. Non-science stream students are those who do not intend to pursue further education in science or science-related areas (IPST, 2002; Boonklurb, 2004).

Implementation of the science curriculum was devolved to schools, and schools are required to develop a school-based science curriculum and materials, taking into consideration the guidelines in the Science Education Framework, the learning needs of society, local wisdom, and desirable attributes of members of the family, community, society and nation. School-based curriculum prescription related to the school environment and society comprises 70% from Science Education Framework and 30% local curriculum (called the 70:30 model). So, schools are responsible for coordinating and cooperating with individuals, families, communities, community organizations, local administration organizations, private persons, private organizations, professional bodies, religious institutions, enterprises and other social institutions.

IMPLEMENTATION OF A NEW SCIENCE CURRICULUM

After the science curriculum was implemented in Thai schools, much research was done to see how this implementation had occurred. Research suggests there have been some successes and, as might be expected, with such sweeping reforms, some difficulties in implementation. The successes and difficulties associated with the implementation of the Science Curriculum Framework from the perspectives of student achievement, changes to pedagogy, and the community are now described.

The Science Curriculum Framework

An evaluation of implementing the national science curriculum at the first and second levels (i.e., primary Grades 1-6) suggested that the strengths of the curriculum are that it emphasizes science process skills, develops students' critical thinking, reasoning, analyzing, criticizing and creative skills; the learning process emphasizes 'learner centered' and 'hands-on activities'; and the learning resources place emphasis on local knowledge (IPST, 2006). Problems and obstacles in implementing curriculum are reported to be that the content and learning standards are still too broad in scope; there are not enough curriculum documents; and, there is not enough time to develop a school-based curriculum (Chuleekan, 2006; Department of Curriculum and Instruction Development, 2003; IPST, 2006). It also seems that students from different schools achieve differently according to the nature of the school-based curriculum (ONEC, 2005). For example, students studying at urban schools generally achieve higher grades than those studying at rural schools. Some small rural schools were very concerned about how to develop and implement a school-based curriculum, feeling that the science content is too broad in scope, and not consistent with the learning standards proposed in the Science Curriculum Framework. Some parts of science content were seen as too difficult for students, with insufficient learning time meaning few did experiments (IPST, 2006; Kanklang, 2005). Finally, it seems as if the school curriculum still places too much emphasis on subject content relative to science process skills (ONEC, 2005).

Student Achievement

Student learning outcomes after implementing the new science curriculum point to improvements in students' thinking skills, science process skills, and scientific attitudes (Regional Office for Educational, Religious & Cultural Development: Educational Region 1, 2004). Evaluation of the learning outcomes of Grade 1 and Grade 4 students indicates that more than 90% of the students at Grades 1 and 4 achieved the learning standards set in science substances (Department of Curriculum and Instruction Development, 2003). It also seems that students become more responsible for their own learning, are more creative, make sacrifices, are more sympathetic, and use various sources and methods to search for knowledge (Department of Curriculum and Instruction Development, 2003). Some three years after implementing the new curriculum, evaluation of the learning outcomes of students at the basic education level, likewise indicated higher achievement, with 40% gaining a 'high' level of science achievement and about 25% 'satisfactory' level (55-64%) (Kanklang, 2005). Looking in greater detail, it seems that science students were better at creative thinking; management, communication, and technology skills, and science process skills. Further research suggests that students now have a more positive attitude toward science (IPST, 2006), and in particular, students who learned using IPST teacher manuals (developed to help teachers change their pedagogy to be more learner-centered in nature) had more positive attitudes when compared with students who did not, especially with respect to things such as reasoning, being curious and open-minded, and engaging in creative thinking.

Interestingly, however, some research suggests that students were less capable in terms of science content knowledge after the reforms. For example, evaluation of student achievement at Grades 3 and 6 using a standard test for content knowledge developed by IPST, showed only around a 46% pass rate for these grades (IPST, 2006).

Changes to Pedagogy

One intention of the Thai educational reforms was to help science teachers to become more aware of ways to improve student learning, and how to modify teaching to meet new learning standards. It does seem that that after the education reforms, teachers are more engaged in classroom research in order to improve teaching and learning in their classroom (Vihokto et al., 2003). However, because Thai science teachers are more familiar with the traditional national curriculum, many struggle to develop and implement a school-based science curriculum. Teachers have been provided with teacher and student manuals developed by IPST, and they generally follow the learning activities proposed in the manuals. But overall, it appears administrators, teachers, parents, and communities all lack an understanding of how to develop a school-based curriculum (Boontim, 2004; Chuleekarn, 2003; Dangyai, 2004; IPST, 2006, Regional Office for Educational, Religious & Cultural Development: Educational Region 1-12, 2004; Regional Office for Educational, Religious & Cultural Development: Educational Region 5, 2004; Regional Office for Educational, Religious & Cultural Development: Educational Region 9, 2004). In particular, it seems that teachers in many schools, especially small schools, have difficulty in interpreting the learning standards from the curriculum documents, and achieving these learning standards via a school-based curriculum (Faikhamta & Roadrangka, 2005; ONEC, 2005). A factor here is that there are few teachers with knowledge of curriculum development.

Probably the key limiting factor in implementing Thai education reforms concerns shifting from a teacher-centered to student-centered teaching approach. A learner-centered teaching approach is seen by many Thai teachers to be radically new, and many teachers are suspicious of its effectiveness. Narot (2004) and Pillay (2002) argue that Thai teachers lack an understanding of the concepts, principles, and processes involved in a learner-centered teaching approach - as a consequence they commonly have negative attitudes towards such a change. As an example, Fry (2002) argues that Thai teachers do not understand that activity-based learning is only one of many methods that could be used to promote active, learner-centered learning, and also that they assume that learner-centered learning rejects all memorization. A further factor is that many Thai science teachers are not science graduates, and concerns with their own science content knowledge drives them to retain a teacher-centered approach in which it is less likely their shortcomings will be exposed (Faikhamta, Bunsawansong & Roadrangka, 2006; IPST, 2006; Rukpolmaung, 2005).

Community

One of the intentions, and indeed perceived strengths, of a school-based curriculum, is that it allows for input from parents and communities in curriculum development and implementation. In the case of the Thai educational reforms, the intention was that communities would work within the national guidelines, but develop a local schooling system that best meets the need of local communities. This notion is that teachers would strive to integrate subject content with the local society and draw upon local wisdom to develop science, technology and society (STS) notions. So, students could learn within the local context, and this might help them become aware of the importance of interrelationships between science, technology, humanity and the natural environment, and learn to apply knowledge in daily life and society. Research suggests that since the reforms there is, indeed, greater cooperation between schools and communities (Kanklang, 2005; IPST, 2006). However, a particular limiting factor is a lack of an understanding about the curriculum and school system by parents and communities (Boontim, 2004; Dangyai, 2004; Regional Office for Educational, Religious & Cultural Development: Educational Region 1-12, 2004; Regional Office for Educational, Religious & Cultural Development: Educational Region 5, 2004; Regional Office for Educational, Religious & Cultural Development: Educational Region 9, 2004). Vihokto et al. (2003) suggests that many Thai communities believe that the school should set the curriculum, because many in the communities have little formal education.

CONCLUSIONS AND IMPLICATIONS

Since its instigation in 1999, science curriculum implementation and learning reform has been an ongoing process in Thailand. It does seem that whilst some progress has been made, there is a long way to go if the nation is to realize the benefits of curriculum reform in the way intended by the Thai government and educational officials. A vital factor influencing successful curriculum implementation in Thailand is the nature of the Science Curriculum Framework which, it seems, is too broad in scope. As a consequence, when teachers try to use it as a basis to develop a school-based curriculum, they struggle to organize learning standards and science content (ONEC, 2005). A second important factor inhibiting effective curriculum reform in Thailand is teachers' lack of understanding about school-based curriculum development and constructivist-based or learner-centered teaching approaches. Prior to the education reforms, teachers used learning materials such as student and teacher manuals developed by IPST. They follow the teaching and learning activities in these manuals consistently, rather than developing their own school-based learning activities.

Thai teachers, and their students, are very familiar with a highly traditional, didactic, teaching approach. It is thus hardly surprising that it seems it will take a long time for teachers to shift their understanding and teaching practice towards constructivist-based views of teaching and learning (Dahsah, 2007; Faikhamta, 2007).

In order to enhance the success of science curriculum development implementation, all stakeholders in Thai science education have a crucial role to play. The following suggestions are made to help elucidate the roles a stakeholder might play.

- All stakeholders concerned with the development of school-based curriculum need to develop an understanding of curriculum development and implementation. Teacher training institutes, for example, should have curriculum development as a compulsory course, and schools and national institutes should be exposed to workshops about curriculum development;
- The Science Curriculum Framework should retain its flexibility, but give more specific detail about content and standards, making it easier for schools to develop their own curriculum to match such standards. The framework could, for example, give examples or exemplars of a school-based curriculum;
- Further national, regional, and school level research about curriculum implementation is needed to evaluate curriculum implementation at all levels;
- Schools should work to inform their community members about the importance of participation between communities and schools in school-based curriculum development;
- Schools need to provide community stakeholders with guidelines for educational management, and suggestions as to how their community members might participate in the schooling system, and in curriculum development and implementation;
- Schools and educational institutions should work together to identify and help science teachers who did not graduate in science to promote their understanding of both content and pedagogy; and
- Teacher educational training institutions need to evaluate and, where necessary, revise their curricula and teaching methods to prepare science teachers capable of delivering the education reforms signaled by the Thai Government.

REFERENCES AND BIBLIOGRAPHY

Bell, B. (1993). *Children's science: Constructivism and learning in science*. Australia: Deakin University Press.

Boonklurb, N. (2004). *Current trend and main concern as regards science curriculum in Thailand*. Retrieved 14 September 2004, from http://www.ibe.unesco.org/National/China/NewChinaPdf/Ilthailand.pdf

Boontim, S. (2004). Evaluation of school curriculum management, basic education, Education Region 6. In Bureau of Inspection and Evaluation, Office of the Permanent Secretary (Ed.), *Report for education evaluation and development (B.E. 2541-2546)*. Bangkok: The Express Transportation Organization of Thailand.

Chuleekan, V. (2006). *Research summary: Evaluation of school curriculum management, Education Region 1-12*. Retrieved 13 November 2006, from http://www.inspect6.moe.go.th/curriculum/vijai.htm.

Dahsah, C. (2007). *Teaching and learning using conceptual change to promote Grade 10 student understanding and problem-solving skills in stoichiometry*. Unpublished PhD thesis, Kasetsart University, Bangkok, Thailand.

Dangyai, N. (2004). Factors that effect to school curriculum of school in Education Region 1. In Bureau of Inspection and Evaluation, Office of the Permanent Secretary (Ed.), *Report for education evaluation and development (B.E. 2541-2546)*. Bangkok: The Express Transportation Organization of Thailand.

Department of Curriculum and Instruction Development. (2003). *Report of follow up and evaluation of school-based curriculum implementation: Primary education level*. Bangkok: Education Research Division, Department of Curriculum and Instruction Development, Ministry of Education.

Faikhamta, C. (2007). *The development of pre-service chemistry teachers' pedagogical content knowledge: From a methods course to field experience*. Unpublished PhD thesis, Kasetsart University, Bangkok, Thailand.

Faikhamta, C., Bunsawansong, P., & Roadrangka, V. (2006). Exploring chemistry content knowledge of pre-service science teachers. *Kasetsart Journal (Social Sciences)*, *27*(1), 39-50.

Faikhamta, C. & Roadrangka, V. (2005). Problems in professional experience training of student teachers in the Project for Promotion of Science and Mathematics Talented Teachers (PSMT). *Songklanakarin Journal of Social Sciences and Humanities*, *11*(2), 151-163.

Fry, G.W. (2002). *Synthesis report: Crisis to opportunity, the challenges of educational reform in Thailand*. Bangkok: Office of the National Education Commission (ONEC).

Institute for the Promotion of Teaching Science and Technology. (2006). *Evaluation of implementing basic education curriculum: Science substance*. Bangkok: IPST.

Institute for the Promotion of Teaching Science and Technology. (2002). *Thai science teachers' standards*. Bangkok: IPST.

Kanklang, V. (2005). *Research report: A study of perceptions of education administrators on trend of effectiveness of education management according to the 2001 Basic Education Curriculum*. Bangkok: Bureau of Academic and Educational Standard Office, Office of the Basic Education Commission, Ministry of Education.

Klainnin, S. (2002). *Research report: Thai science education compare with international study*. Bangkok: The Institute for the Promotion of Teaching Science and Technology.

Ministry of Education. (2002). *Basic Education Curriculum B.E. 2544 (A.D.2001)*. Bangkok: The Express Transportation Organization of Thailand.

Narot, P. (2004). *A synthesis of education research and trends of future research*. Bangkok: Office of the National Education Commission.

National Economic and Social Development Board. (2002). *The Ninth National Economic and Social Development Plan*. Bangkok: Suksapan Panitch.

Nimanong, S., et al. (2004). Evaluation research of educational reform school, Education Region 1. In Bureau of Inspection and Evaluation, Office of the Permanent Secretary (Ed.) *Report for education evaluation and development (B.E. 2541-2546)*. Bangkok: The Express Transportation Organization of Thailand.

Office of the Council of State. (2005). *The 1997 Constitution of the Kingdom of Thailand*. Retrieved 24 December 2006, from http://www.krisdika.go.th/searchShowFile.jsp?path=%2 Fexport%2Fnew_htdocs%2Fdata%2Flaw%2Flaw1%2F%25c306%2F%25c306-10-2540-a0001.pdf

Office of the Education Council. (2004). *Education in Thailand 2004*. Bangkok: Amarin Printing and Publishing.

Office of the National Education Commission. (1999). *National Education Act 1999*. Bangkok: ONEC.

Office of the National Education Commission. (2001). *Report of research and development of science education reform in Thailand.* Bangkok: ONEC.
Office of the National Education Commission. (2005). *Report summary: Evaluation of the development of education reform in Thailand.* Bangkok: ONEC.
Pillay, H. (2002). *Teacher development for quality learning: The Thailand Education Reform Project.* Bangkok: Office of National Education Commission.
Regional Office for Educational, Religious and Cultural Development: Educational Region 1-12. (2004). Evaluation of school curriculum management, basic education, education region 1-12. In Bureau of Inspection and Evaluation, Office of the Permanent Secretary (Ed.), *Report for education evaluation and development (B.E. 2541-2546).* Bangkok: The Express Transportation Organization of Thailand.
Regional Office for Educational, Religious and Cultural Development: Educational Region 5. (2004). Evaluation of School Curriculum Management, Basic Education, Education Region 5. In Bureau of Inspection and Evaluation, Office of the Permanent Secretary (Ed.), *Report for education evaluation and development (B.E. 2541-2546).* Bangkok: The Express Transportation Organization of Thailand.
Regional Office for Educational, Religious and Cultural Development: Educational Region 9. (2004). Evaluation of school curriculum management, basic education, education region 9. In Bureau of Inspection and Evaluation, Office of the Permanent Secretary (Ed.) *Report for education evaluation and development (B.E. 2541-2546).* Bangkok: The Express Transportation Organization of Thailand.
Rukpolmaung, C. (2005). *Situation of insufficiency of Thai teachers at basic education level.* Bangkok: Office of National Education Commission.
Silpabanlaeng, R., Roadrangka, V., Chaiso, P., Yutakom, N., & Phanwichien, K. (2006). The trend of pre-service science teacher education in the fifth years period (2002-2016). *Kasetsart Journal (Social Sciences)*, *27*(1), 39-50.
Tobin, K., Tippins, D.J., & Gallard, A.J. (1994). Research on instructional strategies for teaching science. In D.L. Gabel (Ed.). *Handbook of research on science teaching and learning* (pp. 45-93). New York: Macmillan
Vihokto, P. et al. (2003). *Report of follow up and evaluation of school-based curriculum implementation: Primary education level.* Bangkok: Education Research Division, Department of Curriculum and Instruction Development, Ministry of Education.

AFFILIATIONS

Chanyah Dahsah
Institute for Innovation and Development of Learning Process
Mahidol University
Thailand

Chatree Faikhamta
Department of Education
Faculty of Education
Kasetsart University
Thailand

ISHAK AND ZULKIFLI

TEACHER TRAINING OF SECONDARY SCHOOL PHYSICS TEACHERS IN MALAYSIA: CRITICAL ISSUES AND A NEW DIRECTION

INTRODUCTION

The training of science teachers in Malaysia, as in many similar nations, represents a formidable challenge, especially in the case of problematic subjects such as physics. If Malaysia is to achieve developed nation status as outlined in its Vision 2020, then it needs a skilled workforce, and thus the training of science teachers is of utmost importance. Critical issues associated with training of physics teachers in a Malaysian tertiary educational institution are explored in this chapter. Details of a current innovative teacher training intervention, based on the Central and Northern European tradition of 'didaktik analysis', with the intention of developing a new direction for physics teacher training in Malaysia are presented.

Malaysia, located in South East Asia, is struggling to achieve developed nation status as outlined in Vision 2020. One of the challenges in this vision is greater emphasis on science and technology. The National Education Blueprint 2006-2010 launched by the Malaysian Prime Minister in January 2007 outlines six strategies, one of which stresses the importance of teacher training. Science teacher education is thus seen to play a key role in assisting the Malaysian government to realize the Vision. Hence, it is important to focus on critical issues for physics teacher training in Malaysia - local issues and reality - in order to map the ways in which physics education could and should develop. The authors argue here that traditional teacher training regimes that focus almost entirely on enhancing trainee teachers' pedagogy or pedagogical content knowledge, lack the capacity to meet Malaysian teacher training needs. What is needed is a teacher training regime that takes cognizance of local needs and context, and produces new physics teachers with the ability to produce learning activities that will motivate students to engage meaningfully with relevant physics content.

CONCEPTIONS OF PHYSICS AND PHYSICS TEACHING

If we are to develop an understanding of physics content needs in school teaching and in teacher training, we first need to consider physics itself as a scientific discipline. Physics education involves the learning and teaching of specific physics content, most of which is derived from physicists. However, the physics content as presented by teachers of physics might not be in accordance to the views of physicists. One way this potential mismatch might be addressed is for physicists

Richard K. Coll, Neil Taylor (eds.), Science Education in Context: An International Examination of the Influence of Context on Science Curricula Development and Implementation, 301–312.

engaged in research and development to become engaged in developing 'teaching experiments' which involve physics teachers working in the research of physics instruction in secondary schools (Lijnse, 2000). Such scaffolding might support teachers who are required to teach physics, and who lack physics knowledge, for example, graduates who major in other sciences (chemistry, or biology) or other branches of science disciplines. This latter scenario is often encountered in Malaysian secondary schools, particularly in rural areas, and an increasing lack of qualified physics teachers in secondary schools contributes to this issue.

A key role physics teachers play, is to ensure that their students gain scientific knowledge similar to that accepted by physicists. However, research suggests that students commonly develop conceptions of physics (and other science disciplines) which are different from scientific explanations, particularly when the topics are outside their personal experience. An example of this comes from the National Olympiad Physics Competition for secondary physics students held annually in Malaysia, for which students are often found to be unable to apply physics principles despite being exposed frequently to textbook problems and physics tests. Research reported in the international science education literature suggests that many students worldwide hold ideas or beliefs in various science content domains that are contrary to the intended outcomes of school science teaching (Duit, 2004; Duit & Treagust, 2003; Fensham, 2001). These student 'alternative conceptions' are reported to be robust and difficult to extinguish through teaching, although a variety of conceptual change approaches have been proposed and evaluated (Duit & Treagust, 2003; Harrison & Treagust, 2000). Duit and Treagust (1998) claim that students' conceptual progress towards understanding and learning science concepts and principles after instruction quite frequently appear to be limited. This is in many ways an extraordinary observation, given that there have been literally thousands of studies investigating and attempting to remediate students' alternative conceptions (Duit, 2004). Fensham (2004) and Lijnse (2000) argue that one fundamental reason science concepts are poorly learned may be that the problematic nature of the content itself is ignored when teaching and researching in science education to develop new pedagogy. In other words, physics as a discipline of knowledge is 'foreign', or outside everyday experiences of many secondary school students *and their teachers*. In brief, the consequence of this is that teachers seldom think about what content to include and the relationship between science as practiced by scientists and science that can and should be taught in the classroom.

Wellington and Osborne (2001) comment on the role the teachers of science play in mediating the conceptions of science among secondary school students, and argue that "as teachers of science … our primary skills lie not in our ability to do science, but in our ability to interpret and convey a complex and fascinating subject" (p. 138). This is an accord with what Fensham has written of many times (see, e.g., Fensham, 2000, 2001, 2004). Fensham points to a need to develop content from primary sources of scientific knowledge into the context of school knowledge. In other words, teachers need to look at the original work of the specific content, and transform this into the intended school science curriculum. Teachers then need to have knowledge of specific content - each content area of which requires different teaching methods or instructional approaches. It is important to note that an individual teacher achieving understanding of specific content, does not necessarily mean he or she has the ability to teach it. Colomb (1999) describes this as a process in which scientific knowledge is transformed into school knowledge, and then into taught knowledge (Figure 1).

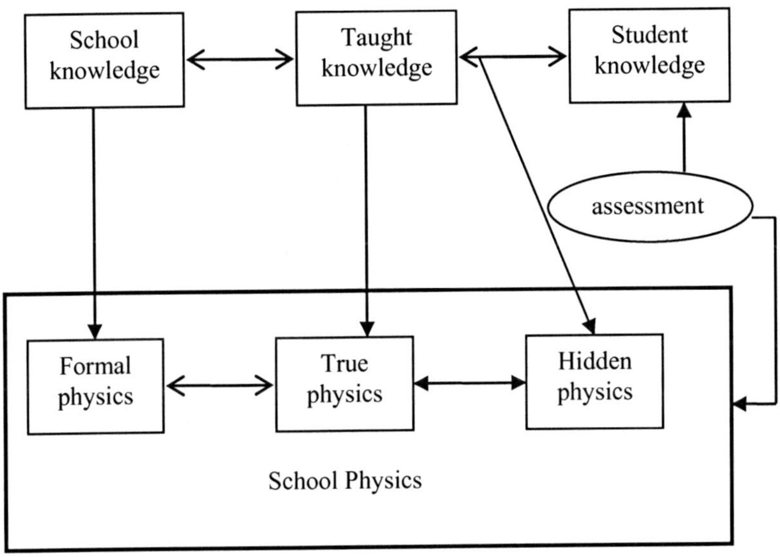

Figure 1. The functioning of knowledge in school (modified from Colomb, 1999)

PHYSICS IN SECONDARY SCHOOLS

In Malaysia, physics as a discipline is seen as abstract, difficult to understand and to conceptualize. As such, physics is regarded as 'high status' by the employment sector. This high status of physics employment-wise makes teaching physics unattractive to prospective physics teachers. The Malaysian Government in 1994 introduced a 'master teacher' scheme. This scheme enables teachers in secondary schools who have been teaching continuously for at least 12 years to be promoted to become a master teacher. Such a teacher is regarded as expert in his or her subject, and the position is similar in status to that of a principal at a secondary school. The master teacher is still heavily involved in the teaching and learning process in the classroom and is expected to assist other schools within the region, for example, conducting in-service courses for colleagues coordinated by the Training Division of the State Education Department.

REFORM OF THE MALAYSIAN SECONDARY SCHOOL PHYSICS CURRICULUM

Physics learning in Malaysia from the 1970s was based on the Malaysian New Modern Physics Curriculum of 1976. Physics learning was traditional and based on the use of textbooks and, to some extent, laboratory work. Emphasis in this version of the curriculum was on 'discovery learning' with students engaging in a variety of activities in order to develop an understanding of physics concepts through independent investigation. Through students' own efforts, it was argued, their native ability and intellectual capability, would enable them to learn the desired physics concepts. However, in most cases, physics concepts were in fact learnt via rote memorization. The teacher transferred physics knowledge to the students; 'filling them up' with facts, concepts, principles and laws. In other words, teachers strived to equip their students with the physics content knowledge needed to pass examinations.

A new physics curriculum was introduced in 1992. Other than focusing on conceptual physics, the new curriculum introduced 12 science process skills and a Science as a Process Approach - an adaptation from the USA and UK science curricula (Table 1).

Table 1. Scientific skills in the 1992 Malaysian Physics Curriculum

Process Skills	*Manipulative Skills*
Identifying problems	Presenting information systematically
Formulating hypotheses	Applying
Planning and conducting studies	Managing laboratory apparatus correctly and safely
Making inferences	
Observing	
Recording	
Interpreting data	
Manipulating	
Making conclusions	

For the first time, learner-centered education was introduced, and this was to be based on von Glasersfeld's (2002) notions of constructivism: that students come into classroom having prior knowledge formed by years of experience and learning. As students grow, their view of prior experiences changes and affects their interpretation of observations; and that challenging, revising, and restructuring students' prior knowledge requires much effort from teachers. The teacher was expected to explore students' prior understanding of specific content and help transform any misconceptions. In order to achieve this, the teachers were expected to challenge students' knowledge or ideas by engaging in activities such as experiments, asking questions and problem-solving. In addition, the teachers were supposed to assume the role of a 'facilitator' of student learning. Although the notions of constructivism were intended to be implemented in Malaysian secondary school classrooms, this view of learning - although advocated widely -

was not extensively practiced by physics teachers, in either rural or urban areas when confronted with classroom realties.

A decade later, another new physics curriculum was introduced. The main aim of this new curriculum was to provide students with the knowledge and skills in science and technology that would enable them to solve problems, and make decisions about everyday life based on good scientific attitudes developed during learning (Kementerian Pendidikan Malaysia, 2002). Other than inquiry and constructivism - both already existing in the curricula - some new approaches such as Science, Technology and Society (STS), contextual, and mastery learning, were also emphasized. At the same time, the then Malaysian Prime Minister announced that science and mathematics were to be taught in English. The Kementerian Pendidikan Malaysia (2004) thus produced another revision of the physics curriculum in English which was first implemented at Form 4 (i.e., Year 10, or 16 years old) in 2006.

TEACHING PHYSICS IN ENGLISH IN MALAYSIA

A shift in medium of instruction for the primary and secondary school science and mathematics curricula in Malaysia from the Malay language to English began in 2003. This change was implemented with pupils in Year 3 at the primary school level, and in Form 1 (Year 7, or 13 years old) at the secondary school level. However, teaching of other school curricula remained in the Malay language. At the beginning of 2006, the Form 4 physics (i.e., Year 10, or 16 years old) curriculum was taught in English for the first time. Many Malaysian students are weak in English, particularly those from the rural areas, and thus faced problems in understanding English language, as well as understanding physics content. Even capable students with the ability and desire to learn physics, took longer to understand physics when English was the medium of instruction. It remains unclear to what extent students' problems with conceptual understanding of physics are due to English language difficulties.

SECONDARY PHYSICS TEACHER TRAINING IN MALAYSIA

Enrolments of pre-service teachers in Malaysian physics teacher training courses nationwide have declined at an alarming rate in recent years. At the same time, teacher training institutions face increasing challenges in training current intakes of trainee teachers. For example, some pre-service teachers have sound physics knowledge, but others, even those with university physics learning experience, lack understanding of physics concepts, and either cohort may experience difficulty teaching physics - especially in English. If trainees do not understand specific physics content, it seems likely that they will have difficulty teaching such content. In any case, the role of the teacher is paramount in helping students move towards the scientific view of a variety of concepts. Hence, teacher education plays an important role in assisting science teacher trainees to develop specific content knowledge, so that their school students learn physics knowledge more easily.

Here it is argued that traditional methods of teacher training have largely failed, and that the training of physics teachers is in need of a renaissance. In Malaysia, as in many other countries, teacher training is based on the 'pedagogical approach' in which trainers strive to develop trainees' pedagogical knowledge, and in some cases their content knowledge. The results from a study of pre-service teachers discussed below support the notion that training based on the Central and North European 'didaktik approach' (not to be confused with 'didactic' see below) can help pre-service teachers come to understand physics as it is understood by physicists, devise ways of teaching *specific* physics content, and as a consequence enhance their confidence or self-efficacy in the teaching physics.

THE 'PEDAGOGY TRADITION'

The 'pedagogy tradition' in science teacher training programs places emphasis on pre-service teachers' ability to transform science content knowledge through planning, preparation, and trial or microteaching under supervision. Shulman's (1987) notion of pedagogical content knowledge (PCK) is a form of the pedagogical tradition in which teaching strategies are developed and incorporate knowledge of subject matter, understanding the specifics of students' learning difficulties, and students' conceptions of specific content subject matter (van Driel, Verloop & de Vos, 1998). This PCK-based approach is the norm in Malaysian teacher training schools, but there is little emphasis on science content compared with teaching methods. Hence, Malaysian teacher training courses have typically been based on developing trainee teachers' pedagogy. In practice in Malaysia, following this tradition, pre-service teachers are taught how students learn, how students acquire knowledge, and how to evaluate students' scientific skills, scientific attitudes and their values.

THE 'DIDAKTIK TRADITION'

Fensham (2001) suggests that there are two key issues that need to be addressed if we wish to improve teacher training. The first is this notion that science content be 'treated as problematic' and the second is that of content analysis. To say a teacher should treat science content as problematic, does not mean that students find the content itself difficult (i.e., they struggle to understand mechanics, electromagnetism, etc.), but instead refers to the relationship between science as viewed by practicing scientists, and what the teacher presents in the classroom; what we might call 'school science' (Jenkins, 2000).

Didaktik analysis (a German term) is very different to didactic teaching, that is, teaching via a transmissive approach. Didaktik analysis has its tradition in Northern and Central Europe (e.g., Finland, Germany, Sweden - see, Uljens, 1997) and refers to the art of teaching or 'how to teach' (Bertrand & Houssaye, 1999; Colomb, 1999); to teach, to be a teacher, to educate or study teaching (Hopmann & Riquarts, 1995 & 2000). It is thus a tradition of thinking about teaching and learning (Westbury, 2000), what should be taught and learned (Gundem, 2000), and a disciplinary subject matter, or any means of formalizing a disciplinary subject matter for learning (Tochon, 1999).

Didaktik analysis in teacher training programs has three elements: the teacher, the students, and the knowledge, in what is described as the 'didaktik triangle' (see Figure 2, Tiberghein, Jossem & Barojas, 1998). This triangle can be used in teacher training to explain to pre-service teachers how to prepare lessons (Hopmann & Riquarts, 1995). In practice what this means is that teaching requires knowing what the content of instruction should be like; where it comes from; and how it is used. Teaching is only possible if instruction takes care of the progress of learning and the development of the student; and teaching has to be take cognizance of both the content and the student.

Didaktik analysis requires the teacher to engage in content analysis; understanding of general student alternative conceptions as reported in the science education literature; knowledge of his/her own students - their alternative conceptions, learning styles and cognitive abilities; and knowledge of appropriate pedagogy based on this learner-centered analysis. Didaktik thus consists of reflection on, and planning about, teaching, with an emphasis on the relationship between the student and knowledge (Bertrand & Houssaye, 1999).

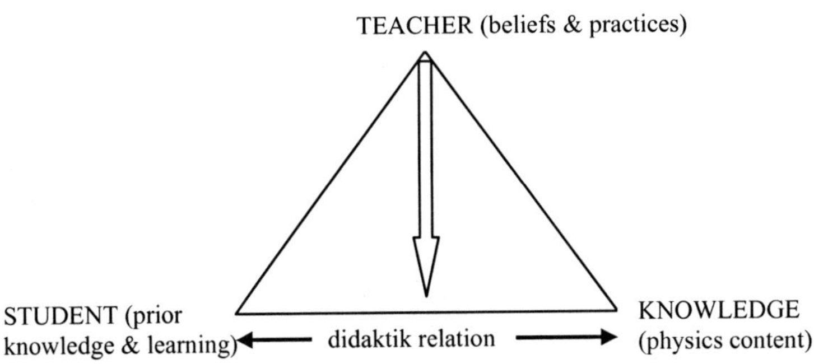

Figure 2. The didaktik triangle (modified from Colomb, 1999 & Kansanen & Meri, 1999)

DIDAKTIK OF PHYSICS

The didaktik tradition provides guidance for some of the core tasks in pre-service teacher training. Questions that arise from the didaktik tradition are: What do teachers know about the content they are teaching?, and what do students see and comprehend when they are confronted with the content? (Hopmann & Riquarts, 1995, 2000;Westbury, 2000). To answer these questions, Marton and Ramsden (1988) note that one needs to have a 'subject' didaktik, because this forms a major component in teacher education and research. The key point here being the focus on *specific* content; hence, the need for a *didaktik of physics* that "defines the analysis and mapping of the different ways students experience and conceptualize various content domains" (Marton & Ramsden, 1988, p. 283), in this case from physics.

Klafki's (2000) model of didaktik analysis is now widely used in teacher education in Central and Northern European countries. It poses five 'sets of questions' to be used in pre-service teacher courses, as a guideline for preparing lesson plans. Gundem (2000) and Kunzli and Kruger (2000) reduced Klafki's (2000) rather complex model of didaktik analysis to a consideration of:

- What is to be taught and learned? (the content aspect);
- How is content to be taught and learned? (the method aspect); and
- Why is content to be taught and learned? (the goal/aims aspect).

According to Fensham (2004), these questions emphasize the importance of students' prior ideas. Therefore, it is important that teachers utilize research findings of student alternative conceptions when analyzing specific science content. The teacher also needs to anticipate the students' future, what might affect them as adults, and is required to undertake a close examination of the structure of content.

DIDAKTIK ANALYSIS IN MALAYSIAN TEACHER TRAINING

A didaktik analysis of physics based on these notions was introduced as an intervention into a teacher education program in a school of education training program in Malaysia; specifically in a physics teaching methods course for third and fourth year pre-service physics teachers in their final year of their undergraduate program. In practical terms, this consisted of helping the pre-service teacher trainees understand how to teach physics by drawing on didaktik analysis. Teaching practices, in the microteaching and during a practicum, engaged in as part of the intervention, helped the pre-service physics teachers to reflect on didaktik analysis and inform their decisions about teaching secondary school physics.

One hundred and thirteen pre-service physics teachers were involved in this study: 35 third years (15 males & 20 females), and 78 fourth years (18 males & 60 females). Data were gathered through the use of written reports of pre-service teachers' prior physics learning experiences at secondary and tertiary levels, two tests of conceptual analysis - one on kinematics (Beichner, 1994), and one on force and motion (Thornton & Sokoloff, 1998). Questionnaires about teaching attitudes and teaching competencies, also were used, as were interviews after the microteaching and practicum. We also examined written assignments of didaktik analysis of physics, daily lesson plans, and triangulated this with observations of microteaching and the practicum to see if their teaching was in fact based on didaktik analysis. Other data came from written reflections about the methods course generally, and the practicum, examination of relevant questions from final examinations, and field notes made by the first author during meetings with the pre-service physics teachers.

The research findings suggest the pre-service teachers' beliefs about, and attitude toward physics teaching were influenced by their prior physics learning experiences. In particular, the physics content knowledge influenced their physics teaching self-efficacy. As a result of their engagement with didakik analysis in the methods course, they gained in conceptual understanding of specific physics content, and this enhanced their physics teaching self-efficacy beliefs, resulting in a more positive attitude toward physics and learning. Observation of the microteaching and practicum suggested the pre-service teachers' attitudes were shaped by their beliefs about their experience of the methods course generally and the didaktik analysis specifically, consistent with the apparent improvement in attitude toward physics and learning, their physics teaching self-efficacy beliefs, and their conceptual understanding of specific physics content.

The research also identified some factors that influenced the pre-service physics teachers' views of didaktik analysis. The pre-service physics teachers' experiences influenced their attitude toward physics and learning and subsequently their beliefs about physics teaching practices in the microteaching and practicum. One fundamental theme to emerge from all these data sources is that they believed in the value of didaktik analysis-based in microteaching, but it was hard to find an example of a pre-service physics teacher who actually implemented this in their lesson plans, and teaching sequence during the practicum. In other words, there was lack of correspondence between pre-service physics teachers' expressed beliefs about didaktik analysis-based teaching, and their teaching practice in the classroom. It would seem that having some experience with didaktik analysis assignments and teaching practice in the microteaching alone, is not enough to ensure that pre-service physics teachers will actually use didaktik analysis-based teaching methods in their classrooms. However, the use of the didaktik analysis seemed to help better prepare pre-service physics teachers for engaging in didaktik analysis-based teaching practice. At the conclusion of the course, many pre-service physics teachers had a more positive learning experience of didaktik analysis.

During the practicum, most of the pre-service physics teachers were bounded by the physics content contained in the curriculum specifications, and this subsequently had an impact on their lesson plans and the teaching sequences used in the classroom. They were typically required to teach directly from the curriculum specifications in order to meet the requirements set out in individual schools, and to try to meet the criteria established by the didaktik analysis thus some caused tension. In these circumstances, the pre-service physics teachers were likely to teach without having had the opportunity to apply fully the didaktik analysis-based teaching practices. In addition, as a result of lack of accessible resources and lack of content knowledge for specific content areas, some pre-service physics teachers were not able to adopt the didaktik analysis-based teaching practices. Thus, the authors think it is important that pre-service physics teachers are given support and opportunities by their mentor teachers to undertake didaktik analysis of other physics content areas of their own.

CONCLUSIONS

From this study, the authors have come to believe that the pre-service teachers themselves, their beliefs and attitudes, their specific content knowledge, their practices with their students in the classroom are the heart of the success of didaktik analysis-based teaching. This study adds to the growing literature suggesting that introducing didaktik analysis into science and physics teaching methods courses requires the identification of pre-service physics teachers' attitude toward, and beliefs about physics teaching, their attitude toward physics and learning, their physics teaching self-efficacy beliefs, and their conceptual understanding of specific physics content. The introduction of the didaktik analysis-based teaching practice in this physics teaching methods course in Malaysia is just the beginning. It does suggest this approach can work in a different education context to the origins of didaktik analysis (i.e., Europe). Thus, our research continues with new cohorts in the physics teaching methods course.

REFERENCES AND BIBLIOGRAPHY

Beichner, R.J. (1994). Testing student interpretation of kinematic graphs. *American Journal of Physics, 62*(8), 750-762.

Bertrand, Y., & Houssaye, J. (1999). Pédagogie and didactique: An incestuous relationship. *Instructional Science, 27*, 33-51.

Colomb, J. (1999). School knowledge and didactic analysis: A research perspective in comparative didactics. *Instructional Science, 27*, 53-71.

Duit, R., & Treagust, D. F. (1998). Learning in science: From behaviourism towards social constructivism and beyond. In. B.J. Fraser & K.G. Tobin (Eds.), *International handbook of science education* (Part I, pp. 3-26). Dordrecht: Kluwer.

Duit, R., & Treagust, D. F. (2003). Conceptual change: A powerful framework for improving science teaching and learning. *International Journal of Science Education, 25*(6), 671-688.

Duit, R. (2004). *Bibliography: Students' and teachers' conceptions and science education*. Retrieved 23 May 2005, from http://www.ipn.uni-kiel.de/aktuell/stcse/stcse.html

Fensham, P.J. (2000). Providing suitable content in the 'science for all' curriculum. In R. Millar, J. Leach & J. Osborne (Eds.), *Improving science education: The contribution of research* (pp. 147-164). Buckingham, UK: Open University Press.

Fensham, P.J. (2001). Science content as problematic: Issues for research. In H. Behrendt, H. Dahncke, R.M. Duit, W. Graber, M. Komorek, A. Kross & P. Reiska (Eds.), *Research in science education: Past, present, and future* (pp. 27-41). London: Kluwer.

Fensham, P.J. (2004). *Defining an identity: The evolution of science education as a field of research*. London: Kluwer.

Gudmundsdottir, S., Reinertsen, A., & Nordtømme, N.P. (2000). Klafki's didaktik analysis as a conceptual framework for research on teaching. In I. Westbury, S. Hopmann & K. Riquarts (Eds.), *Teaching as reflective practice: The German didaktik tradition* (pp. 319-334). London: Lawrence Erlbaum.

Gundem, B.B. (2000). Understanding European didactics. In B. Moon, S. Brown & M. Ben-Peretz (Eds.), *Routledge international companion to education* (pp. 235-262). London: Routledge.

Harrison, A.G., & Treagust, D.F. (2000). Learning about atoms, molecules, and chemical bonds: A case study of multiple model use in grade-11 chemistry. *Science Education, 84*(3), 352-381.

Hopmann, S., & Riquarts, K. (1995). Starting dialogue: Issues in a beginning conversation between didaktik and the curriculum traditions. *Journal of Curriculum Studies, 27*(1), 3-12.

Hopmann, S., & Riquarts, K. (2000). Starting a dialogue: A beginning conversation between didaktik and the curriculum traditions. In I. Westbury, S. Hopmann & K. Riquarts (Eds.), *Teaching as a reflective practice: The German didaktik tradition* (pp. 3-11). London: Lawrence Erlbaum.

Jenkins, E. (2000). 'Science for all': Time for a paradigm shift? In R. Millar, J. Leach & J. Osborne (Eds.), *Improving science education: The contribution of research* (pp. 207-226). Buckingham, UK: Open University Press.

Kansanen, P. (2003). Studying - the realistic bridge between instruction and learning. An attempt to conceptual whole of the teaching-studying-learning process. *Educational Studies, 29*(2/3), 221-232.

Kansanen, P., & Meri, M. (1999). The didactic relation in the teaching-studying-learning process. *The TNTEE Journal, 2*, 107-116.

Kementerian Pendidikan Malaysia (Ministry of Education Malaysia). (2002). *Huraian sukatan pelajaran fizik tingkatan empat* (Form 4 physics curriculum specifications). Kuala Lumpur: Curriculum Development Centre.

Kementerian Pendidikan Malaysia (Ministry of Education Malaysia). (2004). *Integrated curriculum for secondary schools.* Kuala Lumpur: Curriculum Development Centre.

Klafki, W. (2000). Didactic analysis as the core of preparation for instruction. In I. Westbury, S. Hopmann & K. Riquarts (Eds.), *Teaching as a reflective practice: The German didaktik tradition* (pp. 139-159). London: Lawrence Erlbaum.

Kunzli, R., & Kruger, G. H. (2000). German didaktik: Models of re-presentation, of intercourse, and of experience. In I. Westbury, S. Hopmann & K. Riquarts (Eds.), *Teaching as a reflective practice: The German didaktik tradition* (pp. 41-54). London: Lawrence Erlbaum.

Leach, J., & Scott, P. (2000). Children's thinking, learning, teaching and constructivism. In M. Monk & J. Osborne (Eds.), *Good practice in science teaching: What research has to say* (pp. 41-56). Buckingham, UK: Open University Press.

Leach, J., & Scott, P. (2002). Designing and evaluating science teaching sequences: An approach drawing upon the concept of learning demand and a social constructivist perspective on learning. *Studies in Science Education, 38*, 115-142.

Lijnse, P. (2000). Didactics of science: The forgotten dimension in science education research? In R. Millar, J. Leach & J. Osborne (Eds.), *Improving science education: The contribution of research* (pp. 308-326). Buckingham, UK: Open University Press.

Marton, F., & Ramsden, P. (1988). What does it take to improve learning? In P. Ramsen (Ed.), *Improving learning: New perspectives* (pp. 268-286). London: Kogan Page.

Shulman, L.S. (1987). Knowledge and teaching: Foundations of a new reform. *Harvard Educational Review, 57*(1), 1-22.

Thornton, R.K., & Sokoloff, D.R. (1998). Assessing student learning Newton's laws: The force and motion conceptual evaluation and the evaluation of active learning laboratory and lecture curricula. *American Journal of Physics, 66*(4), 338-352.

Tiberghein, A., Jossem, E.L., & Barojas, J. (1998). *Making the results of research in physics education available to teacher educators.* Retrieved 26 July 2005, from http://www.physics.ohio-state/edu/jossem/ICPE/BOOKS.html

Tochon, F.V. (1999). Semiotic foundations for building the new didactics: An introduction to the prototype features of the discipline. *Instructional Science, 27*, 9-32.

Uljens. M. (1997). *School didactics and learning: A school didactic model framing an analysis of pedagogical implications of learning theory.* Hove, UK: Psychology Press.

Van Driel, J. H., Verloop, N., & de Vos, W. (1998). Developing science teachers' pedagogical content knowledge. *Journal of Research in Science Teaching, 35*(6), 673-695.

Vasquez-Levy, D. (2002). Bildung-centred didaktik: A framework for examining the educational potential of subject matter. *Journal of Curriculum Studies, 34*(1), 117-128.

Von Glasersfeld, E. (2002). *Radical constructivism in mathematics education.* Dordrecht: Kluwer.

Warren, B., Ballenger, C., Ogonowski, M., Rosebery, A.S., & Barnes, J.H. (2001). Rethinking diversity in learning science: The logic of everyday sense-making. *Journal of Research in Science Teaching, 38*(5), 529-552.

Wellington, J., & Osborne, J. (2001). *Language and literacy in science education*. Milton Keynes, UK: Open University Press.
Wells, G. (1994). Learning and teaching 'scientific concepts': Vygotsky's ideas revisited. Retrieved 20 September 2005, from http://education.ucsc.edu/faculty/gwells/
Westbury, I. (2000). Teaching as reflective practice: What might didaktik teach curriculum? In I. Westbury, S. Hopmann & K. Riquarts (Eds.), *Teaching as a reflective practice: The German didaktik tradition* (pp. 15-39). London: Lawrence Erlbaum.

AFFILIATIONS

Mohd. Zaki Ishak
School of Education & Social Development
University of Malaysia Sabah
Malaysia

Zulkifli Mohamed
School of Education & Social Development
University of Malaysia Sabah
Malaysia

TENZIN AND MAXWELL

PRIMARY SCIENCE EDUCATION IN BHUTAN: DEVELOPMENT AND CHALLENGES

INTRODUCTION

The land-locked mountainous Himalayan Kingdom of Bhutan, sandwiched between the giant neighbors China to the north and India to the south, has adopted an approach to development that attempts to straddle traditional values and customs but also adopts many of the features of Western development (see Maxwell, 2007). After centuries of self-imposed isolation from outside influences, the process of modernization in Bhutan began in earnest from 1961 with the commencement of the First Five Year Plan. Introduction and then modernization of the secular education system, to complement Buddhist monastic education, was always seen as a crucial part of Bhutan's development. Science education has had an important place in the modernization process over the last five decades because in "many years to come there shall remain enough demands in the kingdom for doctors, scientist and engineers as well as many technical hands for which science and mathematics education are the pre-encouragement" (Royal Government of Bhutan [RGoB], 1989a, p. 17).

One of the earliest statements of purpose of school education in Bhutan was enshrined in its National Education Policy:

> In the process of educational development, the primary need is to prepare the type of educated manpower required for the all round development of our country. … Also we must build our educational system on local foundations, take what is good and useful from other systems and mould the whole into a system of education which is truly Bhutanese. (RGoB, 1976, p. 1)

A 'truly Bhutanese' education had to wait as most early energy was put into creating primary schools. The curriculum, its examination and teachers were introduced from India. The school education structure that has developed in Bhutan is set out in Figure 1.

The purpose of primary education in Bhutan is to provide basic literacy skills, knowledge of the country's history, geography, culture and traditions and the fundamentals of agriculture, health and hygiene, and population education. Moral science, values education, and art and craft are accorded special attention. These knowledge and skills are provided through a nation-wide new system of activity-based learning.

Richard K. Coll, Neil Taylor (eds.), Science Education in Context: An International Examination of the Influence of Context on Science Curricula Development and Implementation, 313–332.

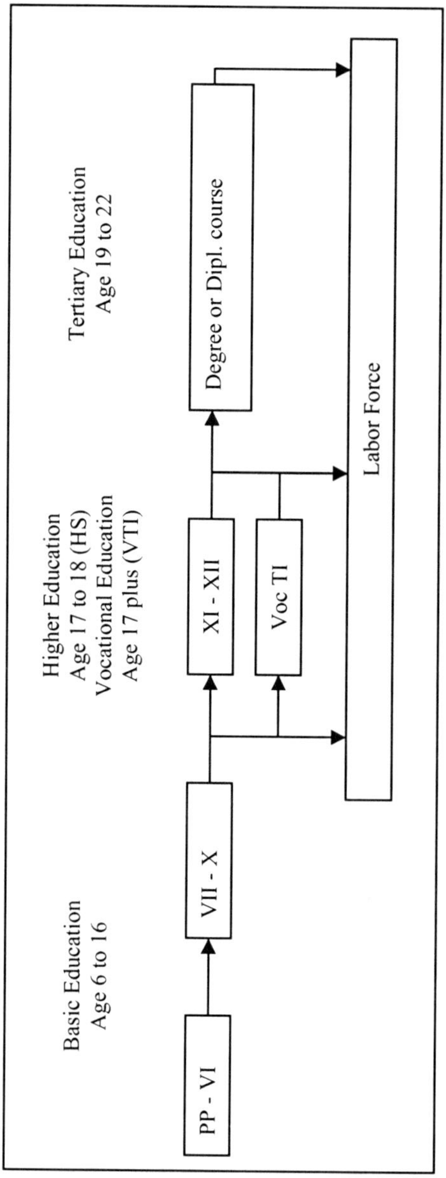

Figure 1. General education structure in Bhutan (RGoB, 2006)

SCIENCE EDUCATION IN THE PAST

The development of a modern education system in Bhutan since 1961 has been nothing short of remarkable. Expansion of schools and elevation of accessibility to education have been consistent national priorities. With the exception of the curriculum for Dzongkha, the national language of the country, the institutionalization of modern education in Bhutan was initiated with curricula borrowed from neighboring countries, particularly India. Understandably, it was compelled to do so due to lack of indigenous expertise to develop curricula and teach them.

Secular enrolments have risen from virtually zero in the 1950s to around 130,000 in all forms of secular education early this century. The school participation rates, although not yet full, nevertheless represent a remarkable achievement in five decades (Maxwell, 2007). The higher urban data reflect relatively easier urban access and other social and economic factors. Provision of secondary school places is still an issue in Bhutan and secondary places are sought after.

Table 1. 2002 School participation rates (%) in Bhutan as reported in Maxwell (2007, p. 55)

	Rural	*Urban*
School Participation Rate (Ages 6-12)	65.3	89.0
School Participation Rate (Ages 13-19)	46.3	77.3

SCIENCE EDUCATION IN PRIMARY SCHOOLS IN THE PAST

Science education has been accorded great importance ever since modern education was instituted in the country. The National Education Policy anticipated that the students leaving primary schools will be able to:

- Speak, read and write simple Dzongkha;
- Speak, read and write simple English;
- Have an acquaintance with a few simple principles of hygiene; and
- Have an acquaintance with a few simple principles of progressive Agriculture, Horticulture and Animal Husbandry, etc. (RoGB, 1976, p. 4)

Thus, in the curriculum formal documentation, 'science' appeared in its more utilitarian forms and was adopted from India. General science from Classes I to VI had wide ranges of topics, heavily influenced by textbooks. Science was de-contextualized. Maxwell remembers a postgraduate student from Bhutan asking him as they were driving across the Australian countryside: "Is that a willow tree?" It was indeed a willow tree.

The learning process was structured without any flexibility since the syllabus 'must be covered'. Science learning was heavily content-based and examination-centered, that is, recitation of scientific facts and principles on demand. Dorji (2003) recollects:

> Teaching was mainly from the front and only the teachers talked. Students were mostly engaged in extensive homework and writing responses to questions that were given at the end of each chapter in the textbooks. Recitation of the meaning of words in the texts was considered important. (p. 6)

Corporal punishment was rampant with the least thought given to how effective the teacher's teaching approaches were or the children's social and intellectual differences. For instance, Tenzin can vividly recollect the times when he was in Class V in 1975. The English teacher stood behind him while he was asked to write an answer to a question on the blackboard. By the time Tenzin had finished writing the answer he had received two or three lashings on his calf.

Eventually the Education Division developed a curriculum section whose earliest tasks were to develop some Bhutanese materials to use. In the early 1990s prior to the New Approach to Primary Education (NAPE)-oriented science, the primary science curriculum was based on the principles that content includes broad-based concepts and principles of science; is related to different age groups and daily life; serves the needs of the community; is closely associated to the environment; and is dealt with within the available time under existing conditions of staff, equipment, and so on (Yadav, 1992). The general science course for the primary school was perceived to have been organized around the following main units: Our Surroundings, Nature and Things, Energy and Work, Life, Health, and Agriculture. It remained heavily content-centered and was dominated by the biological sciences. Consistent with the level of general and professional education of teachers in Bhutan, and consistent with the experiences that the Bhutanese teachers themselves had when they were at school, the curriculum generally, and science in particular, remained teacher-centered and exam-oriented at all levels (Laird, Maxwell, Tenzin & Gyamtso, 1999).

The promotion of students to the next grade was largely determined by a paper and pencil test administered at the end of each year (RGoB, 2003). Though students were assessed mid-year, those achievements had no bearing on their promotion. The Review of Primary Education in Bhutan (RGoB, 1993a) attributes high rates of failure, drop out and repetition in those days to the system of summative assessment. Though the one-shot summative assessment had advantages for teachers, it did not encourage them to make allowances for the social and psychological differences of learners.

INSTITUTIONALIZATION OF THE NEW APPROACH TO PRIMARY EDUCATION

From the 1980s onwards, the national curricula were expanded and consolidated in primary education. New syllabi for primary classrooms were developed to enable Bhutan to have curricula which were uniquely Bhutanese yet incorporating the best principles of modern education from the extensive experience of other nations. As compared to the previous approach, the New Approach to Primary Education, or NAPE, was essentially learner-centered. A significant development along with the introduction of NAPE was the extension of educational accessibility through the establishment of Extended Classrooms (ECRs) which later evolved into Community Schools from 1992 for rural and remote children. These developments were assisted by Bhutan signing the Jomtien Declaration in 1990, which reaffirmed the goal of Universal Primary Education by the Year 2000.

NAPE was the springboard in launching Bhutan's first education system reform to localize its curricula. NAPE was guided by such broad tenets as:

- To provide our students with wholesome education consisting of a package of knowledge and skills that will serve them well and adequately in their adult life;
- To design our education system in such a manner that our children understand, appreciate and support the social and political harmony that exists in the country; and
- To inculcate amongst the students a greater understanding of an appreciation for the predominantly agriculture-based rural lives and, as such, develop in them resourcefulness and dignity of labor as well as the traditional value of "*Phada Bugi Thadamtshi Lay Jumdrey*" (Loyalty and affection between parents and children) (RGoB, 1989a, pp. 9-10).

Secondly, NAPE was based on progressive principles such as:

- Relevant and appropriate syllabi and content in line with the national education policy and responsible to dynamic national needs;
- Appropriate teacher education to enhance the quality of teaching;
- Provision of relevant academic materials;
- Regular supervision and inspection; and
- Standard teacher-pupil ratio (RGoB, 1993a, p. 19).

In 1985, work on NAPE was begun by involving experienced educationists from Bhutan and expatriate consultants, resulting in new learning areas for Bhutanese primary science (see Table 2). The 'core learning areas' were taught in the classroom situation and assessed formally. The personal development learning areas were not assessed and were taught mainly through co-curricular activities. Time was allocated for all the areas.

Pre-primary to Class III was to be composed of three main 'tool' subjects and a new subject called Environmental Studies (EVS). EVS consisted of aspects of geography, history, science, health and hygiene, agriculture, culture and traditions, arts and crafts, values education, physical education, music, dance and drama.

Table 2. Learning areas for primary level (RGoB, 1996)

Class	*Core learning areas*	*Personal development areas*
PP – III	English	Visual Arts
	Dzongkha	Performing Arts
	Mathematics	Socially Useful & Productive Work
	Environmental Studies (EVS)	Agriculture & Social Forestry
		Values Education
IV – VI	English	Visual Arts
	Dzongkha	Performing Arts
	Mathematics	Socially Useful & Productive Work
	Science	Agriculture & Social Forestry
	Social Studies	Values Education

The consolidation of other subjects into EVS was thought to be crucial because:

- The existing curriculum at all levels is grossly inadequate in the learning and teaching of and from the local environment especially teaching and learning of history and geography of Bhutan;
- There are too many subjects and textbooks taught, particularly in the lower classes i.e. up to Class III;
- Learning amongst children is mostly by rote; and
- Classroom scenes are dominated mostly by 'chalk and talk' limiting children's self expression (RGoB, 1989a, p. 9).

From Classes IV to VI, the Environmental Studies base leads to two subjects: Social Studies and Science.

In 1986, NAPE was initially piloted in 13 selected primary schools. Within the period of 1987 – 1992, NAPE was expanded first in Classes PP to IV in 28 schools. After re-drafting, NAPE came to all the 253 primary schools in the country. To achieve this, 63 sets of textbooks, readers, manuals and syllabi for various levels from Classes PP to VI were written by the Education Division's Curriculum and Textbook Development Division (CTDD) now known as Curriculum and Professional Support Division (CAPSD). This was the first and major step towards Bhutanization of the primary curriculum during the Sixth Five Year Plan (1987-92).

Ideally, in PP to Class III, NAPE was based on the class teacher system with children taking active part in their lessons, using the local environment of school, district and country as the context of learning. In the class teacher system of teaching, all the subjects were taught by the same teacher throughout the academic year as this was thought to accentuate the development of a meaningful social bond between the teacher and children (RGoB, 1993a).

The shift in philosophy from traditional to a more progressive approach in 1993 meant that the primary science curriculum had the following underpinning the pedagogical assumptions:

- A balance of knowledge, skills and understanding is necessary for a good science education;
- Children should not begin formal education without knowledge, skills and understanding;
- Science happens all around us and much of what children know is learnt from the environment through interaction with it; and
- Children enjoy learning science through activities - observation, measurement, investigation, designing and making, drawing conclusions, communicating their ideas, drawing and recording.

These principles were dramatically different from those that underpinned teaching prior to NAPE.

The EVS curriculum was developed around themes cutting across different subjects (see Figure 2). The seven subjects in Figure 2 are briefly elaborated below as they are interpreted in Bhutan's EVS curriculum.

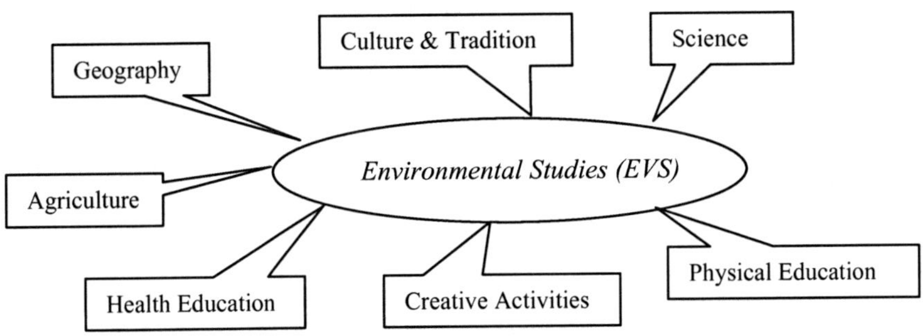

Figure 2. An overview of EVS components (RGoB, 1989b,1989c, 1989d)

History. Children were encouraged to learn history by exploring their living past through their own history, and that of family, friends, class, school, village, monastery, dzong, and so on, by interviewing parents, teachers, friends and other responsible people in their locality. They were also encouraged to learn from stories, songs, dances, customs, rituals and folk legends. While doing so, children would build skills of observing, collecting, recording, comparing and questioning and would learn to respect and feel proud of their heritage, culture and traditions.

Culture and traditions. Children were encouraged to live and work together; provide opportunities to express their feelings and accept others' feelings and beliefs of the society; and deal with other children with compassion and affection.

Geography. Children were encouraged to learn physical geography by imitating the procedures any geographer would follow, starting from measurement, map reading and drawing, drawing objects, and talking about the location and direction.

Science. In NAPE, EVS science was considered vital to provide children with the opportunities to observe, record the observation, ask questions, predict, experiment, and find answers with further observations and experimentation about aspects of the environment.

Health education. Health education was used for the children to learn healthy habits and behavior. Children were provided with the opportunities to practice basic good health and nutrition by encouraging them to wash their hands before and after meals, keep their body clean, keep the surroundings clean, eat at regular times, keep clothes clean, and so on.

Physical education. The EVS ensures that children have time every day for games and play with space and some play materials. Most of the time, everyday materials were used to play games.

Agriculture. Since agriculture is the means of livelihood of the majority of the people of Bhutan, positive attitudes towards agriculture were instilled in Bhutanese young children. The topics included the basic practical skills that the students can use in their farms and in their school's garden.

Creative activities. Children learn about rhythm, music and movement through which they are able to express their feelings. In fact, various components of EVS were taught through the use of rhythm, music and movement.

In other words, ideally, the approach of teaching EVS was through a child-friendly environment. However, with the lack of resources and little teacher education and training in science, little straight science was taught in EVS. On the whole, the overarching tenet of teaching EVS was to develop the basic skills of learning different subjects which they would soon confront as they progressed through their education. To achieve this, EVS was also organized around intellectual processes, for example, planning and science experimentation, social processes such as working together and behaving with respect, and physical skills, for example, development of fine and gross motor skills.

These processes and skills were considered basics to be developed throughout each year, providing ample opportunities to reinforce the practice of the skills. For example, Table 3 illustrates a sequence of EVS science lesson objectives in Class III. To achieve the objectives, children were involved in series of activities to observe, draw, demonstrate record and infer about birds and their habits. Since there was no student textbook, the Teacher's Manual provided guidelines to the teachers on the organization of materials according to the objectives. The manual explains the concepts, while teachers are required to organize supplementary materials for an in-depth study through research and collection of information (Dorji, 2003). But the critical question was: *How could teachers develop resources when most schools had very few (science) books?*

Table 3. An anecdotal text of EVS science lesson in Class III (pp. 107-115)

Birds *(Topic 5)* Main aim: The children will be able to know and talk about birds and their habits. *Local birds and their habits* Objective A: The children can relate a bird's beak to the food the bird eats. Objective B: The children can relate a bird's feet to its food or way of living. Objective C: The children can record similarities and differences between birds and insects. Objective D: The children can record and describe the flight patterns of birds, insects. Objective E: The children can perform physical activities related to flight patterns. Objective F: The children can make pictures and patterns using feathers. Objective G: The children can complete a life cycle of an egg / hen / cock. Each objective is intended to be achieved through a minimum of one to five learning activities. For instance, the objectives A to G are targeted through twenty two activities (to be found in the Teacher Manuals).

The localization of the school curriculum (RGoB, 1976) received a boost when the Government decided that EVS should be taught in the national language, Dzongkha (RGoB, 1994):

> The teaching of EVS in Dzongkha is also expected to end the present tendency to differentiate our teachers into two broad groups, namely, Dzongkha language teachers (ZLT) and the general teachers (at the primary level), as both these groups of teachers are henceforth expected to take up teaching of the EVS in Dzongkha. (p. 5)

It is understood that the decision also supported the development of Dzongkha learning. So, in the same year, the teaching of EVS in Dzongkha was piloted in 26 schools. The English versions of EVS manuals for Class PP to III were translated into Dzongkha. The pilot teaching of Dzongkha EVS through the class teacher system was found to be more effective than teaching through the subject teacher system (RGoB, 1995). Following an evaluation, Dzongkha EVS was implemented nationwide in PP in 1997 and progressively to Class III in 2001 (RGoB, 2001)[i].

NAPE SCIENCE EDUCATION FROM CLASSES IV TO VI

In Bhutan, science is perceived both as a body of knowledge and a process (Yadav, 1992). As a body of knowledge, its facts, figures, laws and principles facilitate learners to provide meaning to the phenomena happening around them. The process component is seen to develop the art of learning science knowledge through the practice of scientific methods such as observation, measurement, hypothesizing and prediction, interpretation, recording and communicating. Through the experiential learning of science, learners come to appreciate the contribution of science towards human endeavor and to value the happy co-existence of living and non-living things. In the preface to the Teachers Manuals, the Ministry of Education indicated the key ideas that influenced the thinking behind the new NAPE Science curriculum:

> The Education Division believes that science education will help our children: understand scientific ideas, develop scientific methods of investigation, relate science to other areas of knowledge, understand the contribution science makes to society, and appreciate the nature of scientific knowledge and by contributing to personal development. (RGoB, 1993b, p. ii)

Furthermore, in the Bhutanese context, it is vital that science education reinforces the Buddhist belief system. For instance, killing is seen as a sin which is scientifically explained as a destruction of the ecosystem that may lead to irreversible natural calamities; and prostration in the temple can be equated to physical exercises essential for sound health and mind.

The science curriculum for Classes IV to VI was developed around a series of themes that are relevant to school life. Knowledge, skills and understanding are developed in a number of appropriate contexts with a set of values and attitudes underlying all the learning and teaching taking place. It is a spiral curriculum in which a concept is introduced and returned to at different times as the children's thinking develops. The primary science curriculum is concerned with contexts as well as giving a broad understanding in science. These contexts relate to the main social and economic activities in Bhutan. Their inclusion develops children's appreciation of the complex relationship between science, technology and society. Based on the Bhutanese contexts, Table 4 sets out the 10 themes, topics and skills identified for the primary science education.

Each topic is delivered through 15 learning activities, via the Teachers' Manual, and review exercises arranged thematically. For instance, Table 5 below illustrates the varieties of learning activities identified thematically for teaching the topic 'Measurement' in the Class IV student text. The main thrust is making children do science so that children learn the epistemology of science rather than the learning more about the science. The purpose is for the learners to generate or construct their own scientific ideas through their own responsible learning from sensory input; for example, sights, sounds, smell and so on (Osborne & Freyberg, 1986). Most of the learning activities teach investigation through testing the environment.

Table 4. Themes, topics and skills of primary science education (RGoB,1993b,1993c, 1993d)

Themes	Topics			Skills focus
	Class IV	Class V	Class VI	
1. Diversity and ecology 2. Life processes 3. Human Influences on the Earth 4. Materials 5. The Earth in space 6. Forces 7. Electricity and Magnetism 8. Energy transfers 9. Sound 10. Light	1. Measuring 2. Water and soil 3. Food and health 4. Sight and sound 5. Trees and wood 6. Cloths and dyes	1. Sorting and separating 2. Farming and gardening 3. Energy 4. Keeping health 5. Pushing and pulling 6. Air	1. Materials and properties 2. Electricity and magnetism 3. Structure and forces 4. Living and growing 5. Changes and reactions 6. Planet earth	1. Observation 2. Identifying patterns and relationships 3. Sorting and classifying 4. Ordering and sequencing 5. Measuring 6. Communication 7. Data handling 8. Interpreting and drawing conclusion 9. Prediction and making hypotheses 10. Experimenting 11. Fair tests

Table 5. Thematic activities for Measurement in Class IV (RGoB, 1993b)

	Topic 1 *Measurement*
1.1	Why Measure?
1.2	Looking at length
1.3	How big are our hands?
1.4	How tall are we?
1.5	Time yourself
1.6	How many heart beats?
1.7	Pendulum time
1.8	Using a measuring cylinder
1.9	Making a measuring container
1.10	Will water change shape?
1.11	Measuring a stone
1.12	Making a balance
1.13	Stretching a spring
1.14	Weighing
1.15	Making a wall display
1.16	Think Again

Pedagogically, this makes sense since most young children of this cohort cannot cope with the learning of abstract concepts; rather, they relish direct interaction with objects. The science curriculum implementation experience shows that pushing hard too early with the learning of abstract concepts alienates children and proportionately encourages rote memorization. However, the way that science was taught was also heavily influenced by the ways in which it was assessed. In 1996, the Ministry of Education published *The Purpose of School Education in Bhutan* which provides basic information on the learning areas with a rationale, aims, learning experiences, and learning outcomes for each subject. This was prepared with the intention to guide the teachers in preparing their lessons at different levels of school with appropriate learning experiences and learning outcomes (Dorji, 2003).

HOW ARE STUDENTS ASSESSED IN NAPE SCIENCE?

The weaknesses of assessment practiced in Bhutan prior to NAPE were, firstly, assessment gave very little attention to the understanding of ideas and mastery of the skills. Reproduction from memory encouraged rote learning. Secondly, although assessment activities of various forms like dictation, class work, homework, weekly and monthly tests and project work were carried out by the teachers, they did not have any weighting in the overall assessment record. Only summative assessments carried out at the end of the terms were valued. Much more highly valued were the formal, end-of-year tests, at all levels of the primary school in all subjects (carried out by the Bhutan Board of Examinations (BBE) established in 1986). Thirdly, all tests, apart from those constructed by the BBE, did not test what they were supposed to, indicating that the teachers did not have an adequate understanding, and skills in the construction, of tests (later rectified by BBE, see below). Consequently, teachers could not rely on the results (RGoB, 2003). In NAPE science, a more diverse form of assessment that ascertains the student's performance in terms of both content and process was needed.

In the light of these shortfalls in the assessment, the new curriculum instituted two significant changes in the system of assessment:

- Diversification of summative assessment with the inclusion of Continuous Assessment (CA)/ongoing evaluation with weighting; and
- A mandate to develop test blueprints based on the Bloom's taxonomy of learning objectives prior to the setting of questions.

These ideas were promulgated through the greater mass of primary teachers via various educational forums and in-service courses spearheaded by the BBE.

Continuous assessment (CA) is generally construed in Bhutan as an ongoing process that uses observations and other techniques to gather information on the progress of student's learning in terms of content and the scientific processes. It is carried out through the academic year and uses a variety of activities to gather a picture of a student's performance. Ideally, CA consists of formative, summative and prognostic evaluation and is focused on individual students as well as on the performance level of whole class. The formative assessment (FA) component is carried out with the use of specific FA tools (such as checklists and rubrics) and techniques (such as observation). The summative assessment (SA) component uses tests, both at the school and national levels. Prognostic assessment (PA) is based on formative and summative assessment results in order to assist future learning. The information gathered from FA has bearing on the student's Progress Report Form (PRF) and is used for the Prognostic Assessment of students (RGoB, 2003). In fact, CA is still not well understood by teachers in Bhutan (Dukpa, 2000)

The other milestone in the change of assessment practices associated with NAPE and the introduction of CA is the use of a Test Blueprint. This has been developed by teachers based on the principles of Bloom's taxonomy of learning objectives, and all the schools are required to use the tool for setting questions for any examinations. This innovation has improved the validity and reliability of many school-made tests in Bhutan.

A test blueprint is a guide to test construction and use. Teachers are advised to set the question papers by including questions from all the levels so that it helps to test all the skills in science. Therefore, teachers should prepare a blueprint for each question paper they are setting. This will also help the students of various difficulty levels to perform well in the examinations. In the test blueprint, the six levels of learning objectives are placed across the top with the subject topics down the left hand side (see Table 7). Different questions framed from all the topics designed to test the varieties of learning objectives are mapped out. This greatly facilitates the test setter to have an overview of the learning objectives, topics, and type of questions with marks. It ensures that all the topics are covered as per the weighting of the syllabus. Tables 6 and 7 illustrate the types of test items and their distribution, and an example of test blueprint for Class VI science examination respectively.

Table 6. Types of questions (RGoB, 2002)

Question variety	*Items*	*Marks*	*Remarks*
I Multiple Choice Questions (MCQ)	25 items	25 marks	
II Completion Variety / Fill in the blanks (Cm)	5 items	5 marks	
III Match the following variety (Mat)	5 items	5 marks	
IV Correction Variety (Cr)	5 items	5 marks	
V Questions Variety (QV)	31 items	60 marks	

In Bhutan, the year-end examination was the tool for a summative assessment of the achievements of students. It was essential to grade and rank pupils since they were promoted based upon the results. The examinations continue to be particularly important at the end of Class VI where the number of places available in Class VII is often restricted. At the end of the academic year, a progress report of a student is prepared for promotion to the next grade, taking into consideration the cumulative marks obtained. This system of assessment reduced the detention and drop out rates in the schools. In very recent times, the BBE has dropped external examinations for Classes VI and VIII altogether, but not for Class X where they remain since they mark the end of basic education for students.

Table 7. An example of Table of Classification (Blueprint) (RGoB, 2002)

OBJECTIVES / SKILLS TOPIC/CONTENT	*KNOWLEDGE*	*COMPREHENSION*	*APPLICATION*	*ANALYSIS*	*SYNTHESIS*	*EVALUATION*	*TOTAL*
1. Materials & Properties	*4*	*4*	*7*	*1*			*16*
1.1 Everyday materials			MCQ QV(3)				2 items 4 marks
1.2 Paper							
1.3 Investigating with paper		MCQ					1 item 1 mark
1.4 What is matter?	QV(2)						1 item 2 marks
1.5 Solids							
1.6 Elastic bands			MCQ	MCQ			2 items 2 marks
1.7 Bouncing balls							
1.8 Solid materials							
1.9 Looking at buildings	QV(2)						1 item 2 marks
1.10 Are solids and liquids alike?		QV(2)					1 item 2 marks
1.11 Gases							
1.12 Soluble materials & solutions							
1.13 Processes			QV(2)				2 items 2 marks
1.14 A load of rubbish		MCQ					1 item 1 mark
1.15 Design a rubbish bin							

MAJOR IMPLICATIONS FOR TEACHERS

The NAPE curriculum resulted in considerable implications for teachers, not to mention the system, as it strove to create materials and classrooms to house students. Dorji (2003) observed that the new curriculum emphasized that children learn at their own pace rather than making all of them achieve the objectives at the same time. The new curriculum demanded a paradigm shift from lecturing to an experience of the individual child exploring and learning.

Firstly, and crucial to making NAPE EVS and General Science a success story, teachers needed to modify their perspective of student science learning, change their teaching strategies and thereby their roles. The process of teaching/learning in science was no longer construed as transmission of scientific facts and figures, but of providing opportunities for learners to generate and construct their scientific ideas with a series of learning activities, such as observation, measurement, investigation, recording, communicating, and so on. The NAPE curriculum completely changed the paradigm of teaching/learning. Teachers needed to embrace new child-centered teaching strategies in lieu of the traditional chalk and talk. Since the learning is predominantly by investigation through testing the environment, or the children's systematic interaction with the environment, teaching/learning materials must be pre-arranged and students' group activities identified. Group activities, whether through field trips, cooperative learning or group circus or individually, must be laid out to break the monotony of the learning process. Consequently, the role of the teacher was largely re-construed as a motivator whereby the teacher encouraged children to be active learners; worked as a diagnostician to discover the strength and weakness of learners and plan the lesson accordingly; acted as a guide to usher children through the learning activities gearing towards the learning objectives; was an innovator using their own imagination to stimulate learning in themselves as well as children; became an experimenter to explore new ways of learning that were appealing for children; and finally, shared findings of experiments on teaching with others to further improve the ideas developed (Osborne & Freyberg, 1985).

These changes in the teacher role in NAPE generally, and science in particular, were not automatic and training with support and scaffolding was crucial (Fullan 1991). These were major demands, especially for teachers whose own knowledge of science was not well developed and whose professional history was steeped in didactic methods.

Secondly, students were provided with a science textbook each while teachers were nationally provided with student's textbooks, Teacher's Manuals and a general teacher's guide *Teaching of Primary Science for Classes IV to VI*. Primary schools were also generally equipped with basic science equipment and chemicals such as test tubes, beakers, measuring cylinders, magnets, measuring scales, mild acid and alkali. This resource support to schools lessened the burden to improvise equipment in the limited time teachers generally have due to large class sizes, and the heavy workload due to shortage of teachers, especially in remote schools and allowed more time for planning the upcoming lessons. However, whether the science teachers have the capacity to use effectively the basic science equipment, that continues to be provided to schools, remains a concern. Currently, with UNICEF assistance, primary science teacher training on the use of the basic science equipment has begun through the cluster-based in-service programs.

SOME IMPLICATIONS OF EVS, GENERAL SCIENCE AND NAPE MORE GENERALLY

The novelty of teaching EVS in Dzongkha and General Science in English, however, was fraught with some problems in teaching as well as in terminologies. The Dzongkha Development Commission (DDC) attempted to translate the scientific terms as and when requested but it proved a difficult task. For instance, in Dzongkha anything that grows underground is termed as '*rato*', meaning 'root'. Therefore, the concept of 'potato' as a stem might be taught as a root, which would conflict with the scientific concept of the potato. This applies to other crops like ginger. Further, since there are no equivalent Dzongkha words for many scientific terms, the translation of the words was generally based on utility or characteristics which make Dzongkha equivalent words very cumbersome to learn and remember. Furthermore, the Dzongkha equivalent words are not popular amongst the people, including the teachers and their students.

In general, people from different walks of life prefer to use English to Dzongkha for writing. This is largely due to the complex nature of the Dzongkha script. An evaluation of (Dzongkha) EVS teaching indicated that teachers, particularly the non-Dzongkha specialists, faced difficulties in spelling in Dzongkha (RGoB, 2000). This is a hugely limiting factor for teaching EVS. Fearful of exposing their weaknesses, such teachers are generally very reluctant to deviate from the Teacher Manuals. On the other hand, though, Dzongkha specialists (ZLTs) ranked higher in language competency, generally have poor EVS content knowledge, and process skills meaning they also rely on the manuals. The report also adds that some teachers, particularly those not trained in science, do not find the manuals user friendly. Furthermore, field observation raised doubts about teaching geography, science, culture and health through EVS in Dzongkha since teachers generally focus on language rather than the content and skills of the EVS constituting subjects. On the other hand, the desired objective of improving Dzongkha may well be being achieved. Also, the coupling of environmental science with Dzongkha does afford particular advantages at a cultural level. The gaps between the intended curriculum and curriculum practices indicate the need to strengthen the teacher preparation process, including the in-service upgrade BEd by distance education modules (Maxwell et al., 2006) and also to improve the curriculum materials.

The launching of NAPE was also the dawn of in-service training of teachers. National Based In-service Programs (NBIP) and later District Based In-service Programs (DBIP) eventually resulted in 1,500 teachers trained on the use of NAPE in the 1990s. However, it is important to note that 60% of primary teachers had only Class X or less as their formal general education credentials supported by two years in teacher education (Laird et al., 1999). Since then, in-service education, including in EVS and General Science, has been further modified and strengthened by instituting the system of the cascading mode of training through the conduct of Training of Trainers (TOT), Cluster Based In-service Program (CBIP) and School Based In-service Program (SBIP), as well as NBIPs and DBIPs for better impact and judicious use of training resources. This was further strengthened with the institutionalization of the concept of education resource centers (RC) with the mandate to facilitate professional development programs, especially for the remote school teachers.

The introduction of resource centers is a strategy to bring teachers together from time to time to learn from each other's practices and experiences, and from visiting experts; to develop teaching materials (Dorji, 2003); and to practice their innovation and the ideas gained through in-service programs. By 2006, Bhutan had established 35 RCs across the country under the assistance of various donors such as UNICEF.

The NAPE work of the 1990s introduced the Bhutanization of the curriculum and also heralded the removal of formal end-of-year examinations. Previous reliance upon India curriculum materials and examinations was gradually replaced by documents, practices and institutions on the part of the Royal Government of Bhutan. In 2006, the control of the final Class XII examinations was taken over by the BBE.

Notwithstanding the gains made by NAPE, there have been substantial criticisms. There is a clear view, articulated by a senior Bhutanese educator (Dorji, 2003), that NAPE was essentially before its time, especially due to the teaching cadre's lack of ability to handle the student-centered approach that NAPE demanded. By experience and training, despite the influences of the NBIPs, teachers were used to the more traditional teacher-centered approach and found it hard to change. To make things worse, there were not always sufficient materials for science experimentation or lighthouse models of best practices. The general view is that NAPE has to some extent hindered the development of education in Bhutan which is being addressed at the start of the new millennium.

Bhutan's resources have been stretched as it has created an education system from scratch in five decades. A huge 10% of GDP has been spent on education during that time (Maxwell, 2007), indicative of the commitment that Bhutan has made to education. Resources have always been, and continue to be, a major constraint on what has been possible. The international community through grants and loans by governments and the assistance of NGOs such as UNICEF, Save the Children (US), DANIDA and HELVITAS, have been crucial. With the projected income from hydro power, it is likely that Bhutan will become less dependent upon such assistance (Maxwell, 2007).

LOOKING FORWARD

Only time will judge the utility and strength of science education in creating the kinds of people Bhutan needs. Education leaders and the promulgation of critical thinking within the system via overseas study and exposure to the good practices of curriculum elsewhere have constantly led to questions about the quality of science curriculum for Classes IV to VIII. Already this has been happening in the bigger educational forums like Annual Education Conferences and CAPSD Board Meetings.

The biggest concern is the lack of knowledge of scientific concepts observed in the primary science textbooks and manuals which certainly do not facilitate a student self-study. Revisions have been undertaken from time to time. For example, the science textbooks and manuals for Classes IV to VIII were revised through a series of writers' workshops with the science teachers of different categories of schools in 2000 to 2001. In 2002, Dr Jack Holbrook, a UNESCO science consultant, reviewed the revised science curriculum from Classes IV to X by considering the relative contributions of biology, physics and chemistry. He found that:

> The course tries to be true to its intention and relates to the environment. In this it tries to build on the learning in Classes IV to VI and interrelate the Biology, Physics and Chemistry through the environment; adopt an investigative approach; promote student centered learning; make the study of science learning purposeful, interesting and enjoyable; and develop values and show concern to the environment. (Holbrook, 2002, pp. 2-5)

He commended the courses for Classes VII to VIII as suitable for Bhutan, and able to meet the goals of science education. He recommended that the text be redesigned to have a more investigatory approach; include more activities; create a balance of biology, physics and chemistry; and strengthen the objectives related to skills. He also recommended in relation to the teacher's guide the inclusion of student competencies and hence a move towards a criterion reference approach; a strengthening of the end of chapter questions; an expectation for students to undertake some form of student-based work every lesson and that hands-on activities are present, on an average, at least one in a week; and, finally, the revision of Classes IX and X courses to better relate to the science in Classes VII and VIII (Holbrook, 2002). Further, he added that although teachers have attended workshops on science in Classes VII to VIII, this has not proved sufficient and there is an important need for more workshops and to make this on-going feature.

Such changes to the science education in Bhutan will take time to implement. To revitalize the utility of the curriculum in the contemporary world of technology, a major science curriculum reform from Classes PP to XII is being earmarked in the 10th Five Year Plan commencing from June 2008. Unlike in the past, when the textbook/content directed the learning, the reform will be initiated with an extensive needs analysis leading to drafting of a science curriculum framework followed by students' performance achievement level for different classes. These documents should underpin the development of textbooks and teacher's guides. To ensure the effective implementation of the new science curriculum, all teachers will be trained through cluster-based workshops, scaffolded with school visits by the facilitators from CAPSD and the Colleges of Education. No doubt, this major revision will create challenges for the Ministry in developing and supplying materials and equipment for the teachers as they come to understand the reforms and take part in the various forms of in-service training. Additionally, the two Colleges of Education will need to make adjustments to pre- and in-service programs.

CONCLUSION

Primary science education in Bhutan is relatively new, as is its modern education system. Initially, heavily based on rote memorization of science facts and examined in that way also, the NAPE science education curriculum attempted to change this. The more progressive ideas built into NAPE were clearly a challenge to many in the teaching cadre. Now, more child-centered teaching is to be found in Bhutan, including in EVS and General Science. Moreover, NAPE has been crucial in making the science taught in Bhutan clearly related to the environment in which Bhutanese children live.

There are many questions that can still be asked about science education in Bhutan. Concerns exist about the teachers' level of confidence and ability in handling the science curriculum. To what extent is science taught emphasizing science processes in schools? How do children learn science against the backdrop of a non-technological environment, particularly in the remote places? How does the teaching of science through EVS impact on children's foundation knowledge of science? These and many more questions can be asked of the science curriculum and its implementation in Bhutan. Nevertheless, science education in its two present guises, Environmental Studies (PP-III) and General Science (IV-VI), has been an important part of the curriculum in Bhutanese primary schools since the formal beginning of secular education in Bhutan around 50 years ago. It is certain to remain so, albeit in a different form.

NOTES

[i] Bolstered by this success, the (now) Ministry of Education launched the translation and piloting of Social Studies for Class IV in Dzongkha for the period of 2005 to 2007. This change is scheduled to go nationwide in 2008. History for Class VII taught in Dzongkha was piloted in 2006 and 2007, and is earmarked to go nationwide in 2008 The Dzongkha Bhutan History for Class VIII will be piloted in the same year (RGoB 2005, p III).

REFERENCES AND BIBLIOGRAPHY

Dorji, J. (2003). *Quality of education in Bhutan*. Thimphu, Bhutan: KMT Press.

Dukpa, P.R. (2000). *Continuous assessment as a method of student evaluation in primary schools in Bhutan: An analysis of policy and practices in English curriculum 1986-1998*. Unpublished EdD dissertation, University of New England, Armidale, Australia.

Fullan, M.G. (1991). *The new meaning of educational change*. London: Cassell.

Holbrook, J. (2002). *Some insights into Class VII and VIII science in Bhutan. Unpublished presentation*. Bhutan: Division of Education, Royal Government of Bhutan.

Laird, D.J., Maxwell, T.W., Tenzin, W., & Jamtsho, S. (1999). *INSET within Bhutan 1996-1998 and the INSET Framework within Bhutan, 2000-2004/6*. Thimpu, Bhutan: Royal Government of Bhutan.

Maxwell, T.W. (2007). The important issues facing children in the Kingdom of Bhutan. In I. Epstein & J. Pattnaik (Eds.), *The Greenwood encyclopedia of children's issues worldwide: Asia and Oceania* (pp. 53-77). New York: Greenwood.

Maxwell, T.W., Laird, D.J., Namgay, Reid, J.M., Laird, J., & Gyamtso, D. (2006). *Focus on student learning outcome: Report of the review of the Bachelor of Education (primary) in distance education mode in-service program to the Ministry of Education and the Royal University of Bhutan*. Samtse, Bhutan: Samtse College of Education.
Osborne, R., & Freyberg, P. (1986). *Learning in science: Implication of children's science*. New York: Heinemann.
Royal Government of Bhutan. (1976). *National education policy*. Thimpu, Bhutan: Bhutan Division of Education.
Royal Government of Bhutan. (1988). *Syllabi for Classes 1 - 6*. Thimphu: Division of Education.
Royal Government of Bhutan. (1989a). *Second quarterly policy guidelines and instructions (March)*. Thimphu, Bhutan: Division of Education.
Royal Government of Bhutan. (1989b). *EVS Teachers' manual for Class PP*. Thimphu, Bhutan: Division of Education.
Royal Government of Bhutan. (1989c). *EVS Teachers' manual for Class II*. Thimphu, Bhutan: Division of Education.
Royal Government of Bhutan. (1989d). *EVS Teachers' manual for Class III*. Thimphu, Bhutan: Division of Education.
Royal Government of Bhutan. (1993a). *Review of primary education in Bhutan - Final report*. Thimphu: Division of Education.
Royal Government of Bhutan. (1993b). *Primary science: Class IV - Activity guide*. Thimphu, Bhutan: Division of Education.
Royal Government of Bhutan. (1993c). *Primary science: Class V - Activity guide*. Thimphu, Bhutan: Division of Education.
Royal Government of Bhutan. (1993d). *Primary science: Class VI - Activity guide*. Thimphu, Bhutan: Division of Education.
Royal Government of Bhutan. (1994). *Thirteenth quarterly policy guidelines and instructions (October)*. Thimphu, Bhutan: Division of Education.
Royal Government of Bhutan. (1995). *Fifteenth quarterly policy guidelines and instructions (November)*. Thimphu, Bhutan: Division of Education.
Royal Government of Bhutan. (1996). *The purpose of school education in Bhutan*. Thimphu, Bhutan: Division of Education.
Royal Government of Bhutan. (2000). *Evaluation of Dzongkha EVS*. Unpublished report prepared for the Curriculum and Professional Support Division. Thimphu, Bhutan: Division of Education.
Royal Government of Bhutan. (2001). *Twentieth education policy guidelines and instructions (EPGI), (August 2001)*. Thimphu, Bhutan: Division of Education.
Royal Government of Bhutan. (2002). *Teaching of primary science (Revised): Unpublished report*. Division of Education.
Royal Government of Bhutan. (2003). *A guide to continuous assessment*. Thimphu, Bhutan: Ministry of Education.
Royal Government of Bhutan. (2005). *Good governance plus*. Thimphu, Bhutan: KMT Printing Press.
Royal Government of Bhutan. (2006). *General statistics*. Thimphu, Bhutan: Ministry of Education.
Yadav, M.S. (1992). *Teaching of science*. New Delhi: Anmol Publications.

AFFILIATIONS

Wangpo Tenzin
Curriculum and Professional Support Division
Ministry of Education
Bhutan

T. W. Maxwell
School of Education
University of New England
Australia

CHENG

ASSESSMENT REFORM IN SCIENCE IN HONG KONG: POLICY, TEACHER BELIEFS AND PRACTICE

INTRODUCTION

Science education in Hong Kong, as in many other places worldwide, has undergone a number of reforms as the nation attempts to develop and implement curricula that takes account local context and needs. This reform occurred in two waves; the first being the abolishment of advanced level practical papers in science subjects. The second reform consisted of reform of the assessment regime; striving to better match assessment with the curriculum. Here the author analyses changes in the science assessment policy in Hong Kong, and summarizes findings of a number of local research studies into secondary science teachers' beliefs and practices in the assessment of their students' science learning.

The first wave of assessment reform in Hong Kong science started with the abolishment of practical papers in biology and chemistry at the Hong Kong Advanced Level Examination (HKAL) in 1995 and 1997, respectively. A teacher assessment scheme (TAS) was subsequently established and for the first time school-based practical work was assessed at a school level, and it constituted 20% of the HKAL score. Recently, the Government has decided to restructure secondary education from a two-year senior secondary and two-year HKAL preparation period, to a three-year senior secondary period. This is seen as the 'second wave' of assessment reform in which the new senior secondary level curriculum encourages alternative forms of assessment, and de-emphasizes public examinations as a measure of student learning. Starting from 2009, the public examination will include a one-off written examination, and a school-based assessment (SBA) component. It is intended that the SBA compose 20% of the total score of the public assessment, consisting of a variety of assessment tasks - both practical and non-practical related tasks - that reflect students' skills in the performance of practical work, their abilities in carrying out scientific investigations, and their generic skills (e.g., creativity, critical thinking). Recently, local science educators have expressed concerns as to whether or not teachers are prepared for such assessment reform.

BACKGROUND – EDUCATION AND ASSESSMENT REFORM IN HONG KONG

Education reform was launched in Hong Kong in 2001 after the framework of learning was redefined by the Education Commission in 2000 (Education Commission, 2000). Traditional subjects like history, biology, and physics were

Richard K. Coll, Neil Taylor (eds.), Science Education in Context: An International Examination of the Influence of Context on Science Curricula Development and Implementation, 333–341.

categorized into different 'key learning areas' (KLAs), with all the science subjects classified under the science KLA. Learning was no longer defined merely by the subject matter, but emphasis was also placed on generic learning skills. The development of nine generic learning skills (collaboration, communication, creativity, critical thinking, use of information technology, problem-solving, numeracy, self-management, and study skills) was advocated. These changes redefined learning, and teaching could no longer be conducted by strictly following the textbook. Teachers were encouraged to design tailor-made school-based curricula, and provide opportunities to enhance the development of the students' generic skills. Forms of assessment alternative to the pen and paper tests were required and needed to be integrated into the public examinations. Education reform, therefore, meant a shift in the culture of both learning, and teaching and assessing science.

Though education reform was launched in 2001, Hong Kong secondary teachers did not feel the impact at that moment, as no changes to the public examination system were announced. Primary schools, however, were the first to respond to the education reform. The primary science curriculum was abolished in 1996, and science topics were taught as part of an integrated subject named 'general studies', which includes science, social sciences, health education, environmental education, and information technology topics. While primary teachers were adapting to the teaching of the new subject, they also started to implement school-based projects in the late 1990s, intended to provide opportunities for pupils to develop the nine generic skills. These two changes were related, as many schools implemented their projects in relation to the subject general studies, given its multidisciplinary nature. For primary teachers, there were more challenges to face when the Government announced the implementation of a system-wide assessment (system assessment) in the form of pen and paper tests for all Primary 6, Secondary 1 and Secondary 3 students starting in 2004, in which pupils' performance in English, Chinese and mathematics was assessed (Education & Manpower Bureau, 2006a). This seemed something of a contradictory move, because primary teachers who had just mastered the skills of guiding pupils to conduct project learning, now identified drilling pupils to prepare for the system assessment as their new 'mission'.

Secondary teachers did not start to feel the impact of education reform until 2005 (Education & Manpower Bureau, 2006b) when the Government announced the introduction of a new secondary school system. Secondary education was restructured from a two-year senior secondary and two-year HKAL preparation period, to a three-year senior secondary period (Education & Manpower Bureau, 2004). With this restructuring, the curricula for all the secondary subjects were rewritten along with the mode of public examinations. Instead of having a one-off pen and paper examination as the sole determinant of the grade, 20% would be allocated to school-based assessment (SBA) for all subjects. This may not have been unfamiliar to the science teachers, as the teacher assessment scheme (TAS) was implemented in science subjects from 1995. The TAS and SBA in science subjects are comparable, as they both involve the assessment of school-based tasks assigned to students during the lessons, which can include practical work and field study reports. While the students are completing these tasks, the teachers can assess their attitude in science learning, their science process skills, and their understanding of science concepts. Despite the similarities, there were new practices that science teachers needed to prepare for. In its ideal form, the SBA should include different forms of alternative assessment tasks that may or may not be practical activities, and the use of assessment rubrics in assessing tasks that do not have a definite response. The assignment of different forms of assessment tasks is consistent with recent trends in science education, meaning the adoption of tasks that can cover a variety of skills such as problem-solving ability, and evaluate different aspects of cognition via say concept mapping, creative assessments, journals, oral interviews as well as self-perception questionnaires to compare students' understanding before and after science lessons (Dana, Lorsback, Hook & Briscoe, 1991; Heinze-Fry, 1992; Torff, 1997). Daws and Singh (1999) report that self-assessment can deepen science learning.

FACING THE CHALLENGES

In order to support teachers in meeting the challenges of the assessment reform, Paterno (2001) proposes the notion of 'assessment literacy', and suggests that teachers need to be familiar with alternatives to traditional measurements of learning. The choice of assessment methods is emphasized by Mertler (2005) who defined assessment literacy as being able to "understand which assessment methods to use to gather dependable information on student achievement, communicate assessment results effectively, and … use assessment to maximize student motivation and learning" (p. 78). Facing the challenges of the assessment reform is not easy, and Hargreaves and Earl (2002) describe a number of difficulties that teachers face. These difficulties can be categorized as: *technological perspectives*, which include the necessary skills to implement innovative assessment practices, the issue of time and ways to write feedback that stimulates student learning; *cultural perspectives*, which include increasing the involvement of students in the assessment process, an emphasis on self-assessment, reflection on learning and new ways of reporting student learning to parents; the *political perspective*, which includes inconsistencies in assessment policies; and finally, the *postmodern perspective*, which includes discerning if students' work is their own.

Identifying Practices and Beliefs in Science Assessment among Primary and Secondary Teachers

In order to identify the prevailing assessment practices and beliefs among primary and secondary teachers, the author conducted the first local quantitative study in 2000 as part of an international study called Schools Around the World[i] (Cheng & Cheung, 2001), and a follow-up study in 2003 (Cheng & Cheung, 2005) which aimed to identify changes (if any) for teachers two years after implementation. Forty-two secondary school teachers and 35 primary teachers - representing those who teach science (i.e., general studies) at primary levels, integrated science at Secondary 1 to 3 levels, and biology at Secondary 4 to 7 levels - were invited to participate in the study in 2000. They came from 14 primary and 23 secondary schools in Hong Kong. Questionnaires were administered before the first teacher professional development workshop of the SAW project and these included three questions aimed at discovering their current practices in science and biology assessment, their beliefs about the assessment, and their perceptions of their competence in carrying it out. The same group of teachers was to complete the same questionnaire again in 2003, two years after the project was implemented.

Assignment of Students' Work before, during and after a Lesson

The findings suggest that both the primary and the secondary teachers assigned a wider range of different types of students' work in the post-test (Cheng & Cheung, 2005). The change among the primary teachers was more significant in terms of the assignment of student work before the lessons, which may reflect a change in the fundamental belief that students have prior concepts before they come to lessons. Another observation was that the teachers assigned both the traditional forms of homework as well as more innovative tasks to students after the lessons. The teachers were less innovative in assigning different types of tasks to students during the lessons.

Teachers' Beliefs and Perceived Competence in Science Assessment

There was little change in the beliefs and self-perceived competence in science assessment among the teachers. In contrast with other studies which have found primary teachers to be a 'weak point' in science education (Cobern & Loving, 2002; Gardner & Cochran, 1993; McDuffie, 2001; Skamp & Mueller, 2001; Yates & Chandler, 2001), the primary teachers here felt more competent in assessing students' science learning and, in particular, in assessing their strengths and weaknesses, in using the assessment results, in making their teaching more effective, as well as in communicating frequently with parents about student performance. The secondary teachers, in contrast, were more worried about setting high but realistic objectives for individual students, as well as communicating with parents about student performance.

UNRESOLVED ISSUES

The findings revealed two unresolved issues among the secondary teachers (Cheng & Cheung, 2005). The first relates to their beliefs about implementing assessment to improve student learning and their emphasis on preparing students for the public examination. The second relates to their perception of competence in science assessment.

The Dilemma of Assessment for Learning and the Public Examination

The findings suggest that the primary teachers out-performed their secondary counterparts, mostly because they did not have to worry about public or system-wide assessment until June 2003, when the system assessment was implemented. The secondary teachers, however, found themselves in conflict between education reform directions, and the requirements of the public examination. The teachers' views or beliefs were contradictory, and their conceptions of assessment quite confused. They felt that spending less time on examinations and having more time to attempt alternative assessment tasks was good, and did not think that examinations are the most effective way of motivating students to learn, or that exam results accurately represent students' performance. Their practice suggests that they were assigning both the traditional homework and the different types of assessment tasks after science lessons, but their assignment of science assessment tasks was less innovative during the lessons. This suggests that their method of assessment was not consistent with their beliefs or the education reforms, but were merely actions taken to meet the requirements of public examinations. This reflects the reality of current practices in many schools in which assessments are conducted basically to fulfill summative functions. Despite the fact that examinations taking the form of standardized tests have been reported as deficient by a number of researchers (see, e.g., Baron, 1991; Kulm, 1991; Haetrel, 1991; Raizen & Kaser, 1989), this is likely to continue to account for a major part (80%) of the senior secondary public assessment. This sends a message to teachers that although examinations have their problems and limitations, they are still a prominent determinant of a student's final grade.

Perception of Low Competence in Science Assessment

It seems Hong Kong secondary school teachers doubt their competence in responding to the changing demands of science assessment, and do not consider themselves as very competent. In particular, teachers seem to have difficulty understanding and using a variety of methods for assessing students and setting high but realistic objectives for individual students. This provides direction for future professional development programs.

Besides raising teachers' confidence and competence in alternative and formative assessment, an implementation plan also needs to be developed. Teachers need to be able to visualize how new and old assessment practices can be put in place and how formative functions of assessment are to be re-emphasized, how different types of student learning outcomes are to be addressed, how to implement alternative assessment forms while also preparing students for examinations, and finally, how to provide feedback, set assessment criteria and encourage self and peer assessment. Moreover, McTighe and Emberger (2006) suggest that teachers collaborate to develop new assessment strategies. They remind us that teachers can collaboratively design assessment tasks, have peers review assessments for feedback on designs, and conduct group evaluation of student work elicited by the assessment tasks.

PREPARING TEACHERS TO FACE THE CHALLENGES OF THE ASSESSMENT REFORM

In order to address the new demands for SBA, the author collaborated with nine teachers in four secondary schools to design scientific investigation tasks with the use of assessment rubrics to enhance student performance.[ii] As well as helping junior secondary students to develop their science process skills, they also were guided to understand the requirements of an assessment rubric so that they could conduct self-assessment and improve their own performance. The study analyzed science teachers' concerns in implementing scientific investigation tasks, helping students to understand assessment rubrics, and identifying ways that can support the development of scientific investigation skills among junior secondary students. The project included an intervention that took the form of teacher professional development workshops and school-based support. School-based support activities included working with teachers in designing the scientific investigation tasks, deciding the forms of assessment to be implemented (e.g., peer, self, formative), determining how the tasks address the demands of the assessment rubrics, and fine-tuning the assessment rubrics that address the ability of the students in the schools. The teaching resources that addressed the needs of individual schools were jointly designed and developed by the project team and the teachers in the schools. When the teachers implemented the suggested scientific investigation tasks in their lessons, classroom observations were made, and the process of implementation was videotaped. Interviews with the teachers were targeted to find out teachers' concerns in designing and assessing scientific investigation tasks, as well as their experience in using assessment rubrics in the project.

The results suggest that the teachers observed among students a better understanding of the requirements of the scientific investigation tasks (e.g., in defining what is good observation and design) and could follow the requirements in the rubrics to improve their performance. However, they were concerned about whether the students could continue to apply the same criteria in future scientific investigation tasks. One of the teachers also reported that the performance of her Secondary 1 students (aged 12 years) was greatly enhanced to a level comparable to Secondary 5 students (aged 17 years). Furthermore, the teachers found the assessment rubrics useful in grading scientific investigation tasks, as they provided clear criteria, they also reported their concerns. Two of the teachers encountered some difficulty in understanding the criteria themselves, and three of them found it hard to explain to students the meaning of these criteria. They reported a lack of teaching time, difficulties in helping students to apply their understanding of the criteria to the tasks they were working on, and difficulties in helping students to distinguish differences between the requirements of different levels.

The experience of this project suggests that school-based support for and collaboration between teachers can provide successful experiences for implementing new science assessment practices. Teachers were able to visualize how old scientific investigation tasks and new assessment practices, that is, the use of rubrics, can be put in place at the same time. They could see how formative functions of assessment are re-emphasized by explaining to students the requirements of the tasks and encouraging self-assessment. They also understood how different types of student learning outcomes can be addressed, for example, science process skills, how to implement alternative assessment like engaging students in science projects while they also prepare students for examinations, and how to provide feedback, set assessment criteria and encourage self and peer assessments. Continual support or repeated experience is required so as to provide opportunities for students to apply their learning in new tasks, and for teachers to consolidate their newly acquired pedagogical skills. These experiences are essential for teachers to build up their confidence in adopting alternative assessment practices.

CONCLUSIONS AND RECOMMENDATIONS

In light of the persistent emphasis on the recall and comprehension of science knowledge in previous public examinations, the perceived lack of competence in assessment among teachers, and the continual adherence to summative functions of assessment or examinations (as 80% of the score will be accounted for by the one-off pen-and-paper test), the changes in assessment practice seem to be unrealistic, if not impossible. It is not surprising to find calls for delaying or reducing the percentage of SBA among teachers and school principals in local newspapers. These requests reflect a lack of confidence in their ability to implement SBA. Without having accumulated successful experiences and witnessed improved student learning, the voices of opposition among teachers will continue to be strong. The experience from the research projects described in this chapter suggests that teacher development opportunities are essential to increasing teachers' professional competence and confidence in implementing alternative science assessments. These opportunities may take the form of workshops, but the impact of school-based projects seems to be stronger. In school-based projects, teachers can experience the integration of existing science teaching pedagogy and new assessment methods. They can witness improvement in student learning and learn how to strike a balance between the demands of the public examinations and the SBA. Moreover, the government needs to develop a scenario for future directions of development and share it with the teachers, for example, whether the proportion of SBA is to increase or remain unchanged in the next 10 years. With a vision for the future, teachers will be able to balance their efforts in preparing students for SBA and the pen and paper tests in the public examinations.

NOTES

i This project originated from the Schools Around the World Project (SAW) which was initiated in the United States by an education research foundation, the Council for Basic Education (CBE). Apart from Hong Kong and the United States, there are seven other participating countries: Japan, the United Kingdom, Australia, the Czech Republic, France, Germany, and Portugal. The project aims to build up a professional network of teachers internationally such that teachers can work together to share student work samples and their understandings of science assessment.

ii This project was funded by the Research Committee of the Hong Kong Institute of Education.

REFERENCES AND BIBLIOGRAPHY

Baron, J.B. (1991). Performance assessment: Blurring the edges of assessment, curriculum, and instruction. In G. Kulm & S.M. Malcolm (Eds.), *Science assessment in the service of reform* (pp. 247-266). Washington, DC: American Association for the Advancement of Science.

Cheng M.H., & Cheung, W.M. (2001). Science and biology assessment in relation to the recently proposed education reform in Hong Kong. *Journal of Biology Education, 35*(4), 170-177.

Cheng, M.H., & Cheung, W.M. (2005). Science and biology assessment in Hong Kong: Progress and developments. *Journal of Biological Education, 40*(1), 11-16.

Cobern, W.W., & Loving, C.C. (2002). Investigation of preservice elementary teachers' thinking about science. *Journal of Research in Science Teaching, 39*(10), 1016-1031.

Dana, T.M., Lorsback, A.W., Hook, K., & Briscoe, C. (1991). Students showing what they know: A look at alternative assessments. In G. Kulm & S.M. Malcolm (Eds.), *Science assessment in the service of reform* (pp. 331-337). Washington, DC: American Association for the Advancement of Science.

Daws, N., & Singh, B. (1999). Formative assessment strategies in secondary science. *School Science Review, 80*(293), 71-78.

Education Commission. (2000). *Reform proposals for the education system in Hong Kong*. Hong Kong Special Administrative Region of the People's Republic of China: Printing Department.

Education and Manpower Bureau. (2004). *Reforming the academic structure for senior secondary education and higher education: Actions for investing in the future*. Retrieved 8 November 2004, from http://www.emb.gov.hk/index.aspx?langno=1&nodeID=3998

Education and Manpower Bureau. (2006a). *Assessment for learning resource bank: Basic competency assessment*. Retrieved 8 November 2004, from http://www.emb.gov.hk/index.aspx?nodeID =2412&langno=1

Education and Manpower Bureau. (2006b). *New senior secondary curriculum and assessment guides: Provisional final draft of curriculum parts*. Retrieved 8 November 2004, from http://www.emb.gov.hk/index.aspx?langno=1&nodeid=5149

Gardner, A.L., & Cochran, K.F. (Eds.). (1993). *Conference proceedings: Critical issues in reforming elementary teacher preparation in mathematics and science*. Greeley, CO: Center for Research on Teaching and Learning.

Haetrel, E.H. (1991). Form and function in assessing science education. In G. Kulm & S.M. Malcolm (Eds.), *Science assessment in the service of reform* (pp. 233-246). Washington, DC: American Association for the Advancement of Science.

Hargreaves, A., & Earl, L. (2002). Perspectives on alternative assessment reform. *American Educational Research Journal, 39*(1), 69-95.

Heinze-Fry, J.A. (1992). Using student self-assessment of biological concepts in an introductory biology course. *Journal of College Science Teaching, 22*, 39-43.

Kulm, G. (1991). The control of assessment. In G. Kulm & S.M. Malcolm (Eds.), *Science assessment in the service of reform* (pp. 55-68). Washington, DC: American Association for the Advancement of Science.

McDuffie, T.E. Jr. (2001). Scientists - geeks and nerds? *Science and Children*, *38*, 6-19.

McTighe, J., & Emberger, M. (2006). Teamwork on assessments creates powerful professional development. *Journal of Staff Development*, *27*(1), 38-44.

Mertler, C.A. (2005). Secondary teachers' assessment literacy: Does classroom experience make a difference? *American Secondary Education*, *33*(2), 76-92.

Paterno, J. (2001). *Measuring success: A glossary of assessment terms. Building cathedrals: Compassion for the 21st century*. Retrieved 8 March 2005, from http://www.angelfire.com/wa2/buildingcathedrals/measuringsuccess.html

Raizen, S.A.. & Kaser, J.S. (1989). Assessing science learning in elementary school: Why, what, and how? *Phi Delta Kappan*, *70*(9), 718-722.

Skamp, K., & Mueller, A. (2001). Student teachers' conceptions about effective primary science teaching: A longitudinal study. *International Journal of Science Education*, *23*, 331-351.

Torff, B. (1997). *Multiple intelligence and assessment*. Arlington Heights, IL: SkyLight Training and Publishing.

Yates, G.C.R. & Chandler, M. (2001). Where have all the skeptics gone? Patterns of new age beliefs and anti-scientific attitudes in preservice primary teachers. *Research in Science Education*, *30*, 377-387.

AFFILIATION

May Hung Cheng
Hong Kong Institute of Education
Hong Kong

LEARNING FROM OTHER COUNTRIES: A CRITICAL EXAMINATION OF THE CURRENT PRIMARY SCIENCE CURRICULUM REFORMS IN MAINLAND CHINA

INTRODUCTION

The current primary science curriculum was reformed in mainland China around the turn of the new millennium, and a new curriculum was published in July 2001 by the Ministry of Education. The curriculum documents describe the nine years of compulsory schooling, the science curriculum for Grades 3-6, and contain detailed curriculum standards for primary schools. The curriculum was developed by a panel of science educators, and is separate from the Grades 7 to 9 science curriculum standards for junior secondary schools. The new primary science curriculum[i] is very different in nature from its predecessor - the 'nature' syllabus for primary school. The old curriculum ostensibly aims to:

> Direct students to gain some shallow and basic knowledge of natural sciences, meanwhile developing their interest in and aspirations towards science, as well as the ability to learn and apply science, so that they will receive an education in such aspects as views of scientific nature, scientific attitudes, love for their hometown, love for their motherland, and love of nature, and thus promote the healthy development of their body and mind (State Education Commission, 1992, p. 1)

The new science curriculum standard, in contrast emphasizes 'scientific literacy' as its fundamental principle. It claims, among other things, that science learning should be inquiry-based; that its content should meet the needs both of individual students as well as the society; and that it should enable students to understand the nature of science (Ministry of Education [MoE], 2001). The differences between the two documents are quite striking, at least in terms of wording and the concepts used. Although the early nature study syllabus from the 1990s was also influenced by foreign ideas, the new science curriculum standard (Grades 3-6) was more directly affected by international science education reform trends of the 1980s. The question is, to what extent have the core ideas inherent in current science curriculum reforms in other countries been adopted in the Chinese curriculum development and implementation? Additionally, has this adoption of foreign ideas considered the Chinese national context? This chapter is, therefore, a critical reflection and review of this question, and is intended to inform the ongoing reform of the primary science curriculum in China, and provide some lessons for other developing countries

Richard K. Coll, Neil Taylor (eds.), Science Education in Context: An International Examination of the Influence of Context on Science Curricula Development and Implementation, 343–352.

experiencing similar curriculum change in primary science.

First, some background is provided about the reform of the primary science curriculum, and its transition from 'nature' to 'science'. This is followed by a consideration of the Chinese context through the discussion of issues and disputes that have arisen from the reform of the school curriculum. Illustrations are provided of successful projects that have adopted international science curricula and instructional ideas in school science. The chapter then advocates a systemic approach to primary science curriculum reform, and provides suggestions for science education research accompanying the development and implementation of science curriculum, and science teacher education and the development needed for its support.

FROM 'NATURE STUDY' TO 'SCIENCE': A TRANSITION TO A MORE BALANCED PRIMARY SCIENCE CURRICULUM

The primary science curriculum in mainland China is now undergoing a major transformation and the current reform of school curriculum and instruction has been under way since the beginning of the new millennium (Hao, 2006; MoE, 2001). Prior to the reform, primary science teaching had occurred rather sporadically, with science taking the form of 'nature study' in the middle and upper years of primary schools for almost the entire twentieth century - similar to the situation in other countries. There were some differences due to the Chinese intellectual tradition that emphasizes acquisition of book knowledge on the part of pupils during instruction. In contrast, the reform of the current primary science curriculum and instruction places priority on inquiry learning and teaching, that is, 'doing science', a somewhat alien concept for Chinese primary science teachers (Ding & Hu, 2005). As a consequence of the shift in focus there also was a shift in the name of the subject from 'nature' to 'science' to reflect the true character of both the changed content, and the new approach to primary science learning/teaching.

Owing to the Chinese cultural traditions characterized by instilling in students knowledge from books and neglecting process skills and hands-on activities in learning, it was natural that nature study as a primary school subject also reflected such values. The assessment system was also congruent with this traditional learning/teaching approach, in that it focused on the recall of knowledge in testing and examinations. Besides 'basic knowledge' transferred from the book to the student's head, it was (and to a great extent still is) emphasized that 'basic competence and skills' should be developed in students as well. This, among other things, means mainly the ability to conduct observations, the ability to do experiments, and logical thinking - all of which were embedded in the curriculum. Teaching meant, hopefully, that these things would be copied through instruction by students (Li, 1992). The so called 'dual basics', a discourse prevalent in mainland Chinese school curriculum policy and theory for more than half a century, represented the major theoretical underpinnings of the school curriculum in practice in nearly all the subjects, including primary science.

Despite the rhetoric of giving priority to the 'basic competence and skills', the old nature study curriculum was not balanced in terms of scientific literacy, that is, it prioritized *learning science* over *learning about science* and *doing science* (Hodson,

1998). Looking from the perspective of the implemented or enacted curriculum and the student-experienced curriculum, for most primary science teachers and their students, what they taught and learned arguably concentrated on some natural and/or scientific facts, learning terms, or pieces of knowledge that could be readily tested in paper and pen assessment.

From the 1980s through to the 1990s, the primary science curriculum took the form of nature study as described above, but was not uninfluenced by foreign ideas and practices in science education, although the influence was indeed limited in scope. During this era, China was somewhat 'cut off' from the outside world, particularly during the Cultural Revolution that lasted from 1966 to 1976. It was only when the opening up and reform era came in the late 1970s, that Chinese educators begin to have opportunities to 'look through the windows' of comparative education and foreign educational reforms and theories, and to engage in international academic exchanges with their foreign counterparts. So 1980 resulted in a translation of a primary science education monograph, *New Directions in Elementary Science Teaching* (1969) that originated with the US science educators, Paul DeHart Hurd of Stanford University, and James Joseph Gallagher of the University Extension Madison, Wisconsin. This was translated by Mogeng Liu, a senior editor and primary science educator with the People's Education Press, who was later to lead the reform of Chinese primary science curriculum until his death in 2000. In 1982 and 1984, a Harvard science educator, Professor Brenda Lansdown, was invited by Liu to visit China, and she had an enormous influence on Chinese primary science curriculum development and subsequent implementation. Her book, *Teaching Elementary Science through Investigation and Colloquium*, first published in 1971, was translated into Chinese and published in 1983. In 1984 when she was 80 years old, she gave a science lesson on the topic 'snail' to Grade 3 students in a primary school in Beijing, thus forming solid friendships with Liu and other Chinese science educators and primary science teachers. The primary science curriculum reform during the period from 1980 to 2000, promoted by the visits of Professor Brenda Lansdown and the translation of her book into Chinese, in a sense, made up for the loss that China had made in the 1960s when science curriculum reform in many Western countries was in full swing. For instance, some of the famous primary science teachers, in this case, those teachers with the special (and the highest) titles (*Teji Jiaoshi*), because of their recognized excellence in teaching, established themselves professionally just after they had learned and benefited from the ideas of Brenda Lansdown. This was recognized by some of these teacher themselves (see, for example, Lu, 1996), and for the first time in the history of primary science teaching/learning in China, inquiry or investigation received the attention of Chinese science teachers and was seen as an important approach to teaching and learning. But the idea of inquiry only permeated the teaching practice of those few primary science teachers who embraced with enthusiasm the reforming of their science classrooms, and who aspired to become a renowned teacher. For the majority of primary science teachers, their concerns remained with testing and examinations, and therefore, inquiry/investigation approaches to teaching and learning science were beyond them.

An analysis of the primary science curriculum documents and textbooks from this period shows that although some ideas prevalent in the Western science education community were made accessible to Chinese education policy makers and primary science educators, the discourses used in China were still traditional in nature. Heavy emphasis was still placed on the acquisition of basic knowledge and skills, and the way they were obtained was through teachers lecturing passive students, thus neglecting the idea of the student making sense of the content and engaging in meaning-making. For instance, in the nature study syllabus published in 1992, the aims of the course were specified in terms of the following four key terms: knowledge, interest, competence, and character education (Li, 1992; State Education Commission, 1992). Here, by competence, it means specifically "develop[ing] the competence of learning and applying science in it" and by character education, it refers specially to "receive[ing] an education in such aspects as views of scientific nature, scientific attitudes, love for their hometown, love for their motherland, and love of nature" (Li, 1992, p. 13). No one argues that these aims in primary science education are unimportant, but the problem is that the discourses throughout the documents sound as if they come from the late nineteenth or early twentieth century, when nature study was first introduced into the elementary school curriculum.

Unlike the reform of nature study as a school subject in the 1980s and 1990s, the current primary science curriculum has taken in many more ideas and experiences from other countries and/or other areas than mainland China on board (e.g., Taiwan & Hong Kong). For example, several American science education documents have been translated into Chinese, such as the *Science for All Americans* (1989) and *Benchmarks for Scientific Literacy* (1993) produced by the AAAS, and the *National Science Education Standards* (1996) produced by the USA National Academy of Science. The translation of these documents into Chinese provides Chinese science educators and teachers with a wealth of new ideas about international science education reform. Around 2000 when the science curriculum standards document for Grades 3 to 6 was in the making, science education curriculum documents of such other countries and areas such as England, Australia, New Zealand, Japan, Canada, Russia, Hong Kong and Taiwan, as well as those of the USA, were also translated into Chinese - mainly for reference (Cai, 2003). Building on these 'borrowed' ideas and previous experiences of Chinese science teachers in the teaching of nature, a new primary science curriculum standard was established, which was couched in much more internationalized science education discourses than had been the case in the nature study syllabus. For the first time in the history of Chinese science education, a series of interrelated new concepts such as *views of science*, *scientific literacy*, *scientific inquiry*, *the nature of science*, and *process skills*, and so on, entered the science education curriculum documents. This led to further internationalization or globalization of Chinese science education.

FOLLOWING THE GOLDEN MEAN: THE CHINESE NATIONAL CONTEXT RECONSIDERED

It goes without saying that any country, while undergoing a curriculum reform, has to take into account its national context (and even the local context) in adopting curricula ideas from other countries. Historical experience time and again shows that any attempt at curriculum reform is doomed to failure if no consideration is given to the local settings in terms of such factors as cultural traditions, educational heritage, social mindsets, teacher preparation and development, and parent and student expectations (Ding, 1996; Sifuna, 1992; Wang, B.L., 2006).

The way the primary science curriculum reform in China has been conducted is progressive in nature, with a few experimental districts first trialing the new curriculum, and then expanding the 'experiment' to more and more districts and provinces by phasing in the new curriculum and phasing out the old one. By September, 2006, all the schools should have adopted the new textbooks based on the new science curriculum standards. One of the problems with the new primary science curriculum (and this is also true of other curricula) that has arisen in the curriculum development and implementation process, is that the new primary science curriculum is perceived to be 'city-oriented' and suited mainly to developed districts within China, while failing to meet the needs of children in rural and backward economic districts (Wang, Y.C., 2006). Like many other developing nations, China is a dualistically-structured nation with relatively developed (culturally and economically) cities and coastal areas on the one hand, and vast rural, mountainous, and backward areas on the other. Given the fact that the rural population amounts more than 70% of the total population, any national curriculum that cannot meet the needs of children in rural communities is surely inappropriate?

In addition to the issues mentioned above, there are huge differences within a given geographical area. For example, schools in the city of Beijing proper have much better facilities, and better trained teachers than those in surrounding rural districts. On the other hand, some schools labeled 'key-point schools' which, in the case of secondary schools have been re-styled as model high schools, have greatly superior school buildings, more advanced facilities and more better-qualified teachers. Because of this, such schools usually are able to attract higher-caliber students, and therefore many more of their graduates can go on to study in prestigious universities or colleges through high stakes competitive entrance examinations.

Another particular problem with the new primary science curriculum lies in the fact that time available for science teaching in Grades 1 and 2 is reduced, although nominally the science content is supposed to be put in another course titled 'moral education and life' (*Pinde yu Shenghuo*). In the previous curriculum nature study was taught in the entire primary stage from Grade 1 through to Grade 6 since 1986 (Zhong & Gao, 2007). Ironically, however, science education is also part of kindergarten learning (i.e., 3-5 year old children). Cutting down science for Grades 1 and 2 children is surprising, given the national modernization program that has been underway from the late 1970s.

Besides these two obvious issues, there are other more subtle aspects about the national context that should be considered if the primary science curriculum reform is to be successful. Before dealing with those aspects, it is necessary to discuss two disputes arising from the curriculum reform within the Chinese education research community.

The National Context and Its Role in Curriculum Reform

Over the past six years, as the curriculum reform of basic education has been ongoing, there has been a strong debate within the Chinese education community about the national context (*Guoqing*) and its role in curriculum reform. Those educators who have been responsible for the design of the reform are inclined to the view that a new school curriculum, based on constructivism, postmodernism and other Western theories, in which the development of the individual student commands central stage, should replace the old view which was based on pedagogy deeply influenced by the former Soviet educator I.A. Kairov, that certainly in the case of China, was very examination-oriented (Zhong, 2005). For these curriculum theorists, the national context should not be espoused as a pretext for holding the back reform, and in this sense, they hold that the national context can be ignored in favor of global trends of the times and truth. In contrast, other educational theorists consider the national context should feature more in thinking about the current curriculum reform in general, and the primary science curriculum in particular. In their opinion, the national context has been historically formulated and is dynamic, meaning curriculum reform should be subjected to considerations of national context, positively taking advantage of it, and therefore modifying, rather than disregarding it (Wang, B.L., 2006). It is palpably clear that the in-depth reason behind the dispute is, in the view of the author, something of a power struggle, although it is acknowledged that neither side is necessarily totally right or wrong. As Michael W. Apple argues, education is "a site of conflict about the kind of knowledge that is and should be taught, about whose knowledge is 'official' and about who has the right to decide both what is to be taught and how teaching and learning are to be evaluated" (2004, p. vii).

Guiding Principles of Curriculum Reform

In the current primary science curriculum reform, as is the case of the reform of the reform of the entire curriculum for compulsory schooling in China, the guiding principles or theoretical foundations underpinning the curriculum reform are explicitly claimed to be constructivism, multiple intelligences theory, post-modern theory, and so on (Zhong, 2005; Zhong, Cui & Zhang, 2001). For other curriculum theorists, however, this means only the transplanting of foreign curriculum theories without considering the Chinese context. Because modern science has not been rooted in 'Chinese soil', the level of scientific literacy of most Chinese people is not as high as that in developed countries. The curriculum theories adopted from other countries, they argue, cannot constitute guiding principles or the theoretical foundations for the new curriculum reform, as they are not suited to the Chinese context. Rather, they point out that the epistemology of Marxism, including Deng Xiaoping and Jiang Zeming's thoughts, should form the theoretical foundation of curriculum (Jin & Ai, 2005).

Identifying the National Context (Guoqing), and How to Adapt the Primary Science Curriculum Development and Implementation to the National Context

The national context (*Guoqing*) in terms of schooling is what we confront every day: millions of students, crowded classrooms of 40-80 students, large numbers of ill-trained science teachers, inadequate facilities and equipment in science laboratories, and above all, competitive and high stake testing and examinations. In such circumstances, science curriculum reform is indeed onerous, but this does not constitute a pretext that advanced curriculum and educational ideas adopted from other countries cannot be applied in Chinese classrooms. The successful experimentation of the French-introduced 'learning by doing' project of primary science education that has been implemented in many cities in China shows that inquiry learning/teaching in primary science can take place, if, among other things, teachers are well trained. In Beijing, our initial efforts at assessment for learning in primary science classrooms also have proved successful, in that the teachers involved in our project have engaged in collaborative action research together with us, and this enabled them to comprehend what is meant by, for example, constructivism, assessment for learning, teacher professional development, and so on. They 'live' these ideas by practicing them in their daily work creatively. For example, one experienced specialist science teacher in our team found that the use of assessment for learning in her classroom made her focus more on the key questions she had in her lesson plan, and these newly-designed questions were of considerable help in bringing in learning for understanding for her Grade 6 students. Her students were more active in answering questions, their interest in learning and doing science dramatically increased, and their performance in science tests improved. She also tried other strategies successfully, such as 'wait time', peer self-assessment in the classroom, making use of summative assessment for the purpose of formative assessment, and so on. The preliminary results were striking. In an interview with her, she remarked:

> Reflecting on the previous teaching that I had done, although I had students in mind and wanted to start teaching where they were, it had to be put aside when it came to teaching. In fact, what I did in class was more from my own perspective than from the children's. Now, I am really considering their points of view while in teaching.

ADVOCATING A SYSTEMIC APPROACH TO THE REFORM OF PRIMARY SCIENCE CURRICULUM

As noted above, primary science curriculum reform is currently underway in mainland China, and this has met with various problems and issues, some of which may have been avoided if a more systemic approach to the reform had been made. The experiences of the past years have led us to believe that merely developing a new science curriculum is not adequate. Curriculum is only one link in the whole 'educational chain', and if we want to change the curriculum, the other links of the chain have to be changed as well. Some authors have found that in quite a number of schools throughout China, the new science curriculum has been in place, yet with the old pedagogy remains in practice (e.g., Hao, 2006). Because of lack of funding for continuous professional development, most primary science teachers have had little or no chance of engaging in the further learning required of them by the curriculum reform. In a survey of the current status of the implementation of the new science curriculum in 2006 it was found that 58% of the teachers surveyed responded that they had little knowledge of the new science curriculum standard; 80% of teachers teaching primary science have no science background; 60% of teachers reported having few or no facilities or apparatus in teaching primary science; and 51% of teachers found difficulty in trying to use inquiry-based teaching methods (Hu, Han & Liu, 2007). In such circumstances, it is understandable that the gap between the new theories, such as constructivism, and hard reality of China classrooms can not easily be bridged.

A systemic approach to the reform of primary science reform, therefore, first demands that teacher education and teacher development are commensurate with the new curriculum. Without teachers who really understand the rationale behind the curriculum, the best possible curriculum cannot be implemented to good effect.

Second, the assessment practice of student performance should be appropriately modified to suit the new curriculum development and implementation. So far, although developmental assessment has been encouraged for teachers to use in their classrooms, few teachers know what it is or how to apply it in their classrooms. Assessment for learning or formative assessment appears in the reform document and various journal articles, but is seldom put to use by Chinese science teachers as they are, for the most part, ignorant of how to apply it. So the problem does not rest on the national context not allowing for the innovation, but lies in the fact that most teachers have no access to the new ideas and little academic support in trialing them. On the other hand, some of the national context should be modified to ensure that innovations can be effected. For example, the 'teach to test' culture should be modified somewhat in order that inquiry-based curriculum and instruction has a chance of success. Otherwise, no reform will make any difference.

Last, but certainly not least, primary science curriculum reform, as a highly complicated and complex professional endeavor, calls for research-evidenced support as well as participation of curriculum experts, science researchers and science educators. As a new field of education, science education research, in most countries other than the USA, has emerged, institutionalized, and developed from the science curriculum reforms since the 1960s (Fensham, 2004). However, for historical reasons, science curriculum reform in China did not happen until the beginning of reform and opening up of the nation politically and economically in the late 1970s. Unlike many other countries, science education research has not been seen as important in the science curriculum reforms, and therefore in China there is a dearth of science educators in general, and science curriculum specialists in particular who are active in research. One of the reasons why there exists a gap between the adopted theory and practice, in the view of the author, is that few academic and intellectual support staff have been sent to support school science teachers since the country is short of domestic science education researchers. It is high time that Chinese educational authorities recognized the need for the training of such a professional body in preparation for the future reform of science education.

IN CONCLUSION

In the final analysis, we come to the conclusion that learning from other countries in the primary science curriculum reform is absolutely essential, for two key reasons. The first is that, unlike other subjects such as history and the mother tongue, which are peculiarly different in different cultures, science teaching and learning is much more universal, and therefore there is more common ground in science curriculum development and implementation between China and other countries. The second is that other countries, like the USA, the UK, Australia, France and Germany, and so on, have accumulated much experience in science curriculum reform since the 1960s, and much of their reform has been research-based. These countries can thus provide developing countries like China, that lack research capability in science education, with both needed experience and insights. While learning from other countries is emphasized, it does not mean that a consideration of the national context (and equally important, the local situations) in adopting and adapting theories from abroad is not significant; rather, as Atkin and Black (2003) point out:

> Research results … could not in any simple sense be applied to practice. The context and person of the teacher were all important, so we called for patient development work in which teachers could draw upon the research lessons as stimuli or provocations that would both encourage and support them in inventing new knowledge about pedagogy as they transformed the ideas into workable practices. (p. 142)

ACKNOWLEDGEMENT

This chapter is part of the findings of a research project titled Primary Science Curriculum, Teaching/Learning, and Assessment: A Theoretical and Practical Research Study from the Perspective of International Comparison, funded by the Foundation for the author as a National Excellent Doctoral Dissertation of the Peoples Republic of China (FANEDD) (Project No.: 200308).

NOTE

[i] This in the form presented in 2001 is termed an 'experimental edition' indicative of the on-going nature of the reform of primary science in China.

REFERENCES AND BIBLIOGRAPHY

Apple, M.W. (2004). *Ideology and curriculum* (3rd ed.). New York: Routledge-Farmer.

Atkin, J.M., & Black, P. (2003). *Inside science education reform: A history of curriculum and policy change*. Buckingham, UK: Open University Press.

Brandwein, P.F., Lansdown, B., & Blackwood, P.E. (1971). *Teaching elementary science through investigation and colloquium*. New York: Harcourt Brace Jovanovich.

Cai, H.J. (2003). The development of science courses in Chinese primary schools. *Journal of Educational Science of Hunan Normal University*, *2*(3), 70-73.

Ding, B. (1996). A critical review of curriculum diversification in African countries. *Comparative Education Review*, *4*, 6-11 [in Chinese].

Ding, B., & Hu, J. (2005). Some issues in China's school science education reform: A constructivist perspective. *Comparative Education Review*, *7*, 1-6 [in Chinese].

Fensham, P.J. (2004). *Defining an identity: The evolution of science education as a field of research*. Dordrecht: Kluwer.

Hao, D.Y. (2006). New curriculum reform: Problems and solutions. *Educational Research, 316*(5), 25-29 [in Chinese].
Hodson, D. (1998). *Teaching and learning science: Towards a personalized approach*. Buckingham, UK: Open University Press.
Hu, W., Han, Q., & Liu, J. (2007). Investigation into and thoughts of the current situation of the implementation of new curriculum of science in the primary schools. *Theory and Practice of Education, 27*(3), 58-63 [in Chinese].
Hurd, P.D., & Gallagher, J.J. (1969). *New directions in elementary science teaching*. Belmont, CA: Wadsworth.
Jin, L.Y., & Ai, X. (2005). What is the theoretical foundation for the new curriculum reform anyway? *Daily of Chinese Education*, May 28, pp. 4-5.
Li, P.S. (1992). Understanding deeply the nature and task of 'nature' and knowing accurately its aims and demands. In Department of Basic Education, State Education Commission (Ed.), *The learning guidance for the syllabus of nature teaching (Provisionary)* (pp. 10-28). Beijing: People's Education Press.
Lu, P.Q. (1996). The practice and exploration in the teaching of 'nature'. In B. Liu, L. Ma, S.S. Zhao & P.S. Li (Eds.), *Records of thoughts about teaching for Chinese famous teachers with the special title* (pp. 101-134). Nanjing: Jiangsu Education Press.
Ministry of Education. (2001). *Science (for grades 3-6) curriculum standard (Experimental edition*). Beijing Normal University Press.
Ministry of Education. (2001). *Science (7-9) curriculum standard (Experimental edition)*. Beijing: Normal University Press.
National Academy of Science. (1996). *National science education standards*. Washington, DC: National Academy Press.
Sifuna, D.N. (1992). Diversifying the secondary curriculum: The African experience. *International Review of Education, 38*(1), 5-18.
State Education Commission. (1992). *The syllabus of nature teaching (Provisionary)*. Beijing: People's Education Press.
Wang, B.L. (2006a). Curriculum reform in the Chinese context. *Journal of Beijing Normal University (Social Sciences Edition*), *4*(196), 18-27 [in Chinese].
Wang, Y.C. (2006b). The new curriculum reform: Dilemmas and the way forward. *Educational Research and Experiment, 6*, 26-30 [in Chinese].
Zhong, Q. (2005). Curriculum reform in China: Challenge and reform. *Comparative Education Review 187*(12), 18-23 [in Chinese].
Zhong, M., & Gao, L. (2007). Problems and analysis on primary science curriculum reform. *Curriculum, Teaching Material and Method, 27*(6), 77-81 [in Chinese].
Zhong, Q., Cui, Y., & Zhang, H. (2001). *An explanation of the guiding outlines for the curriculum reform of basic education (Draft)*. Shanghai: East China Normal University Press [in Chinese].

AFFILIATION

Bangping Ding
School of Education
Capital Normal University
Beijing
China

CONCLUSIONS AND IMPLICATIONS

COLL AND TAYLOR

THE INFLUENCE OF CONTEXT ON SCIENCE CURRICULA: OBSERVATIONS, CONCLUSIONS AND SOME RECOMMENDATIONS FOR CURRICULUM DEVELOPMENT AND IMPLEMENTATION

INTRODUCTION

The genesis of this project and book was our experiences of teaching science and science education at various levels in developing countries; in the Pacific and the Caribbean. These experiences along with numerous conversations with other teachers and educators who had worked in Africa and elsewhere left us with something of a sense of despair. We constantly confronted Western or foreign science curricula which were plainly alien to science learners in non-Western contexts. We witnessed numerous curricula reforms and professional development initiatives, many of which seemed doomed to failure. In fact Helu-Thaman (1991) referred to the 'wreckage' of aid-funded curricula initiatives all around the Pacific. Probably the most alarming aspect in all of this was the role of the foreign expert. Someone, normally 'aid-funded', who turned up for a short period of time to tell the locals what they should be doing! The naiveté of some of these people was truly remarkable (or perhaps they just didn't care?). Failure of the program or reforms was generally attributed to the locals not 'seeing it through' or not quite understanding the new curriculum initiatives. There was little effort made to take into account local conditions or the views of local experts, especially teachers.

The experiences of the authors of this book provide us with a unique opportunity to consider the nature of science curriculum development and its implementation in an extraordinary variety of educational contexts. The most cursory examination reinforces the central theme of this book: that context exerts a significant influence on the development and implementation of science curricula. As noted in our opening chapter, we as editors endeavored to remain 'hands off' and let the authors' own stories come through. Here we consider what all these fascinating stories have to tell us, and what can we learn from the myriad of experiences and experiments detailed in this book.

First we present a thematic analysis of the stories in terms of 'observations', and essentially try to answer the question; *what has happened and why?* From this we attempt to draw some conclusions, and try to answer the question; *what we can learn from these experiences?* Finally, we present a synthesis of our responses to the first two questions, and develop a framework model for curricula development and implementation that takes into account the complexity of the educational context.

Richard K. Coll, Neil Taylor (eds.), Science Education in Context: An International Examination of the Influence of Context on Science Curricula Development and Implementation, 355–362.

What we present here is inevitably our own picture of the influence of context on curricula development and implementation. It is up to the reader to judge whether or not this captures the sense of the stories told by these authors.

OBSERVATIONS

The first and most compelling observation is the *extent and pace of science curriculum reform* in many countries. Turkey might represent something of the extreme with four main phases and 11 versions of the science curriculum to date (Çalik & Ayas, chapter 13), and the sheer pace of Bhutan's reforms is truly remarkable (Maxwell & Tenzin, chapter 26). But most countries have undergone numerous reforms, and seldom has this appeared to be a measured, incremental development designed to build on past experience (locally or international), or based on sound evaluation of the impact of prior reform.

Curriculum reform does not come cheap, so why have so many countries engaged in such huge reforms? The main reason seems to be the *economic imperative*. Not surprisingly, developing countries want to improve their economic health, and governments worldwide have very much bought into the argument that enhancing science and technology education is crucial in coping with the effects of globalization (Gilli, 2000). This has also been driven somewhat by international agencies such as the World Bank. Even countries that are in fairly good economic health, seek to maintain their status and remain economically competitive, or like Singapore, want to 'future-proof' themselves against potential economic hardship (Tan, Koh & Cheah, chapter 23). A secondary reason, and one perhaps of greater significance in more developed countries, is to *enhance scientific literacy*; taken here to mean the development of citizens who are capable in engaging meaningful debate about science and technological matters (Jenkins, 1990; Laugksch, 2000). This is not unrelated to the economic imperative, but is a little loftier in ambition, and somewhat less directly pragmatic in nature. Prinsloo in chapter 22 suggests that each science curriculum contains important, albeit subliminal messages or codes about what we see science as, and we need coherence between our curricular aims and the messages we send learners about science and the nature of science.

The *political dimension* is a key driver in science curriculum development and reform. This is often related to the economic imperative (in many cases at least initially driven by Sputnik-era concerns of falling behind technologically), and a desire for a scientifically literate society. In other words, science curricula development and reform has been a 'top down' rather than 'bottom up' process. Consistent with this, reform has inevitably been enacted via a *centre-periphery model*, such as that described by McGee (1997). This has often resulted in considerable debate about the purpose and value of reform. In more politically-sensitive countries like China, this results in tension between the various stakeholders (Ding, chapter 28).

The majority of the science curricula reforms reported in this book have drawn upon Western thinking, and new science curricula are inevitably based on some form of foreign or imported Western science curricula. In some cases this has involved the direct use of science curricula from developed countries deemed economically successful. The relationship here to the economic imperative is plain; they are successful, so we should emulate their models of science teaching and learning. Related to this use of an imported model of curricula, is the importation of the theoretical basis to science learning and teaching. So, if not constructivism, certainly a *learner-centered* approach to teaching and learning is advocated in all curriculum documents. Other ideas such as Science-Technology-Society mentioned by Engida and Areaya in chapter 17 feature, and inquiry-based learning, is also popular. The intention seems to be to shift away from science learning consisting of the accumulation of scientific facts via rote memorization. This might at first sight seem a positive thing - if one buys into our thesis that the educational context is a crucial aspect of learning. A learner-centered approach to science education is the necessary yet not sufficient enabler of allowing for educational context in teaching and learning. However, the use of a learner-centered approach to science teaching and learning has mostly consisted of rhetoric, with a learner-centered approach mentioned routinely in documents, but in practice seldom woven through the whole process of curriculum reform - especially in terms of the assessment regime. So despite this apparent beacon of hope, there has mostly been little recognition of the importance of educational context in learner-centered education.

A second factor inhibiting the implementation of reformed science curricula is lack of coherence between what is purported to be a learner-centered approach to education as detailed in various curricular documents and supporting teacher manuals, and the *assessment regime*. Most authors point to the inherent tension between the development of learner-centered curricula and one or more gate-keeping, external, summative examinations, which almost inevitably reward the rote memorization of scientific facts (Hume & Coll, 2005).

Every author comments on the unrealistic expectations of their governments or ministries of education in the frequent introduction of sweeping reforms, without the provision of adequate *teacher professional development*. In most cases any professional development has been of the 'one shot' or 'pit-stop' variety (Grundy, 1995), and rather ad hoc in nature. As Dahsah and Faikhamta note in chapter 24, many teachers, despite being able to articulate these ideas in a fairly convincing fashion, in fact do not know what learner-centered education is actually all about. It takes time for them to clearly understand and get used to such radical changes. If this does not occur, then teachers simply carry on using the same pedagogies they have found effective in the past, as noted by De Beer in chapter 21. As many authors note, professional development on this scale represents a formidable challenge. In China (chapter 28), Pakistan (chapter 9) and India (chapter 10), for example, the sheer scale of the problem is daunting. How can you provide teacher professional development for literally millions of science teachers? Even in better-off countries, this remains problematic.

Remarkably there is *little evidence of evaluation* of these extensive curricula reforms, certainly in terms of enhanced learning outcomes. Evaluation has more commonly involved ascertaining how teachers have coped with the reforms, and whether or not, say learner-centered education actually occurs in the classroom (which frequently it does not). Considering the amount of money and human resources spent on such reforms this is an extraordinary observation given the prevalence of student alternative conceptions reported worldwide (Duit, 2000). Surely one expectation of science curricular reform would be enhanced conceptual understanding? This lack of evaluation of student outcomes does have parallels in other countries, with, for example, New Zealand despite decades of dramatic reforms, never having seriously evaluated student learning outcomes. Related to this, is the observation that comparatively little educational research has been conducted in many of the countries described in this book.

CONCLUSIONS AND RECOMMENDATIONS

Given the above observations it would be easy to become demoralized. Without being too negative, it certainly seems a lot of time and money have been wasted on rather pointless science curriculum reforms, most of which had little hope of success. However, we believe this book presents evidence for hope. We first draw our conclusions and then make some recommendations. We draw our hope from the stories provided by the authors. Our model for curricula development is detailed in Figure 1, and we now describe each of its features in turn.

First, science curriculum reform should draw upon international - including Western – experiences, both successes and failures. But it must take cognizance of local contextual factors. To illustrate, Malawi, as Chikasanda and Mbender note in chapter 18, like many developing countries, wants to reduce poverty and improve the standard of living of its citizens. However, given its educational context, namely that there are few jobs in the technical trades related to science and engineering, it doesn't make much sense to 'invest' vast amounts of scarce resources in producing technically-trained people who have little hope of employment. The needs of a country, including its economic needs should, we believe, dictate the science curriculum that is appropriate. Consider the situation of Bouganville as described by Jenkins in chapter 4. The needs of the 'lost generation' he speaks of are plainly different from those in say Singapore, described by Tan, Koh and Cheah in chapter 23. Finally, as noted by Tsaparlis in chapter 12, the curriculum needs to be balanced in terms of its treatment of science content areas.

> *Recommendation 1.* Curriculum developers and those involved in curriculum reform need to do a detailed and careful 'needs analysis', which locates science curricular aims with respect to actual local, contextual needs, and that is balanced in terms of science content.

Second, curriculum development and reform has too frequently resulted in importation of foreign curricula. It is possible to develop a match between local and international ideas and still address tensions such as those between science and religion as noted by Ambusaidi and Elzain in chapter 8 in the case of Oman, and in Pakistan as noted by Halai in chapter 10. In the case of very large countries, with

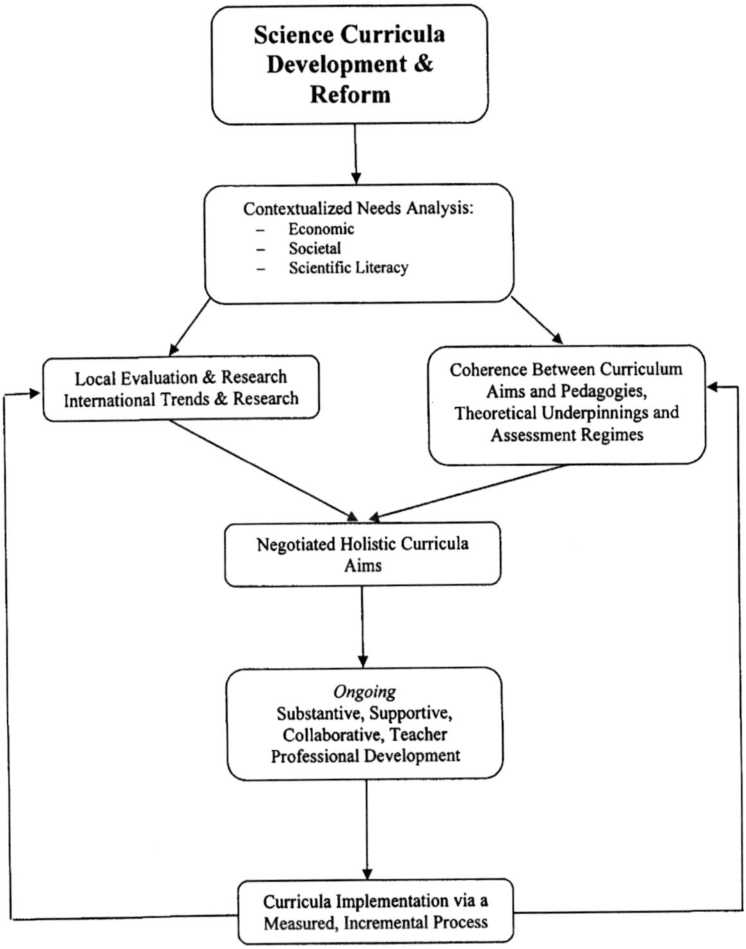

Figure 1. Proposed Model for Curricula Development and Reform

very diverse populations like India (see chapter 9 by Ranade), we think a national science curriculum is unlikely to be workable. We suggest here that what works in one part of India, may not necessarily work in the rest of India, or what works in the USA or UK, may not work in Thailand, China, or Nigeria. So we need to know the strengths and weaknesses of our current science curricula, and what might be done to address any shortcomings, and use this to drive us towards meeting our aims. How else might the science curriculum inform students about important health issues such as HIV/AIDS, which De Beer in chapter 21 and Matsoga in chapter 14 point out is of crucial importance in many African countries, or move towards self-reliance as noted by O-saki in chapter 19? It is this approach, we suggest, that may allow contextual factors, local and regional (see Akinsola, chapter 20), to be taken into consideration in curricula reform.

Recommendation 2. Curriculum development and reform needs to build upon substantive evaluation of past local experiences, and take cognizance of, but be not dominated by, international experiences and research.

Third, any new science curriculum needs to be accompanied by careful consideration of related issues, particularly assessment, whether the theoretical basis to reform is constructivism as occurs most often, or, say Bronfenbrenner's Ecological Systems Theory (1993) as utilized in Papua New Guinea (see Rombo - chapter 3). The experiences described in this book suggest that assessment regimes are the main drivers of teacher behavior. There is little likelihood of teacher change in terms of pedagogy (even with substantive professional development) if the assessment regime remains rooted in the past. In order to be consistent with the learner-centered education, such assessment regimes will almost certainly require a shift away from external summative examinations, and the incorporation of at least a component of school-based assessment such as that described in Sri Lanka by Selvaruby, O'Sullivan and Watts in chapter 11. Fiji, as described by Taylor, Taloga and Ali in chapter 6 shows that a drive to achieve consistency between assessment and curricula aims is now on the agenda at least in some countries. Likewise in chapter 27 Cheng describes how we can incorporate school-based assessment of practical work as part of the assessment regime.

Recommendation 3. Curriculum development and reform needs to comprise a holistic process in which there is coherence between the aims of the curriculum, the theoretical underpinnings, and the assessment regime.

Fourth, curriculum development and reform may not be successful even with changes to assessment regimes, if teacher professional development is inadequate. The authors here have all complained of inadequate professional development. We need coherence between the societal and educational aims of the science curriculum, the assessment regime, and strong support for teachers in terms of professional development if we are to have any hope that these aims will be realized. As Sade notes in chapter 5, we cannot expect - in his case technology teachers - to teach technology in the way we want if they have little understanding of how we view technology in a contextually-appropriate fashion. There needs to be recognition that this professional development is not a cost, but an essential part of curricula reform, as noted by Kasanda in chapter 16. There also needs to be a judicious use of international expertise along the lines proposed by Atweh, Bernardo and Balagtas in chapter 2, which recognizes the value of the teacher as the principal mediator of curriculum implementation. Thus, as argued by these authors, professional development needs to be a collaborative endeavor, and not revisit say the cascade model Koosimile and Prophet (chapter 15) argue has failed in Botswana. The model of professional development described by Mohd. Zaki and

Zulkifli in chapter 25 shows another way we can consider such processes, drawing as it does on Fensham's (2000, 2001) content analysis of science content.

> *Recommendation 4*. Curriculum development and reform needs to be accompanied by substantive, supportive, collaborative teacher professional development that is on-going in nature.

Fifth, curriculum development and reform takes time to do well. In several cases, this has been rushed (e.g., in Turkey, chapter 13, and in Bhutan - chapter 26). So reform needs to be linked to the past, and move forward at a rate that students, teachers and the communities they serve, can cope with. Sadler-McKnight and Rainford in chapter 7 comment that it takes time to throw off the influence of the colonial past; we need to allow such time if we are to bring about meaningful change.

> *Recommendation 5*. Curriculum development and reform needs to be enacted in a slow, measured, manner, than involves a feedback loop based on regular monitoring of successes and failures.

SOME CONCLUDING THOUGHTS

First, readers might feel we have avoided some issues in our summary and recommendations here. For example, we have not recommended for or against a centre-periphery model of curriculum development and reform. Our stance is that if all of the above recommendations are indeed met, then this probably doesn't matter. In other words, implicit in our recommendations is genuine collaboration between officialdom and teachers.

Second, others may well think many of our recommendations are obvious or straight-out commonsense. We certainly hope so! Our position here is that they are, but nonetheless, they have seldom been taken into consideration.

Third, others might say such recommendations are not feasible, given the economic circumstances present in many of the countries described in this book. Maybe. But these countries have already spent vast amounts of money and human capital in reforms that often with very little success. How do we judge such failure? Our position is that we should recognize such failures (and our successes), and learn from them.

We are 'true believers' in the notion that science whilst is not a panacea for economic and other woes in developing and developed countries, it is a crucial part of the equation. Thus we view money spent on the development and effective implementation of a science curriculum that genuinely meets local needs, as an investment, not as a cost.

REFERENCES AND BIBLIOGRAPHY

Bronfenbrenner, U. (1993). The ecology of cognitive development: Research models and fugitive findings. In R.H.Wozniak & K.W. Fischer (Eds.), *Development in context: Acting and thinking in specific environments* (pp. 3-44). Hillsdale, NJ: Erlbaum.

Duit, R. (2000). *Bibliography: Student's alternative frameworks and science education* (5th ed.). Kiel, Germany: University of Kiel.
Fensham, P.J. (2000). Providing suitable content in the 'science for all' curriculum. In R. Millar, J. Leach & J. Osborne (Eds.), *Improving science education: The contribution of research* (pp. 147-164). Buckingham, UK: Open University Press.
Fensham, P.J. (2001). Science content as problematic: Issues for research. In H. Behrendt, H. Dahncke, R.M. Duit, W. Graber, M. Komorek, A. Kross & P. Reiska (Eds.), *Research in science education: Past, present, and future* (pp. 27-41). London: Kluwer.
Gilli, D. (2000). Science education and economic developments: Trends relationships and research agenda. *Studies in Science Education, 35*, 27-58.
Grundy, S. (1995) *Action research as professional development.* Occasional Paper #1, Innovative Links Project, Canberra: Australian Government Publishing Service.
Helu-Thaman, K. (1991). A letter from a curriculum officer. In C. Benson (Ed.), *Report of the Pacific Curriculum Conference* (pp. 98-105). Suva, Fiji: Institute of Education, University of the South Pacific.
Hume, A., & Coll, R.K. (2005, July). *The impact of a new assessment regime on science learning.* Paper presented at the 34th Annual Conference of the Australasian Science Education Research Association. Hamilton, New Zealand.
McGee, C. (1997). *Teachers and curriculum-decision-making.* Palmerston North, New Zealand: Dunmore.
Jenkins, E. (1990). Scientific literacy and school science education. *School Science Review, 71*, 256.
Laugksch, R.C. (2000). Scientific literacy: A conceptual overview. *Science Education, 84*, 71-94.

AFFILIATIONS

Richard K. Coll
Centre for Science & Technology Education Research
University of Waikato
New Zealand

Neil Taylor
School of Education
University of New England
Australia

Other books/book series by Sense Publishers

CULTURAL PERPECTIVES IN SCIENCE EDUCATION
Distinguished Contributors

Series Editors

Kenneth Tobin *The Graduate Center, City University of New York, USA*
Wolff-Michael Roth University of Victoria, Canada

Scope

Cultural Perspectives in Science Education: Distinguished Contributors features a profile of scholarly products selected from across the career of an outstanding science education researcher. Although there are several variants in regards to what is included in the volumes of the series the most basic form consists of republication of 8-10 of the scholar's most significant publications along with a critical review and commentary of these pieces in terms of the field at the time of doing the work, the theories underpinning the research and the methods employed, and the extent to which the work made an impact in science education and beyond. Another genre of Key Works republishes the most influential research in a selected area of interest to science educators. Examples of the areas we will feature include science teacher education, science teaching, language in science, equity, the social nature of scientific knowledge, and conceptions and conceptual change. Collections of articles are placed in an historical context and the rationale for changing perspectives is provided and analyzed in relation to advances and changing priorities in science education. Each volume shows how individuals shaped and were shaped by the cultural context of science education, including its historical unfolding.

THE RE-EMERGENCE OF VALUES IN SCIENCE EDUCATION

Deborah Corrigan *Monash University, Clayton, Australia*
Justin Dillon *King's College, London, UK*
Richard Gunstone (Eds.) *Monash University, Clayton, Australia*

Issues relating to values have always had a place in the school science curriculum. Sometimes this has been only in terms of the inclusion of topics such as 'the nature of science' and/or 'scientific method' and/or particular intentions for laboratory work that relate to 'scientific method.' Sometimes it has been much broader, for example in curricula with STS emphases. Of importance to aspects of this proposal is that different countries/cultures have had different traditions in terms of the place of values in the school [science] curriculum. One obvious very broad difference of this form is the central place in [science] education thinking in many European countries of bildung, and the complete absence of this construct from most [science] curriculum thinking in English speaking contexts. There are numbers of such country/cultural differences.

In the 1990s many countries moved towards various conceptualizations of Outcomes Based Education - OBE (sometimes so labelled and sometimes not). It was usual (but not universal) for OBE focused science curricula to have constrained views of the values that should be implicit and explicit in curriculum; that is views concerned only with 'the nature of science' and 'scientific method' (both usually seen as quite unproblematic).

Currently there are a number of education systems that are changing again, and choosing to move away from Outcomes Based Education (for example, South Africa and several Australian states). One of the most interesting features of many of these movements is the re-embracing of a wider view of the science curriculum, including a reconsideration of the nature and place of the values associated with science in the purposes for and approaches to science education.

Diverse
ISBN 978-90-8790-036-6 hardback
ISBN 978-90-8790-035-9 paperback
287 pages

THE CULTURE OF SCIENCE EDUCATION
Its History in Person

Kenneth Tobin
City University of New York

Wolff-Michael Roth (Eds.)
University of Victoria, Canada

The Culture of Science Education: Its History in Person features the auto/biographies of the professional lives of 22 science educators from 11 countries situated in different places along the career ladder within an ongoing narrative of the cultural history of the field. Many contributors began to identify as science educators at about the time Sputnik was launched but others were not yet born. Hence the book articulates the making of a field with its twists and turns that define a career as a scholar in science education.

Through the eyes of the contributing scholars, the development of science education is seen in the United States and its spread to all parts of the world is tracked, leading to a current situation where some universities from overseas are exporting science education to the United States through graduate programs—especially doctoral degrees. Other key issues addressed are the conceptual personae, such as Jean Piaget and Lev Vygotsky, who have shaped the field of science education and how publishing in English in high-impact journals and obtaining external funds from private and governmental agencies have become driving forces in science education.

The Culture of Science Education: Its History in Person was written for science educators with an interest in the history of science education as it is experienced as lived culture. The book is intended as a reference book for scholars and as a text for graduate students involved in science education.

With Contributions by:
Gilberto Alfaro Varela, Hanna Arzi, Eileen Parsons, Justin Dillon, Lynn Dierking, Reinders Duit, Mariona Espinet, Barry Fraser, James J.Gallagher, Penny Gilmer, Peter Hewson, Chao-Ti Hsiung, Doris Jorde, Jane Butler Kahle, Elizabeth McKinley, Okhee Lee, Lilia Reyes, Svein Sjøberg, Maria Varelas, Manuela Welzel

NEW DIRECTIONS IN MATHEMATICS AND SCIENCE EDUCATION
Volume 3
ISBN 978-90-77874-35-6 hardback ISBN 978-90-77874-33-2 paperback

KEY WORKS IN RADICAL CONSTRUCTIVISM
(edited by Marie Larochelle)

Ernst von Glasersfeld

Key works on radical constructivism brings together a number of essays by Ernst von Glasersfeld that illustrate the application of a radical constructivist way of thinking in the areas of education, language, theory of knowledge, and the analysis of a few concepts that are indispensable in almost everything we think and do. Ernst von Glasersfeld's work opens a window on how we know what we know. The present work grew out of a desire to make more accessible this line of thought, to highlight its originality and consistency, and to illustrate its fecundity in the domains of cognition and learning.

The first three parts of this book contain texts by Glasersfeld that outline the constructivist approach and explicate the frequently drastic reconceptualizations he has suggested. Both the last part and the postscript consist of commentaries by Edith Ackermann, Jacques Désautels, Gérard Fourez, Leslie P. Steffe and Kenneth Tobin, scholars in the fields that Glasersfeld has been concerned with. They examine a number of critical aspects pertaining to (radical) constructivism's current and future development, often tracing out paths that warrant further exploration and reflection, in particular concerning the sociopolitical dimension of knowledge. Key works on radical constructivism is intended as a reference book for researchers, educators, and students of education—and for anyone interested in grasping, or deepening their grasp of, radical constructivism's tenets, ambitions and concerns. Readers will discover in this collection of firsthand contributions the contours of a bold, contemporary debate about a most compelling current of thought.

Bold Visions in Educational Research volume 18
ISBN 978-90-8790-086-1 (hardback) ISBN 978-90-8790-085-4 (paperback)
340 pages

Printed in the United States
102248LV00009B/4/A

9 789087 902476